RELATIVISTIC MECHANICS

LECTURE NOTES AND SUPPLEMENTS IN PHYSICS
John David Jackson and David Pines, *Editors*

Introduction to Dispersion Techniques in Field Theory — Gabriel Barton (Sussex)

Lectures on Quantum Mechanics — Gordon Baym (Illinois)

Intermediate Quantum Mechanics Second Edition — H. A. Bethe (Cornell) R. W. Jackiw (Harvard)

The Special Theory of Relativity — David Bohm (Birkbeck)

Matrix Methods in Optical Instrument Design — W. Brouwer (Diffraction Limited)

Relativistic Kinematics of Scattering and Spin — R. Hagedorn (CERN)

Mathematics for Quantum Mechanics — J. D. Jackson (Illinois)

Symmetry in the Solid State — Robert S. Knox, Albert Gold (Rochester)

Fields and Particles — K. Nishijima (Tokyo)

Introduction to Strong Interactions — David Park (Williams)

Elementary Excitations in Solids — David Pines (Illinois)

Relativistic Mechanics — R. D. Sard (Illinois)

RELATIVISTIC MECHANICS

Special Relativity and Classical Particle Dynamics

R. D. SARD

University of Illinois

W. A. BENJAMIN, INC.

New York

1970

RELATIVISTIC MECHANICS

Special Relativity and Classical Particle Dynamics

Copyright © 1970 by W. A. Benjamin, Inc.

Standard Book Numbers: 8053-8490-1 (P)
                                8053-8491-X (C)
Library of Congress Catalog Card Number 74-80665
Manufactured in the United States of America
12345 KMRQR 32109

W. A. BENJAMIN, INC.
New York, New York 10016

To the memory of my parents
Frederick N. and Maria Belloch Sard

PREFACE

This book has grown out of half a semester of lectures on special relativity in the intermediate course on mechanics at the University of Illinois. The class consisted of seniors and beginning graduate students plus a few juniors. The original notes served as a supplement to R. A. Becker's *Theoretical Mechanics*[15], or K. Symon's *Mechanics*[16], or A. Sommerfeld's *Mechanics* [11], which cover only prerelativistic mechanics. In the writing, the book has grown to include such topics as Thomas precession and the classical theory of spin motion (Chapter 5), and the principle of equivalence and the effects of weak gravitational fields (Chapter 6). The treatment of the kinematics of particle reactions (Chapter 4) has been considerably expanded. The book should now be useful not only to the original clientele but also to graduate students studying relativity in classical electrodynamics or preparing to do research in particle, nuclear, or atomic physics. Although it is a self-sufficient textbook of special relativity, it does not pretend to give a balanced picture of the whole subject. The emphasis throughout is on the applications to particle dynamics. The student is expected to go elsewhere to learn the equally important field theory side.

Relativistic particle mechanics is the basic language of contemporary physics. I have tried to present it as a working tool and have illustrated the arguments by exercises and examples that refer whenever possible to realistic situations. Thought experiments are also used, but the wonderful progress of accelerator technology and astrophysical research has given us plenty of real examples of "extreme relativistic" situations. It is no longer necessary, as it was in Einstein's time, to imagine trains or rocket ships moving with speeds near that of light in order to illustrate striking relativistic effects. Such effects are now part of our laboratory environment.

My own experience in particle and cosmic-ray research has influenced the choice of topics. I have tried to impart some of the "facts of life" that every worker in these fields needs to know. The topics discussed are, of course, relevant in a broader domain that includes atomic and nuclear physics.

The word "particle" in the subtitle is used in the modern sense: a system, of small size, that is for the time being in a well-defined quantum state. We do not go into the meaning of "well defined," but remark that the constant quantal attributes having classical definitions are proper mass, spin magnitude, electric charge, and other electromagnetic moments. (There are other, new-fangled attributes, such as isospin, baryon number, strangeness, intrinsic parity.) All of the attributes specifying the particle are either Lorentz invariant or defined specifically in the particle's rest frame.

No apology needs to be offered for presenting a classical treatment of the dynamics of electrons and other microscopic particles. Properly understood, the classical analysis is part of the correct quantal treatment, giving the motion of the expectation values of the dynamical variables averaged over identical experiments. None of the classical results need to be unlearned.

The first chapter begins with a survey of basic notions that is partly a review of familiar material, partly a summary of things to come. There follows a discussion of the concept of the inertial reference frame, leading to the modern definition: An inertial frame is one in which a shielded body has zero acceleration. The next topic is coordinate transformations within a reference frame. It introduces the reader to the classification of quantities (scalars, vectors, tensors, pseudoquantities) according to their behavior under a particular transformation (rotation, inversion). Then we consider transformations between coordinate systems in different inertial frames. The principle of special relativity of Galileo is stated: In every inertial frame the laws of physics have the same form with the same numerical values of the constants. This principle stands unaltered as one of the basic assumptions of relativistic physics. What have changed are the transformation laws. The Galileo transformation equations are presented, expressing the common-sense assumption that relative motion has no effect on measurements of distance and time. We then consider the behavior of the other dynamical variables under this transformation and find that Newtonian mechanics, with its assumption of invariable inertial mass and of forces depending only on instantaneous relative position or velocity, has the required Galilean covariance. The simple case of the interaction of two charged particles moving with equal velocity shows the breakdown of Galilean covariance when electromagnetism is involved. It is stated that the experimentally established principle of special relativity can be saved by replacing the Galileo transformation by the Lorentz transformation. The equations are presented, without proof at this stage, and it is shown that they contain the Galileo transformation in the small-velocity limit. The new transformation equations will entail a new momentum–velocity relation (variable inertial mass) but will not alter electrodynamics. The last major topic covered in this chapter is the relationship between earthbound reference frames and inertial frames. Newtonian mechanics and gravitation theory provide an adequate approximation for this discussion. Inertial and gravitational forces are compared, and a careful analysis is given of weightlessness in a freely falling frame. We find that a laboratory attached to the earth's crust can be treated as inertial by simply adding to a particle's acceleration a constant term g_{eff} representing residual gravitational–inertial effects.

Chapter 2 deals with the Lorentz transformation and the kinematics of a particle. The theory that we refer to as "special relativity" is based on three postulates: Galileo's principle of special relativity, the assumption that light spreads in vacuum isotropically with the same speed in every inertial frame, and the assumption that the Galileo transformation and Newtonian dynamics are valid in the zero-velocity limit. The experimental evidence bearing on the first two postulates is presented. There follows a presentation of the conventions involved in specifying the position and time of an event. This leads to the derivation of the equations of the Lorentz transformation. We show first that the principle of special relativity and the homogeneity and isotropy of inertial frames restrict the transformation equations to a linear two-parameter form [Eq. (2–16)]. If we require absolute simultaneity, the parameters take the values characterizing the Galileo transformation. If instead we impose the second postulate, they take the values specifying the Lorentz transformation. We then discuss intervals and introduce the concepts of proper time and proper length. The Einstein time dilation and the Fitzgerald–Lorentz contraction are discussed. The latter involves changes of orientation that are just as jolting to common sense as the changes in length. Minkowski's hyperbolic graph [(ct, x) plane] is then presented. It does seem to clarify for the student the relativity of temporal and spatial coincidences, and prepares the way for four-dimensional spacetime. The experimental tests of the time dilation are then discussed. The twin "paradox" is shown to be a useful thought experiment for relating the red shift of a rising photon to the special-relativistic time dilation. Then we study the transformation of velocity, learning the role of the speed of light as a limiting velocity. The collimating effect of a transformation with large relative velocity is brought out in an example of particle decay in flight. We then present four-dimensional spacetime as the appropriate framework for correlating events. Physical quantities are classified as four-scalars, four-vectors, four-tensors, according to their behavior under a Lorentz transformation. Then comes particle motion in spacetime terminology: world line, four-velocity, four-acceleration. The simplest possible examples, uniform longitudinal and uniform transverse acceleration, are worked out.

Chapter 3 presents the dynamics of a particle. We first obtain the equation of motion of a charged particle by using Newtonian mechanics in the rest frame and the relativistic transformation equations. This leads to the relativistic expression relating momentum and velocity. The example of two charged particles moving with the same velocity—which served in the first chapter to demonstrate the incompatibility of the Galileo transformation, Newtonian dynamics, and electromagnetic theory—is worked out in detail, as a check of the consistency of the new equation of motion and the new transformation equations. We then offer another derivation of the equation of motion, based on momentum conservation and the relativistic velocity transformation. It uses the thought experiment of Lewis and Tolman, a highly symmetrical collision of equal particles. This derivation shows very clearly that the new kinematics entails a new dynamics, independent of the type of force under consideration. By writing the equation of motion of a charged particle in tensorial form we find the relativistic velocity–kinetic-energy relation. The relations between energy, momentum, and velocity are then developed in general, and it is shown how in the new dynamics the velocity of light in vacuum plays the role of a limiting particle velocity. Four-momentum is defined, with momentum and energy appearing as its spacetime components and proper mass as its invariant length. The popular equation $E = mc^2$ is a way of describing this relation between energy, momentum, and velocity. Then the four-dimensional form of the equation of motion is presented. The next section (3-8) is a kinematical interlude in which the transformation properties of four-momentum are used to transform distribution functions of angle and energy. The results presented here are basic to particle and nuclear physics experiments. We return to dynamics proper with consideration of the motion of a charged particle in various electromagnetic field configurations: pure magnetic field, constant homogeneous electric field, and Coulomb field (orbit only).

The fourth chapter deals with scattering processes in the most general sense: transitions in which the initial and final state consist of independent systems with constant four-momentum. Relativistic units are finally introduced, to save writing c's. The four-momentum of a complex system is defined, and it is shown that every term in its energy sum contributes to the inertial mass, defined as the ratio of momentum to velocity. The separate laws of conservation of energy and momentum are shown to be one grand law of conservation of four-momentum. Energy never disappears from the mechanical budget; it is always conserved no matter how "inelastic" the process. Relativistic dynamics is a framework suitable for describing creation and annihilation processes. Although energy is conserved, proper mass, corresponding to one term in the sum, is not. There is free trade among all forms of energy. The significance of this mass–energy relation is brought out by several examples. Next we consider the consequences of four-momentum conservation for various reactions: one-photon emission; two-photon final state; general two-body final state; two-body initial state (binary collision), leading to one body, two bodies, many bodies. The equations are readily obtained by forming the squares of sums and differences of four-momenta (invariant masses, invariant four-momentum transfers).

These invariants are just the quantities whose distributions are predicted by the relativistic quantum dynamics of the process. The practical significance of the conservation law is to reduce the number of independent dynamical variables and to set limits on their range of variation. Some interesting thresholds are calculated. This last part of the chapter, dealing with four-momentum conservation in particle reactions, provides the essential "relativistic kinematics" needed for their analysis.

The fifth chapter deals with the composition of Lorentz transformations and the kinematics and dynamics of spin. The matrices for various kinds of Lorentz transformation are worked out and cascaded to correspond to successive transformations. The basic physics is the velocity transformation law. The addition of velocities is no longer a matter of Euclidean plane trigonometry but of Lobachevskian hyperbolic trigonometry. The spherical defect of the velocity triangle determines a rotation of the axes associated with successive transformations in different directions. Spin then is defined, as the angular momentum of a system in its rest frame. When a system undergoes transverse acceleration, the rotation of axes in successive transformations makes the spin turn. This is the Thomas precession, evaluated in this chapter for an atomic electron and for a spinning earth satellite. We then take up the dynamics of spin. The equation of motion is set up in the particle's rest frame, and solved for motion in a transverse magnetic field. The angle between spin and momentum reflects the gyromagnetic ratio g of the particle. If $g = 2$, spin and momentum turn together; this is the basis of the "g-2" experiments recently carried out on electrons and muons. The motion is also worked out for a transverse electric field, giving the spin–orbit interaction of atomic physics. Finally the spin equation of motion is written in a manifestly covariant form involving the four-spin or polarization four-vector. This gives an equation of motion valid for a finite time in an inertial frame.

In Chapter 6 we consider the principle of equivalence and its application to motion in weak gravitational fields like the one prevailing on earth. Special relativity, as completed by the principle of equivalence, provides a consistent and adequate formalism. The principle is used to calculate the effects of gravity on clocks and measuring rods. A procedure is set up by which the instruments in a noninertial laboratory are calibrated by reference to comoving inertial frames. The gravitational red-shift formula (free fall of a photon) is derived and its experimental tests are discussed. Finally, the free fall of a material particle is analyzed on the basis of the special-relativity velocity addition formula.

Appendix A (Chapter 1) summarizes transformation properties under rotation and inversion in three-dimensional space. Appendix B (Chapter 1) presents the theory of accelerated reference frames in Galilean kinematics, and Appendix C (Chapter 1) is concerned with detectable effects of the earth's rotation with respect to inertial frames. Appendix D (Chapter 3) gives the transformation properties of electric charge and the electromagnetic field. Appendix E (Chapter 3) presents a formal proof that there is no arbitrary zero level of energy in special relativity. Appendix F (Chapter 6) gives the theory of the Doppler effect.

The bibliography does not list all the books and articles on special relativity that are available to the present-day student. It does aim at historical accuracy in the references, which are selected for relevance to the specific matter of the text. I learned about special relativity from Einstein's first paper [2] and Pauli's *Enzyklopädie* article [87]. The enlightening discussions over the years with colleagues and students cannot be adequately acknowledged.

I have a very special obligation to Giulio Ascoli. His unpublished notes form the basis for the treatment of the Thomas precession and the kinematics and dynamics of spin (Sections 5-3 and 5-4); he has also made available his unpublished result on the free fall of a relativistic particle (Section 6-4). Beyond this, he has served as conscience and touchstone throughout the writing of the book. He has not, however, read any of it, and should not be held responsible for its faults.

The book owes its existence to the initiative of the publishers, W. A. Benjamin, Inc., in arranging to have my Physics 322 lectures transcribed and to their patient support during two complete rewritings. The transcriber, L. G. King, made helpful suggestions. Professor G. M. Almy as head of the physics department has abetted the work by providing time and facilities. My family in all three generations has given constant encouragement. To all I am grateful.

Robert D. Sard

Urbana, Illinois
December 1968

CONTENTS

Preface   vii

Table of Particles   xviii

Notation   xix

Chapter 1. Basic Notions. Coordinate Transformations. Special   **1**
    Relativity. Accelerated Reference Frames

   1-1. Basic Notions   3
        Particle
        Rigid Body
        Reference Frame
        Coordinate System
        Clock
        Momentum
        Mass
        Force
        Energy

   1-2. Inertial Reference Frames   13

   1-3. Coordinate Transformations Within a Reference Frame   16
        Covariance
        Homogeneity and Isotropy of Space

1-4. Transformations Between Inertial Frames: Special Relativity   26
The Principle of Special Relativity
The Galileo Transformation
Covariance of Newtonian Mechanics with Respect to the
  Galileo Transformation Between Inertial Frames
Galilean Invariance of Instantaneous Action at a Distance
  Force
Breakdown of Galilean Invariance in the Electromagnetic
  Interaction
The Lorentz Transformation

1-5. Earthbound Reference Frame and Inertial Frames   38
Inertial Forces
Weightlessness in a Freely Falling Laboratory
Relation Between Earth Frame and Inertial Frames

Problems   46

Chapter 2. The Lorentz Transformation and the Kinematics of a
           Particle   **48**

2-1. The Postulates of Special Relativity   48

2-2. The Location of Events in Space and Time   54

2-3. The Lorentz Transformation   58

2-4. Intervals   75

2-5. Proper Time and Proper Length; Time Dilation and Length
     Contraction   80

2-6. Minkowski's Hyperbolic Graph   86

2-7. Tests of the Einstein Dilation; the Twin "Paradox"   94

2-8. Transformation of Velocity   104

2-9. Spacetime   111

2-10. Spacetime Description of Particle Motion: Four-Velocity,
      Four-Acceleration   118

2-11. Uniform Longitudinal Acceleration   124

2-12. Constant Transverse Acceleration   127

Problems   130

Chapter 3. Dynamics of a Particle   **136**

3-1. The Law of Motion for a Charged Particle   136

3-2. Interaction Between Charged Particles Moving with the Same
Velocity   143

3-3. Dynamics Based on Momentum Conservation   146

3-4. Covariant Equation of Motion of a Charged Particle   152

3-5. Energy, Momentum, and Velocity   155

3-6. Four-Momentum   159

3-7. General Covariant Equation of Motion   166

3-8. Momentum Space and Momentum Distributions. Cross
Sections   168

3-9. Motion of a Charged Particle in a Constant Electromagnetic
Field   182
Charged Particle in Magnetic Field
Charged Particle in a Coulomb Field (Hydrogenic Atom)

Problems   197

Chapter 4. Transitions of a System. Conservation of Four-Momentum. The Mass-Energy Relation. Relativistic Kinematics   **202**

4-1. Scattering Processes   202

4-2. Relativistic Units   204

4-3. Four-Momentum of a Complex System. Inertia of Energy   207

4-4. Conservation of Four-Momentum   214

4-5. Rest Energy. Nonconservation of Proper Mass   217

4-6. Radiative Transitions   218

4-7. Two-Body Final State, both of Zero Proper Mass   225

4-8. Two-Body Final State: General Case   227

4-9. Two-Body Initial State (Binary Collisions)   231
One-Body Final State
Elastic Scattering
Inelastic Scattering, Two-Body Final State

Inelastic Scattering, the General Case. Thresholds   250

4-10. Relativistic Kinematics   250

Problems

Chapter 5. Successive Lorentz Transformations. Motion of Spin   **258**

5-1. Transformation Matrices in Spacetime   258

5-2. Composition of Reference-Frame Transformations   264
      Composition of Parallel Lorentz Transformations
      The Velocity Parameter ("Rapidity") α and its Hyperbolic
          Functions
      Composition of Lorentz Transformations in Different Direc-
          tions

5-3. Spin: Kinematics   281
      Turning of Spin and Velocity in a Pure Lorentz Transformation
      Thomas Precession

5-4. Dynamics of Spin   291
      Motion in a Transverse Magnetic Field. "g–2" Experiments
      Motion in a Transverse Electric Field (Spin–Orbit Interaction)
      Equation of Motion of Four-Spin

Problems   304

Chapter 6. Principle of Equivalence. Motion in a Weak Gravitational
      Field   **309**

6-1. The Principle of Equivalence   309

6-2. The Effect of a Gravitational Field on Clocks and Measuring
      Rods   311

6-3. The Gravitational Red Shift   316

6-4. Free Fall of a Material Particle   320
      Motion Parallel to Field
      Motion Perpendicular to Field
      The General Case

Problems   326

CONTENTS

Appendix A Transformation Properties of Vectors. Tensors. Polar
and Axial Vectors   **327**
Rotation of Cartesian Axes   327
Inversion of Cartesian Axes   333
Polar and Axial Vectors   336

Appendix B Galilean Kinematics of Accelerated Reference Frames   **341**
Analysis of Motion into Translation and Rotation   341
Motion of the Basis Vectors   345
Transformation Equations for the Velocity and Acceleration of a
Particle   346

Appendix C Detectable Effects of the Earth's Rotation with Respect
to Inertial Frames   **351**

Appendix D Transformation Law of Charge and Electromagnetic
Field   **355**

Appendix E The Constant of Integration in the Expression for the
Energy of a Particle   **360**

Appendix F Doppler Effect   **362**

References   **366**

Index   **370**

Table of Particles

This table lists only those particles (and their antiparticles) referred to in the text and the problems; there are many others. Spins and magnetic moments are given in Table 5-1.

Name	Symbol	Proper mass (GeV/c^2)	Electric charge (e_0)	Mean proper life (sec)
Photon	γ	0	0	—
Neutrino(s)	ν	0	0	—
Electron	e	0.5110×10^{-3}	± 1	—
Muon	μ	0.1057	± 1	2.20×10^{-6}
Pion	π^0	0.1350	0	0.9×10^{-16}
	π^{+-}	0.1396	± 1	2.60×10^{-8}
K meson[a]	K^{+-}	0.4938	± 1	1.23×10^{-8}
	K^0 short⎱ long⎰	0.4978	0	⎰0.86×10^{-10}⎱ ⎱5.3×10^{-8}⎰
η meson[b]	η	0.549	0	3×10^{-19}
ρ-meson[b]	ρ	0.765	$0, \pm 1$	0.6×10^{-23}
ω-meson[b]	ω	0.783	0	0.6×10^{-22}
Proton	p	0.9383	± 1	—
Neutron	n	0.9396	0	0.96×10^3
Λ hyperon	Λ	1.116	0	2.5×10^{-10}
Σ hyperon	Σ^+	1.189	± 1	0.81×10^{-10}
	$\Sigma^{0(b)}$	1.193	0	$<10^{-14}$
	Σ^-	1.197	± 1	1.6×10^{-10}
B meson[b]	B	1.22	± 1	0.6×10^{-23}
Ω hyperon[c]	Ω^-	1.672	± 1	1.3×10^{-10}

[a] The K^0 is a mysterious complex, behaving as regards spontaneous decay like a combination of short- and long-lived particles of slightly different proper mass.
[b] Lifetime is not directly measured.
[c] The antiparticle has not yet been observed.

Notation

A letter set in boldface represents a vector in three-dimensional space. A vector $\mathbf{a}$ has magnitude a and components a_x, a_y, a_z. The unit vectors along the x, y, z axes are $\mathbf{i}, \mathbf{j}, \mathbf{k}$, and

$$\mathbf{a} = \mathbf{i}a_x + \mathbf{j}a_y + \mathbf{k}a_z$$

In algebraic manipulations, we replace the subscripts x, y, z by 1, 2, 3, and use lowercase italic letters to represent them. Repeated letter indices in products are summed over the range 1, 2, 3. The scalar product of two vectors $\mathbf{a}$ and $\mathbf{b}$ is written

$$\mathbf{a} \cdot \mathbf{b} = ab\cos(\mathbf{a}, \mathbf{b}) = a_l b_l = a_1 b_1 + a_2 b_2 + a_3 b_3$$

If a letter has a tilde ($\sim$) over it, it represents a four-vector. The four components in a particular coordinate system are labeled by a lowercase Greek subscript. Thus, the four-vector $\tilde{a}$ has the components a_λ. The components

$$a_1 = a_x$$

$$a_2 = a_y$$

$$a_3 = a_z$$

are the components of the vector $\mathbf{a}$ of magnitude a. They are called spacelike because they transform like the space coordinates x, y, z, of an event. We use two conventions for the timelike component, giving either a_4 or a_0, where $a_4 = ia_0$. With t the time coordinate of an event, a_4 transforms like ict and a_0 like ct. The four-vector is denoted in terms of its components as

$$\tilde{a} = (\mathbf{a}, a_4) = \{a_0, \mathbf{a}\}$$

the round brackets going with the x, y, z, ict scheme, the curly brackets (braces) with ct, x, y, z.

The scalar product of two four-vectors $\tilde{a}$ and $\tilde{b}$ is written

$$\tilde{a}\tilde{b} = a_\mu b_\mu = a_1 b_1 + a_2 b_2 + a_3 b_3 + a_4 b_4 \qquad \mu = 1, 2, 3, 4$$
$$= \mathbf{a} \cdot \mathbf{b} + a_4 b_4$$

Repeated letter indices are, again, summed over their entire range. If we use instead the labels 0, 1, 2, 3, the scalar product is

$$\tilde{a}\tilde{b} = a_1 b_1 + a_2 b_2 + a_3 b_3 - a_0 b_0$$
$$= \mathbf{a} \cdot \mathbf{b} - a_0 b_0$$

For manipulating scalar products, the (x, y, z, ict) scheme $(\mu = 1, 2, 3, 4)$ is more convenient, inasmuch as all terms in the scalar product enter the sum with the same sign. Otherwise, we tend to prefer the (ct, x, y, z) scheme, which does not involve the unit imaginary i.

The square or invariant length squared of a four-vector is its scalar product with itself:

$$\tilde{a}^2 = \tilde{a}\tilde{a} = a_\mu a_\mu = a_1^2 + a_2^2 + a_3^2 + a_4^2 = a^2 + a_4^2$$
$$= a_1^2 + a_2^2 + a_3^2 - a_0^2 = a^2 - a_0^2$$

If the norm is positive, the four-vector is spacelike; if negative, it is timelike; if zero, it is lightlike. The four-momentum of a real system is timelike.

A four-tensor of the second rank is denoted by a letter with a double tilde, $\tilde{\tilde{a}}$. Its components are indicated by the plain letter with appropriate lowercase Greek indices, $a_{\mu\nu}$.

A matrix is denoted by a letter in sans serif type, a. Its components are denoted by the same letter in italic type with two subscripts—the first for row, the second for column.

In an effort to reduce confusion about inertial mass and proper mass, we use the italic m for the former and the Greek μ for the latter.

The velocity of reference frame R' with respect to reference frame R is $\mathbf{u}_R(R')$. The magnitude of this velocity relative to that of light in vacuum is variously specified by

$$\beta = \frac{u}{c}$$

or

$$\gamma = \left(1 - \frac{u^2}{c^2}\right)^{-1/2}$$

or

$$\alpha = \tanh^{-1}\left(\frac{u}{c}\right)$$

When possible, we refer to quantities in the rest frame by the corresponding lowercase Greek letter—ρ for r, τ for t, σ for s, ξ for x, η for y, ζ for z, λ for l, θ for h. On other occasions, the dagger superscript [†] labels the rest frame.

In considering transformations among three reference frames R_0, R_1, R_2, we denote, for example, the velocity with respect to R_0 of a point at rest in R_1 by $\mathbf{u}_0(1)$; χ_0 is the angle in R_0 between the velocity of a point at rest in R_1 and the velocity of a point at rest in R_2. We also use the triangular convention of labeling a quantity with two indices by simply using the third; for example,

$$u_2 = |u_0(1)| = |u_1(0)|$$
$$\alpha_0 = |\alpha_1(2)| = |\alpha_2(1)|$$

In discussing collisions or transitions, we use the left superscript position to label the particle, the right superscript position to label the state (initial or final), and the right subscript position to label the coordinate. For example,

$$^{2}v_{x'}^{\text{fin}}$$

is the x' component of the velocity of particle 2 in the final state.

In Chapter 4, we use in addition the left subscript position, to label the reference frame. For example, ^1_2E is the energy of system 1 in the rest frame of system 2; $^1_L\mathbf{P}$ is the momentum of system 1 in the laboratory frame L.

In the dynamics of complex systems we use capital letters for dynamical variables of the system and lowercase letters for those of the component parts. An asterisk identifies the center of mass frame.

In Chapter 6, we reserve Greek letters for noninertial frames and Latin letters for inertial frames.

Chapter 1

Basic Notions. Coordinate Transformations. Special Relativity. Accelerated Reference Frames

Contemporary physics describes nature in terms of particles and fields. The concepts used—such as position, time, momentum, energy, force, mass, angular momentum, charge, and field strength—find their defining examples in the mechanics of particles. In this sense, particle mechanics is the language of all physics.

Nonrelativistic classical mechanics is the work of many men, but it received its definitive formulation, as far as physical content is concerned, in Newton's *Mathematical Principles of Natural Philosophy* (1686) [1]. It cannot be said too often that Newtonian mechanics is the correct mechanics for a vast range of natural phenomena, the range in which Planck's constant (h) can be assumed to be zero and the speed of light in vacuum (c) to be infinitely large. It covers almost all of the behavior of matter in bulk under terrestrial conditions. It would be foolish to attempt to formulate the motion of a man-sized object—such as a machine or a man—in quantal or relativistic terms.

Special relativity (Einstein, 1905) [2] is the culmination of the classical physics of fields and particles. It gives us a formalism suitable for the dynamics of particles at all speeds found in nature, up to and including the speed of light. Nonrelativistic classical mechanics is contained in this formalism as the small-velocity limit. Indeed, what we shall call the third postulate of the theory of special relativity is the requirement that Newtonian mechanics hold in an inertial reference frame in which the particle under consideration is essentially at rest.

In contrast to the quantitative continuity between relativistic and pre-relativistic mechanics, the latter being a limiting case of the former, there is a conceptual jump between the two that gives special relativity a revolutionary character. Space and time intervals that in everyday experience are independent of the reference frame turn out to be markedly different in frames moving with large velocity relative to one another. One can no longer speak without qualification of the simultaneity of two events. There are other strange features of the theory. All of the strangeness stems from the fact that our common sense notions about nature come from a very limited experience, acquired mainly in infancy, while the ingenuity of experimental physicists in exploiting the progress of technology has given us a much larger domain in which to test our theories. Special relativity is a logically consistent theory in precise accord with experiment. We must stretch our minds to fit it.

We nevertheless call special relativity classical because it uses the pre-quantal assumption of a direct one-to-one correspondence between the variables of the theory and the numbers measured in experiments. Position, velocity, momentum, and all other dynamical variables are taken to be real numbers; with sufficient ingenuity one can measure these numbers to arbitrary precision. In quantum mechanics (1926), on the other hand, dynamical variables are operators in an abstract space, and the formulas of the theory determine probability distribution functions for the corresponding real numbers measured in experiments. Although special relativity involves a revolution in our thinking about space and time, it is still a classical theory in the sense just specified. Again, despite the conceptual jump, the old point of view is a mathematical limiting case of the new; the classical approach is valid whenever circumstances permit us to assume that h is essentially zero. There is an important class of such phenomena for which relativistic classical mechanics is the right tool.

In addition, the intimate relationship between quantum and classical mechanics makes the latter an essential tool in using the former. The formulas are the same, with the real numbers of classical theory reinterpreted as operators in quantum theory. One result (Ehrenfest's theorem) is that the value of a dynamical variable, averaged over many identical experiments, is governed by its classical equation of motion. In fact, many of the applications that we shall make of relativistic classical mechanics refer to experimental situations in which quantal effects cannot be neglected.

In our presentation of relativistic classical mechanics, we shall not avoid discussion of the apparent paradoxes of the theory, but the emphasis will be on the "facts of life," as needed for the application of the formalism to realistic physical situations. In this practical approach, we not only acquire valuable technique; we also learn what are the real assumptions of the theory. Thus the physicist is prepared for the continual reappraisal of theory that is required by the progress of experimental and mathematical research.

Relativistic classical mechanics will be presented as a logically consistent method of description. Physics has developed by successive approximations, involving at each new stage redefinitions and extensions of familiar concepts. It is not possible to define these concepts in terms of simpler ones. One simply goes ahead and uses the formalism, thereby learning the meaning of the terms used. This is not to say that we require of a theory only that it "work." We also require logical consistency. And if in the competition of rival theories the situation is unclear, as it usually is, we favor the one that appears simpler, more beautiful, more fruitful. Such a judgment is bound to be conditioned by local traditions and philosophical and religious pressures, and it must be treated with reserve. Scientists believe that experiment eventually forces us to make every correction that is needed.

We shall present only that part of the theory of special relativity that bears on particle mechanics. It will, of course, be necessary to draw on the classical theory of the electromagnetic field, out of which special relativity grew. Our traditional manner of teaching physics—in which the material is divided into mechanics, electricity and magnetism, atomic physics, and so on—results in a fragmented approach to special relativity. It is perhaps just as well, for the relativistic way of thinking should permeate all physics, and multiple passes at relativity may be pedagogically more effective than a treatment in a separate course.

We stop short of general relativity, although we must refer to it in connection with the discussion of accelerated reference frames. Hopefully this book will encourage the student to go on to study general relativity. Current advances in experimental technique are bringing this subject to life as sharper tests become possible.

1-1. BASIC NOTIONS

The following remarks about the basic notions of mechanics are not attempts to explain them in simpler terms, but are rather summary statements of relationships between them. At this stage they will seem more or less vague, depending on the reader's background. They should become clearer as he reads on. The beginner should not be discouraged but should read this section quickly, saving careful study for later. To facilitate skimming, we set off in fine print statements that are previews of results to be obtained later.

Particle

A particle is a system that can, for practical purposes, be localized at a point. Classically such a system is characterized by its inertial mass—determining its response to applied force—and its charge (e.g., electric charge, gravitational charge)—determining its interaction with the rest of the world. Microscopic physics presents us with particles (electron, proton, pion, etc.) that have other parameters as well, such as spin (intrinsic angular momentum), magnetic dipole moment, isospin, baryon number, and so on.

All of the parameters are constants of the motion associated with a well-defined quantum state. If they ever change they do so brusquely, in a "transition." Rather than admit multivalued particle parameters, we describe these inelastic transitions by saying that the particles present in the initial state are destroyed and those present in the final state are created. Instead of being the permanent objects of Newtonian mechanics, particles are merely "more or less transient repositories of conserved quantities" (Primakoff). During its life, a particle is a system specified by constant values of its parameters.

> The parameters are either Lorentz invariant (the same in all inertial reference frames) or defined specifically in the system's rest frame. They have a physical significance independent of the velocity of the particle with respect to anything else, and are in this sense intrinsic.

Rigid Body

The rigid body is basic to our idea of the reference frame in terms of which the motion of a particle is specified. The distance between any two points of a rigid body is constant. But conceptually the distance between two points is determined by applying a measuring rod, itself a rigid body. Thus the statement about constant distance is circular if regarded as a definition of rigidity. We must accept the rigid body as a basic undefined concept.

It is, in principle, impossible to realize physically the ideal rigid measuring rod because the interactions coupling the atoms at one end of the rod to the atoms at the other propagate through the solid at a finite speed. If one end is pushed in by a longitudinal blow, the other end does not move immediately. During the waiting period the length of the rod is different from normal. The speed of the elastic waves cannot even in principle be made infinite because the interatomic forces responsible for elasticity are electromagnetic in nature, and therefore the speed of light in vacuum c is an insurmountable upper limit. But the deformations of the rod can be calibrated and corrected for. Thus the absence of absolute rigidity in nature does not prevent us from using the concept of the ideal rigid body.

Reference Frame

The position of one body can be specified only relative to another. The surveyor's transit must have something to stand on. A reference frame is the space determined by a rigid body regarded as a base for surveying. One can think of the body being extended as far as desired by a lattice of measuring rods (Fig. 1-1). The position of a point is then specified by three numbers serving to label the nearby vertex. (The term "nearby" is used in the sense of the infinitesimal calculus.)

In practice, the surveying is done with light beams assumed to travel in straight lines in Euclidean three-dimensional space. Checks of internal consistency permit correction for possible bending of the light. The agreed unit of length is the meter, now defined as 1,650,763.73 wavelengths in vacuo of light from the unperturbed transition $2p_{10}-5d_5$ in krypton 86. The meter is approximated by the distance between two scratches on a platinum–iridium bar at the International Bureau of Weights and Measures near Paris.

> Relativistically, the concept of the reference frame is extended to include timekeeping as well as position measuring. One imagines identical clocks at the vertices of the lattice. Then the position and time of an event are specified by the threefold label of its adjacent vertex and the reading of the local clock. We shall discuss the synchronization procedure in the next chapter.

Coordinate System

A coordinate system is a particular code for labeling points in a reference frame by three numbers. In one reference frame there are a host of possible coordinate systems: Cartesian, oblique, polar, ellipsoidal, or what you will, with all origins and orientations. The Cartesian systems (three mutually perpendicular axes with equal units of length on each) have a privileged

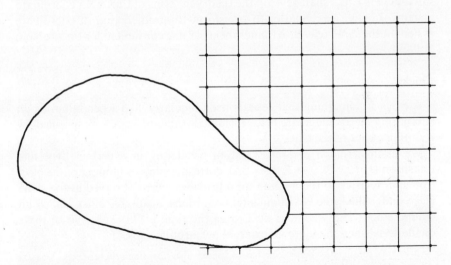

Figure 1-1. A rigid body extended as far as desired by building onto it a lattice of measuring rods.

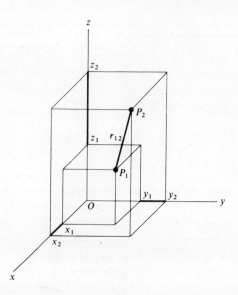

Figure 1-2.

position, because in our three-dimensional Euclidean space the general expression for the distance between two points has the simplest possible form when Cartesian coordinates are used (Fig. 1-2):

$$r_{12} = [(x_1 - x_2)^2 + (y_1 - y_2)^2 + (z_1 - z_2)^2]^{1/2}$$

There are no cross terms, and the coefficient of each coordinate difference squared is $+1$. We shall develop the theory in terms of Cartesian coordinate systems. It is an easy matter to transform to another kind of coordinate system in the same reference frame whenever it is convenient for the problem at hand.

Clock

A clock is something that repeats itself regularly, like a pendulum or an alternating electromagnetic field. Instant of time, or epoch, is specified by counting cycles of the clock.

We assume that a cesium-133 atom oscillating in a certain hyperfine transition ($F, F_z = 4, 0 \leftrightarrow F, F_z = 3, 0$, corresponding to flipping of the electron spin relative to the nuclear spin) provides, when observed at the same place under the same environmental conditions, a suitable clock, called an atomic clock. The second is defined as the time of 9,192,631,770 of these oscillations when there are no external perturbing fields.

> In relativistic theory, one has to specify also how one synchronizes clocks at different places in the reference frame. The theory makes a definite recipe for doing so with electromagnetic signals from a master clock. In Newtonian mechanics, one does not discuss the synhronization of distant clocks; one simply assumes that there is a universal time pervading the frame, and all

frames. Relativistically, each reference frame has its own array of clocks. The transformation between reference frames involves time as well as position coordinates.

Momentum

Momentum, or quantity of motion, or motion *tout court*, played a central role in Newton's formulation of the laws of motion: "the change of motion is proportional to the motive force impressed; and is made in the direction of the right line in which that force is impressed" ([1], Law II). But during the subsequent long period of preoccupation with low-velocity phenomena, the rate of change of motion always reduced to a constant mass times the acceleration, and the momentum concept receded from the foreground. In relativistic and quantal physics, it has regained its prime position.

The momentum of a body is a vector parallel to its velocity; its magnitude is zero when the body is at rest and increases with the magnitude of the velocity. Different bodies with the same velocity have different momenta proportional to their inertial or proper masses. Momentum can be measured by observing the recoil of another body, as in a ballistic pendulum (Fig. 1-3), or, for a particle of known charge, by observing its deviation in a known magnetic field (Fig. 1-4).

Momentum has the remarkable conservation property. It may be traded with other bodies, but the total (vector) momentum of an isolated system is constant in time. This conservation law was formerly regarded as a specific property of certain interactions—the forces occur in equal and opposite pairs ([1], Law III)—but it is now looked on as a manifestation of the homogeneity of physical space.

Relativistically, momentum is the spacelike component of the particle's four-momentum. The timelike component is, to within a constant factor, the particle's energy. For an isolated system, all components of four-momentum are con-

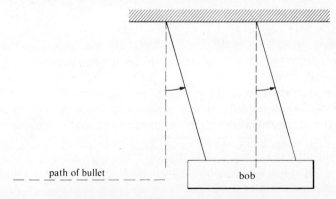

Figure 1-3.   Ballistic pendulum, for measuring the momentum of a bullet.

Figure 1-4. Momentum measured by curvature in a magnetic field. This print shows one of three views of the Brookhaven 80-in. hydrogen bubble chamber, traversed by protons of momentum 7.9 GeV/c. There is a magnetic field of $\mathscr{B} = 18$ kG perpendicular to the print. Charged particles make bubbles in the liquid, here photographed in the light of a flash tube. Measurements of the coordinates of the track images in two or more views permit calculation of the spatial orientation and curvature of the tracks. Then the momentum p is obtained, by using Eq. (3-55b): $p \cos \lambda = (e/c)\mathscr{B}\rho$, where λ is the dip angle of the track from the plane perpendicular to $\mathscr{B}$ and ρ is the radius of curvature. The momentum and dip angle for each track are given on the drawing. The event shown is interpreted as $p + p \rightarrow p + p + \pi^{+} + \pi^{-}$, with the mass assignments as shown on the drawing on the facing page.

served, and conservation of momentum is part of one grand law of conservation of four-momentum.

Mass

Inertial mass, called "mass" in nonrelativistic mechanics, is the scalar coefficient of the velocity in the expression for the momentum:

$$\mathbf{p} = m\mathbf{v} \tag{1-1}$$

It measures the inertia, epitomizing the difference between a Cadillac and a Volkswagen of the same velocity. Measurement of p and v yields the inertial mass of the body. In one mass spectrograph [3], for example (Fig. 1-5), a Wien filter (crossed electric and magnetic fields) determines the velocity without changing it, and a subsequent magnetic deflection determines the momentum.

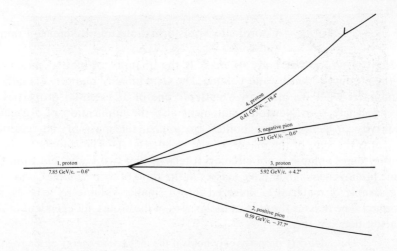

4, proton
0.41 GeV/c, -19.4°

5, negative pion
1.21 GeV/c, -0.6°

1, proton
7.85 GeV/c, -0.6°

3, proton
5.92 GeV/c, +4.2°

2, positive pion
0.59 GeV/c, -37.7°

The gravitational charge (or passive gravitational mass) of a body, measuring its response to gravitational fields, is proportional to its inertial mass. The masses of macroscopic bodies are usually determined by measuring their weights while at rest at the same place on earth.

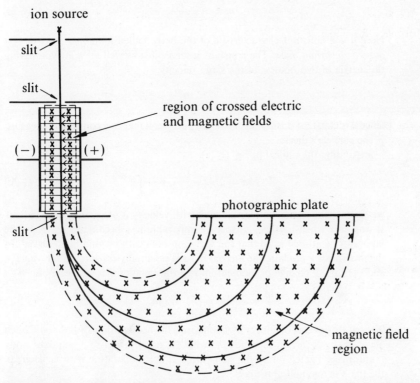

ion source

slit

slit

region of crossed electric
and magnetic fields

(−) (+)

photographic plate

slit

magnetic field
region

Figure 1-5.   Mass spectrograph, consisting of velocity selector followed by magnetic deflection system [3].

9

Newton wrote that body or mass is the quantity of matter, measured by the product of density and volume. The vivid phrase "quantity of matter" illuminates what we mean by "matter": one of its essential properties is inertia. There is in Newton's statement also the implication of a general additivity property. If two bodies are joined, the mass of the resulting system is the sum of the masses of the isolated parts. The statement about density times volume is not circular. It tells us that two bodies cut from the same homogeneous melt have masses in the ratio of their volumes, and that if a sample of material is squeezed into a smaller volume its mass is not changed. The mass is the sum of the masses of the atoms, independent of the spacing between them.

This additivity property only holds if the interaction energy of the parts is negligible. Relativistically, the inertial mass is proportional to the total energy, including that of interaction between the parts. If the heat energy or stress energy of a body is increased, its inertial mass is increased too.

In relativity, we continue to define inertial mass as the ratio of momentum to velocity [Eq. (1-1)]. But there is a new momentum–velocity relation, giving

$$m = \frac{\mu}{\sqrt{1 - v^2/c^2}} \tag{1-2a}$$

Here μ is a constant characteristic of the body, called its *proper mass* or *rest mass* or *invariant mass*. The constant μ serves just as well as m for comparing the inertia of two bodies; at the same velocity,

$$\frac{m_1(v)}{m_2(v)} = \frac{\mu_1}{\mu_2}$$

Since μ is constant during the life of the particle, it makes a convenient measure of the particle's inertia.

Expanding the radical in Eq. (1-2a),

$$m = \mu\left(1 + \frac{1}{2}\frac{v^2}{c^2} + \cdots\right) \tag{1-2b}$$

we see that the change of inertial mass with velocity is of second order in (v/c). It was only when particles of speed comparable to c became available in the laboratory (cathode rays) that the variation of m with v could be observed. In the low velocity domain, the inertial mass is essentially constant at the value μ, and the rate of change of momentum is proper mass times acceleration

$$\frac{d\mathbf{p}}{dt} = \frac{d}{dt}(m\mathbf{v}) \approx \mu\frac{d\mathbf{v}}{dt} = \mu\mathbf{a} \qquad (v \ll c)$$

Then μ or (m) can be found by comparing the acceleration of the body when acted on by a certain force $(d\mathbf{p}/dt)$ with the acceleration of a standard body ("unit mass") acted on by the same force. If the acceleration is twice as great, the mass is one-half unit mass.

Since the inertial mass depends on the velocity of the particle with respect to the reference frame [Eq. (1-2a)], it has different values in different frames. In the reference frame in which the particle is momentarily at rest [$v = 0$ in Eq. (1-2a)],

$$m(0) = \mu$$

The proper mass μ can thus be determined by a measurement of the inertial mass in the rest frame. In any other inertial frame it can be found by measuring p and v and using

$$p = \frac{\mu}{\sqrt{1 - v^2/c^2}} v$$

Thus p and v evidently determine μ as well as m.

The inertial mass is connected with the energy of the particle by the relation

$$E = mc^2 \tag{1-3a}$$

All forms of energy belonging to the particle contribute on an equal footing to its total energy E and thus to its inertial mass m. To within a universal proportionality constant m and E are equal, and are thus physically equivalent. If we know one we know the other. Both are equal within a constant factor to the timelike component of the four-momentum. Conservation of energy and conservation of inertial mass are one law, a projection on the time axis of the broader law of conservation of four-momentum.

In view of the definition [Eq. (1-1)] of m, the physical content of the mass–energy relation [Eq. (1-3a)] for particle motion is simply that

$$\mathbf{p} = \left(\frac{E}{c^2}\right)\mathbf{v} \tag{1-3b}$$

The proper mass is essentially the magnitude, or invariant length, of the four-momentum

$$\mu c = \left[\left(\frac{E}{c}\right)^2 - p^2\right]^{1/2} \tag{1-4}$$

It has the same value in all inertial frames and is one of the parameters characterizing the particle.

In present-day usage, the bare word "mass" more often than not means proper mass. To avoid possible confusion we shall henceforth never use "mass" alone but will always apply a defining adjective, as in "inertial mass" or "proper mass" or "gravitational mass."

Force

Associated psychologically with the muscular stress of a push or a pull, force is defined as the rate of change of momentum.

$$\mathbf{F} = \frac{d\mathbf{p}}{dt} \tag{1-5}$$

In the so-called absolute system of units used in Eq. (1-5), the unit force is one giving unit rate of change of momentum. With this definition force is a vector quantity, and the resultant of two concurrent forces is their vector sum. Because of momentum conservation, forces occur in equal and opposite pairs. In some cases the dependence of force on distance or velocity turns out to have a beautifully simple form [for example, Newton's inverse square law of gravitation, Eq. (1-8)].

When the velocity of a body is small compared with that of light, the rate of change of momentum is proper mass times acceleration, and a force can be measured simply by the acceleration it imparts to a standard body.

The vectorial character of force sometimes enables one to measure a force by balancing it against an opposite known force; with a spring balance, for example, the elastic restoring force of the spring, measured by its elongation, balances the unknown force pulling on the hook. In astronomical and microscopic situations, however, force can only be measured as rate of change of momentum.

> In relativistic mechanics, the force depends on the reference frame, consistent with Eq. (1-5). The transformation law of force has been confirmed experimentally for electromagnetic forces.

Like inertial mass, defined by Eq. (1-1), force is a secondary concept, derived from position, time, and momentum. It is in essence a convenient term for describing how fields affect particle motion.

Energy

The change of energy E of a particle equals the work done by the force $\mathbf{F}$ acting on it (Fig. 1-6).

$$\Delta E = \int dE = \int \mathbf{F} \cdot d\mathbf{r} \tag{1-6}$$

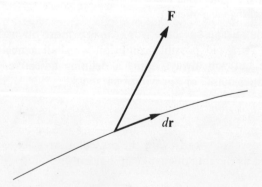

Figure 1-6.

The change of energy of a system of particles is the sum of such terms, one for each particle. The change of energy of a field is defined in such a way as to maintain the energy balance in its interaction with particles. But energy is not a mere bookkeeping term, defined so as always to balance. Energy changes are measurable, because they are associated with very real physical effects, such as heating, dissociation, or luminescence (or explosion!).

Prerelativistic physics is not cognizant of the internal energy (*rest energy*) associated with proper mass, and it permits an arbitrary constant to be added to the total energy of a system without its making any observable difference. In relativity the arbitrariness is removed, with zero energy corresponding to no particles and no fields. The total energy of a system can be determined, for example, by measuring its inertial mass, and using

$$E = mc^2 \tag{1-3a}$$

(There is no arbitrary constant in the ratio of momentum to velocity!)

Energy is tied to momentum, being, to within a constant factor, the timelike component of the four-momentum. It is essentially on an equal footing with momentum. Both are conserved absolutely in all interactions, elastic or inelastic. Position–time (*event*) and momentum–energy (*four-momentum*) are the fundamental objects of relativistic mechanics.

1-2. INERTIAL REFERENCE FRAMES

Deep down in our Newtonian bones we understand by an inertial frame a frame of reference that is at rest or in uniform motion with respect to absolute space. Absolute space is the imagined framework in which bodies move. "Absolute Space, in its own nature, without relation to anything external,... always similar and immoveable" ([1], Scholium to Definitions). It has no direct physical significance. One body can only be located relative to another: the earth relative to the sun; the sun relative to nearby stars; the nearby stars relative to the globular clusters; and so on. In this infinite regress, one never grasps absolute space. (In Newton's words further on in the same discussion, "It may be that there is no body really at rest, to which the places and motions of others may be referred.") Experiment only reveals relative motion.[1]

---

[1] The student should not jump to the conclusion that absolute motion, being undetectable, should for that reason be banished from physics. As a mathematical science, physics makes use of concepts that have only remote connections with laboratory manipulations. All that we have a right to demand is that the concepts fit together in a logical scheme that agrees with all experiments. The profound philosophical objections of Leibniz to the idea of absolute space would have continued to fall on deaf ears were it not for the experimental results demonstrating a breakdown of Newtonian mechanics.

The defining property of an inertial frame is that in it a body moves with constant velocity if and only if the net force on it is zero ([1], Laws I and II). It would appear to be an easy matter to observe the motion of a test body in one's laboratory under this condition. One could arrange for it not to be in contact with anything else and to be free from aerodynamic forces. One could shield it against electromagnetic forces. But there is no shielding against gravity. Going far away from concentrations of matter helps. But the gravitational force is of very long range. For example, our weight is due more to matter deep down in the earth than to nearby buildings. And how can one be sure that there is not far out in space some huge gravitating mass exerting a strong pull on the test body? If this were the case, the laboratory would be subject to this attraction too. If no other outside force acted on it (free fall), its acceleration would be the same as that of the test body, and the test body would show no acceleration with respect to the laboratory. A successful experiment demonstrating zero change of velocity of the test body can be interpreted in either of two ways: (1) the gravitational field in the laboratory is zero, so the net force on the test body is zero. The laboratory is part of an inertial reference frame; (2) the gravitational field in the laboratory is not zero, so the net force on the test body is not zero. The laboratory is in free fall in this external gravitational field.

In practice, one makes comparisons with observations at other times and places and requires of the interpretation that the postulated gravitational field change in a reasonable way. Newton's first law is an affirmation of the existence of inertial frames: Frames exist in which no inconsistency results from saying that a body moving with constant velocity is acted on by zero net force. There remains an ambiguity involving very distant masses: Is our whole set of observable galaxies falling freely in an external gravitational field? The ambiguity is of no practical importance, but it does make it impossible for us to realize physically an inertial frame.

One might hope to identify inertial frames by another property: In inertial frames, the law of motion of a particle [Eq. (1-5)] takes on the simplest possible form, in the sense that the expression for **F** is free from certain additive terms characteristic of frames accelerated with respect to inertial frames. The extra terms, called *inertial forces* (Section 1-5), can be sorted into several due to rotation of the frame, and one due to translational acceleration. The terms due to rotation of the frame can be identified, we shall see, by their characteristic dependence on position and velocity. Thus the centrifugal term gives rise to a radially outward force proportional to the square of the angular velocity and the first power of the distance from the axis; the Coriolis term gives rise to a force proportional to the vector product of angular velocity and particle velocity. These terms have been identified for frames attached to the spinning earth, agreeing with predicted values when the angular velocity of the earth relative to the fixed stars[2] is inserted in the formulas. We conclude that the fixed star frame is not rotating with respect to inertial frames, and that the rotation of an arbitrary frame with respect to inertial frames ("absolute" rotation) can be unambiguously measured. The inertial force term due to the translational acceleration of the noninertial frame is, however, of exactly the same form as the force due to a homogeneous gravitational field, as would be produced by bodies at distances large compared with the dimensions of the region in which the particle moves. Thus the ambiguity persists. We cannot distinguish between a situation in which there is no gravitational field due to distant bodies and the reference frame is inertial, and a situation in which there is a uniform gravitational field and the reference frame is falling freely in it.

This ambiguity serves as the starting point for Einstein's theory of gravitation (general relativity). It is interpreted as the *principle of equivalence* (Chapter 6), emerging as a powerful tool for predicting physical effects.

From the point of view of Newtonian mechanics, the ambiguity is harmless, of concern only to metaphysicians. It does not affect the equations of motion in the fixed star frame, and no contradiction with experiment results from assuming that the fixed star frame is inertial. By the same token, one can perfectly well do without the ghostlike concept of the inertial frame and use instead the fixed star frame.

As an adequate and realistic replacement for the ever-receding inertial frame, we define the *local inertial frame*. A local inertial frame is a reference frame in every point of which a test body shielded from all external influences has zero acceleration. It is, therefore, characterized by weightlessness and the absence of centrifugal and other inertial forces. A local inertial frame is in free fall in the prevailing gravitational field, and the internal gravitational field due to matter belonging to the frame is zero. Furthermore, the frame has zero spin relative to the fixed stars. An orbiting spacecraft without spin and free of drag could serve as the base for a local inertial frame. A local inertial frame has finite extent, its size depending on the specific problem. The boundary is where inhomogeneity of the gravitational field becomes appreciable. For an earth satellite, the boundary is reached where the gravitational field of the earth becomes appreciably different from its value in the cabin. Inside this region, a bullet fired from the cabin moves with constant velocity relative to the cabin.

All but a few physics experiments have been carried out on the surface of a natural satellite of the sun, falling freely in the gravitational field and

---

[2] The fixed stars is a traditional name for the stars regarded as the basis for a frame of reference. Astronomers record the motion of many stars relative to the earth, and compute then a rate of precession and nutation of the earth, a velocity of the solar system, and a rate of rotation of the galaxy, which together make the corrected star motion average to zero. This fit determines the "fixed star" reference frame. The velocity of the stars near the sun with respect to the galactic center is about 200 km sec^{-1} on a radius of about 2.5×10^{22} cm (see [4]). This figure corresponds to a centripetal acceleration (v^2/ρ) on the order of 10^{-8} cm sec^{-2}.

spinning about an axis pointed toward Polaris with an angular velocity of

$$\frac{2\pi}{24 \times 60 \times 60} = 7.27 \times 10^{-5} \text{ rad sec}^{-1}$$

with respect to the fixed stars. The laboratory frame in such cases is not local inertial. A test body in the laboratory does not remain at rest. It is accelerated downward with an "acceleration of gravity" determined by the earth's mass distribution and its rate of spin. Further on (Section 1-5), we shall calculate this acceleration in order to find out how to compare terrestrial experiments with a theory concerned only with local inertial frames.

Inasmuch as the gravitational field varies from place to place in the cosmos, local inertial frames in different regions are not in uniform motion with respect to one another. The spacecraft is accelerated with respect to the center of the earth, the earth with respect to the center of the sun, and so on. In special relativity we concern ourselves exclusively with transformations between members of a given family of local inertial frames, in uniform motion with respect to one another. They belong to a particular region of spacetime. One can think of a host of freely falling laboratories having the same acceleration but different velocities. Hereafter we shall mean by the phrase "inertial frame" a *local* inertial frame in a particular region of the universe at a particular stage in its history.

1-3. COORDINATE TRANSFORMATIONS WITHIN A REFERENCE FRAME

Covariance

In a particular reference frame there are infinitely many possible choices of origin and direction of Cartesian axes. Let us now consider how the terms in our equations change when we go from one such coordinate system as basis to another. If a quantity does not change value under a transformation, it is said to be invariant with respect to that transformation. If two quantities change according to the same law, they are covariant with respect to the transformation. If the two sides of an equation are covariant, the equation is said to be covariant. If it is valid in one coordinate system, it is then valid in every coordinate system reached by the transformation in question. Evidently it is necessary that every additive term in the equation transforms in the same way. We shall shortly consider illustrations of these statements.

At first sight it would seem that the choice of coordinate system is entirely a matter of convenience, and that, therefore, we would be justified in requiring general covariance of our equations with respect to all coordinate transformations that stay in the same reference frame. A little thought shows that this seemingly obvious and trivial requirement is not necessarily valid. It is a matter for experimental test.

Determining whether the position of the origin makes any difference is the same as determining whether our experiments proceed in exactly the same way when carried out at different places. (It is, of course, essential that everything involved in the experiment be shifted, not just part of the material.) A shift of the laboratory, the coordinate system remaining fixed, is equivalent to an opposite shift of the coordinate system with the laboratory remaining fixed.

If the properties of space are different in different places, different locations of the origin will require different forms of the equations. They will not be covariant with respect to displacement of the origin. If the properties of space are different in different directions, space having some built-in directionality, a rotation of the laboratory along with everything affecting the experiment will change the experimental results; correspondingly, different orientations of the coordinate axes in the same laboratory will require different forms of the equations. The equations will not be covariant with respect to rotation of the axes.

Homogeneity and Isotropy of Space

Indifference to location of the origin is called homogeneity of space. Indifference to direction of the axes is called isotropy of space.

> In general relativity space is neither homogeneous nor isotropic, a gravitational field being represented as a curvature of spacetime. This theory differs only minutely in practical consequences from Newton's theory of gravitation, in which the space of an inertial reference frame is homogeneous and isotropic. Only in certain special astronomical situations in which very intense gravitational fields are involved does general relativity predict a detectable departure from Newtonian theory. In this discussion we shall not consider those situations.

Many experiments have been carried out seeking a measurable effect of the earth's velocity relative to the galaxy on the outcome of a terrestrial experiment. These "ether drift" experiments have all given null results. This fact can be considered evidence for homogeneity and isotropy of space over a region of dimensions on the order of the distance traveled by the earth in one day (10^6 km).

A very clear experiment of this type has been carried out by R. W. P. Drever [5]. He looked for fine structure of the nuclear magnetic resonance line for lithium 7 in the ground state as a function of time of day. The line results from the Zeeman effect on the spin $3/2(I = 3/2)$ lithium-7 nucleus in a constant magnetic field; equal spacing of the space-quantized $I_z = 3/2$, $1/2$, $-1/2$, and $-3/2$ levels makes the frequencies of the $|\Delta I_z| = 1$ transitions within the multiplet coincide, giving the narrow line shape observed. As

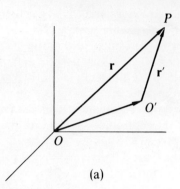

(a)

Figure 1-7a.

these levels correspond to different spatial orientations of the nucleus, an anisotropy of space could shift the $I_z = \pm 3/2$ and the $I_z = \pm 1/2$ levels by different amounts, thereby giving rise to a splitting of the line into a triplet. No evidence for such a splitting was found, at any time of day. It is difficult to put a quantitative interpretation on the negative result—we know too little about the mass distribution in the universe and about the effect of spatial anisotropy on nuclear energy levels— but it is impressive that any level shift, if present, must be less than 2×10^{-16} eV, corresponding to a frequency shift of 0.05 Hz.

The homogeneity of space is a simple requirement mathematically. All but one of the dynamical variables we use (velocity, momentum, mass, and so on) are defined independently of the origin of coordinates. They involve only *changes* of position vectors. The one exception is the position variable itself, the vector **r** drawn from the origin to the particle in question (Fig. 1-7a). Shifting the origin from O to O' changes **r** to the unequal vector **r**′

$$\mathbf{r}' = \mathbf{r} - \overrightarrow{OO'} \tag{1-7}$$

But the difference of two position vectors, $\mathbf{r}_2 - \mathbf{r}_1$, is unaffected by the shift of origin (Fig. 1-7b), for both $\mathbf{r}_1$ and $\mathbf{r}_2$ are changed by the same additive vector $\overrightarrow{OO'}$

$$\mathbf{r}_2' - \mathbf{r}_1' = (\mathbf{r}_2 - \overrightarrow{OO'}) - (\mathbf{r}_1 - \overrightarrow{OO'}) = \mathbf{r}_2 - \mathbf{r}_1$$

In particular,

$$\mathbf{v} = \frac{d\mathbf{r}}{dt} = \lim_{t_2 \to t_1} \frac{\mathbf{r}(t_2) - \mathbf{r}(t_1)}{t_2 - t_1}$$

is invariant with respect to shift of origin. Thus, the momentum $\mathbf{p} = m\mathbf{v}$ [Eq. (1-1)] is independent of the origin. Homogeneity of space requires then that in the equation of motion of a particle

$$\mathbf{F} = \frac{d\mathbf{p}}{dt} \tag{1-5}$$

the force $\mathbf{F}$ involves the position $\mathbf{r}$ only through its difference with some other position vector. The force may depend on position relative to other particles, but not on position relative to the coordinate system. All known

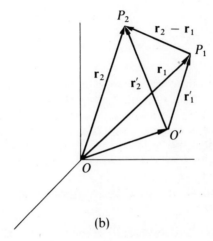

(b)

Figure 1-7b.

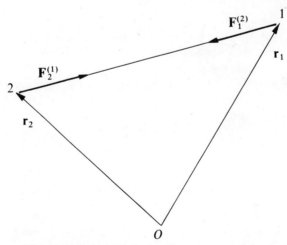

Figure 1-8.   The Newtonian attraction between particles 1 and 2 depends only on their distance apart.

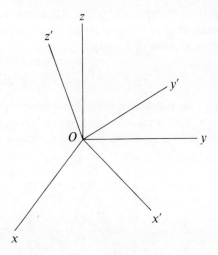

Figure 1-9.

interactions meet this requirement. For example, the Newtonian gravitational force exerted by particle 2 on particle 1 (Fig. 1-8),

$$\mathbf{F}_1^{(2)} = m_1 \cdot G \frac{m_2}{|\mathbf{r}_2 - \mathbf{r}_1|^3}(\mathbf{r}_2 - \mathbf{r}_1) \qquad (1\text{-}8)$$

depends only on the relative positions of the two particles.

The mathematical implications of isotropy are somewhat more involved. Let us consider the way in which different dynamical quantities are affected by a rotation of the axes.[3]

Quantities that are invariant under rotation of the axes are called *scalars*. An example is inertial mass, the numerical value of which is completely independent of the way the axes are oriented. Another is the distance between two points. Although it can be expressed in terms of the coordinates of the points, its value is indifferent to the orientation of the axes. If the primed coordinate system $Ox'y'z'$ is obtainable from the unprimed one $Oxyz$ by a rigid rotation (Fig. 1-9), the distance between any points 1 and 2 is

$$r_{12} = [(x_1 - x_2)^2 + (y_1 - y_2)^2 + (z_1 - z_2)^2]^{1/2}$$

$$= [(x_1' - x_2')^2 + (y_1' - y_2')^2 + (z_1' - z_2')^2]^{1/2} \qquad (1\text{-}9)$$

Vectors are quantities that can be put in one-to-one correspondence with directed line segments (arrows). The projections of a directed line segment

---

[3] Readers unfamiliar with the behavior of vectors and tensors under rotation and inversion should now study Appendix A.

on the axes change in a definite way when the axes are rotated. The components of a vector change in the same way. We can describe the rotation of axes by giving the cosines of the angles between the new and old axes. Shifting from x, y, z to the algebraically more convenient x_1, x_2, x_3 (Fig. 1-10), we denote by a_{ij} the cosine of the angle between the ith primed axis and the jth unprimed axis. Thus,

$$a_{11} = \cos(Ox'_1, Ox_1) = \mathbf{i}'_1 \cdot \mathbf{i}_1 \qquad a_{12} = \cos(Ox'_1, Ox_2) = \mathbf{i}'_1 \cdot \mathbf{i}_2 \qquad \text{etc.}$$

The components A_1, A_2, A_3 of the vector $\mathbf{A}$ in the unprimed coordinate system are related to its components A'_1, A'_2, A'_3 in the primed coordinate system by the transformation law

$$A'_1 = a_{11}A_1 + a_{12}A_2 + a_{13}A_3$$
$$A'_2 = a_{21}A_1 + a_{22}A_2 + a_{23}A_3 \qquad\qquad (1\text{-}10)$$
$$A'_3 = a_{31}A_1 + a_{32}A_2 + a_{33}A_3$$

These equations can be abbreviated by the use of running indices

$$A'_j = \sum_{k=1}^{3} a_{jk}A_k \qquad j = 1, 2, 3 \qquad\qquad (1\text{-}10)$$

and shortened still further by the Einstein summation convention (if an index appears twice in a product, the expression is to be summed over the range of the index) to

$$A'_j = a_{jk}A_k \qquad j, k = 1, 2, 3 \qquad\qquad (1\text{-}10)$$

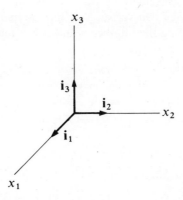

Figure 1-10.

Only if a triple of numbers transforms in this way are they the components of a vector quantity.

The 3×3 array of direction cosines a_{jk} is called the rotation matrix, a.

$$a = \begin{pmatrix} a_{11} & a_{12} & a_{13} \\ a_{21} & a_{22} & a_{23} \\ a_{31} & a_{32} & a_{33} \end{pmatrix}$$

It specifies by Eq. (1-10) how the components of a vector change when the axes are given the indicated rotation. Matrix terminology will prove useful when we consider successive coordinate transformations.

The scalar product of two vectors

$$\mathbf{A} \cdot \mathbf{B} = A_j B_j$$

is invariant

$$A_j B_j = A'_k B'_k$$

and thus is properly named. We call the square root of the scalar product of a vector with itself the *norm* of the vector; it is its length.

Vectors and scalars do not exhaust the list of useful quantities. More complicated geometrical entities called *tensors* are also needed. A tensor is something whose components in a particular coordinate system have two indices and transform as

$$T'_{jk} = a_{jl} a_{km} T_{lm} \qquad (1\text{-}11\text{c})$$

Evidently we have here something transforming like the nine products of the three components of two vectors. We have actually defined a tensor of rank two, the rank being the number of indices. A tensor of third rank has three indices on its components, and

$$T'_{jkl} = a_{jm} a_{kn} a_{lp} T_{mnp} \qquad (1\text{-}11\text{d})$$

A vector is evidently a tensor of rank one

$$T'_j = a_{jk} T_k \qquad (1\text{-}11\text{b})$$

and a scalar is a tensor of rank zero

$$T' = T \qquad (1\text{-}11\text{a})$$

The requirement of isotropy is expressed by requiring covariance of our equations with respect to rotation of the axes. Every additive term in our equations must transform according to the same law ("have the same tensorial character"). For example, a scalar cannot be equated to a component of a vector, nor can one term in a sum be a tensor of second rank while another one is a vector. Such equations could not hold in an arbitrarily rotated coordinate system.

The handedness of a coordinate system is determined by the order in which the axes happen to be labeled. If curling the fingers of the right hand from positive axis 1 to positive axis 2 makes the thumb point in the sense of positive axis 3, the coordinate system is called right handed; otherwise it is left handed.

Reversing one or three axes changes the handedness. The latter transformation

$$x'_1 = -x_1$$

$$x'_2 = -x_2$$

$$x'_3 = -x_3$$

is called *inversion*. An arbitrary transformation from one Cartesian coordinate system to another with the same origin and unit of length can be realized by an inversion (if the handednesses are different) and a rotation.

Inversion is, like rotation, a linear homogeneous transformation [Eq. (1-10)]. The matrix of direction cosines between new and old axes is

$$\mathbf{a} = \begin{pmatrix} -1 & 0 & 0 \\ 0 & -1 & 0 \\ 0 & 0 & -1 \end{pmatrix}$$

or

$$a_{jk} = -\delta_{jk} \tag{1-12}$$

(δ_{jk} is the Kronecker delta, equal to zero if $j \neq k$ and to one if $j = k$.) The definitions of the various ranks of tensor are extended to include inversion as well as rotation.

A scalar is, of course, unchanged in an inversion. The components of a vector are reversed in sign

$$A'_j = -\delta_{jk}A_k = -A_j$$

while those of a tensor of second rank are unchanged because there are two minus signs:

$$B'_{jk} = (-\delta_{jl})(-\delta_{km})B_{lm} = (-)(-)B_{jk} = B_{jk}$$

It is convenient to make use of *pseudotensors*. These are tensors as far as rotation of axes is concerned but under inversion they transform oppositely to tensors. Formally, the transformation law of a pseudotensor has an extra factor, the determinant of the transformation matrix:

$$P'_{jkl} = |a| \cdot a_{jm}a_{kn}a_{lp}P_{mnp} \tag{1-13d}$$

$$P'_{jk} = |a| \cdot a_{jl}a_{km}P_{lm} \tag{1-13c}$$

$$P'_j = |a| \cdot a_{jk}P_k \tag{1-13b}$$

$$P' = |a| \cdot P \tag{1-13a}$$

The determinant is $+1$ for a rotation, -1 for an inversion. A pseudoscalar [Eq. (1-13a)] keeps the same value when the axes are rotated but changes sign when the handedness is changed. A pseudovector [Eq. (1-13b)] has components that transform under rotation like those of a vector [Eqs. (1-11b), (1-10)] but do not change under inversion. A pseudovector can be represented by an arrow but an inversion of axes requires reversal of the arrow (Fig. 1-11). When the distinction between vectors and pseudovectors is significant, they are often referred to as *polar* and *axial* vectors,

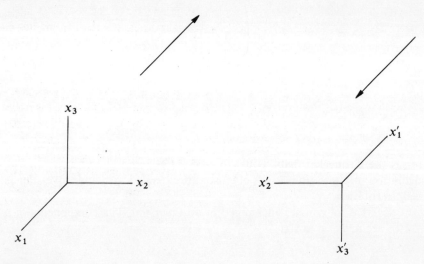

Figure 1-11.

respectively. The components of a pseudotensor of rank two transform like those of a tensor [Eq. (1-11c)] under rotation but change sign under inversion.

An example of a pseudovector is the vector product of two vectors,

$$\mathbf{C} = \mathbf{A} \times \mathbf{B}$$

meaning

$$C_1 = A_2 B_3 - A_3 B_2$$
$$C_2 = A_3 B_1 - A_1 B_3 \qquad (1\text{-}14)$$
$$C_3 = A_1 B_2 - A_2 B_1$$

or, abbreviated,

$$C_j = \varepsilon_{jkl} A_k B_l \qquad (1\text{-}14)$$

where ε_{jkl} is the Levi–Civita symbol, equal to zero whenever any two of the indices are equal and to $+1$ or -1 when jkl is an even or odd permutation of 1, 2, 3. Since inversion reverses the components of both A and B, those of C are unchanged. It can be shown (worked exercise, Appendix A) that the components of $\mathbf{C}$ transform under rotation according to the vector law [Eqs. (1-11b), (1-10)]. It follows from Eq. (1-14) that one must use the right-hand rule (curl the fingers from $\mathbf{A}$ to $\mathbf{B}$, the thumb points along $\mathbf{C}$) to determine the sense of $\mathbf{C}$'s arrow if the coordinate system is right handed, the left-hand rule if the coordinate system is left handed. An important pseudo-vector in mechanics is the *angular momentum*, or moment of momentum, or spin

$$\mathbf{L} = \mathbf{r} \times \mathbf{p}$$

There is unfortunately no notation in general use to identify pseudo-quantities. One is expected just to know whether a particular vector is polar or axial. A sure indication that it is axial is the presence of an odd number of hand rules in the specification of the associated directed line segment.

Screws do exist. Right-handed sugar behaves differently from left-handed sugar. Nevertheless, one can describe the real world just as well with left-handed coordinate axes as with right. There is no reason to prefer one order of labeling axes over another; therefore, we require general covariance of our equations with respect to inversion. One does not add a polar to an axial vector, or a pseudoscalar to a scalar.

The homogeneity and isotropy of the space of an inertial reference frame require that our equations be covariant under shift of origin, rotation of

axes, and inversion of axes. All three transformations leave invariant the
expression [Eq. (1-9)] for the distance between two points. With Δx_1, Δx_2,
and Δx_3 representing the differences of corresponding coordinates, the
line element

$$(\Delta x_1)^2 + (\Delta x_2)^2 + (\Delta x_3)^2$$

is invariant.

1-4. TRANSFORMATIONS BETWEEN INERTIAL FRAMES: SPECIAL RELATIVITY

Hitherto we have compared different coordinate systems belonging to
the same reference frame. This is equivalent to comparing results found in
different laboratories that are rigidly attached to one another. We shall now
consider transformations between coordinate systems belonging to different
reference frames. The reference frames refer to different rigid bodies—such
as, for example, the earth's crust, an orbiting satellite, the fixed stars. We
are now interested in the effect of the relative motion of two rigid bodies on
the description of events in terms of these two as bases.

The Principle of Special Relativity

The simplest relative motion is that of pure translation: Every point of the
body space R' has the same displacement with respect to R in the same R–time

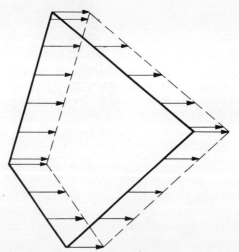

Figure 1-12.   Translation of a rigid body. Each point has the same
displacement.

(Fig. 1-12). When the rate of displacement is constant,

$$\mathbf{u} = \lim \frac{\overrightarrow{PP'}}{\Delta t} = \text{const}$$

we have uniform translation. Then, if R is inertial, R' is inertial, and conversely.

If R and R' are inertial, it is a very important and striking experimental fact that the description of nature is not at all affected by the transformation from one to the other. That is, *the laws of physics have the same form when described in two inertial reference frames. Furthermore, the constants appearing in the equations have the same numerical values. This is the principle of special (or restricted) relativity.*

If we perform an experiment in our laboratory in which we measure, for example, the mass of some ion, and if some other physicists carry out the same experiment in another laboratory that is in uniform motion of translation with respect to ours, they will observe the same deflections and time intervals and derive the same answer as we. Steady motion of a laboratory has no effect on the course of events inside it. This is a fact of everyday experience. When one is in an automobile and is not looking outside, one only notices change of speed or direction. When the motion of the vehicle is really uniform, inside it is just as if it were at rest.

The principle of special relativity was clearly formulated by Galileo, and reaffirmed by Newton. [Newton writes ([1], Book I, Corollary V), "The motions of bodies included in a given space are the same among themselves, whether that space is at rest, or moves uniformly forward in a right line without any circular motion." He then comments, "A clear proof of this we have from the experiment of a ship, where all motions happen after the same manner, whether the ship is at rest, or is carried uniformly forwards in a right line."] Einstein affirmed its validity for all physical phenomena, including those involving the electromagnetic field. As far as particle mechanics itself is concerned, it is present in Newton's formulation.

The experiments have actually involved comparisons between the earth crust frame and frames in uniform motion with respect to it. As we shall see, the noninertial character of these frames does not affect the conclusion stated. It does give rise to observable effects that must be allowed for in the analysis. Experiments inside a laboratory do show effects of its *acceleration* with respect to an inertial frame in the same locality.

Newton carried out a simple experiment to show the intrinsic detectability of rotation ([1], Scholium to Definitions). He hung a bucket of water by a rope and twisted the rope. Before the bucket was released, the water surface was flat (Fig. 1-13a). When it was released, the bucket spun around and the water adopted the characteristic parabolic profile as it took up the motion of

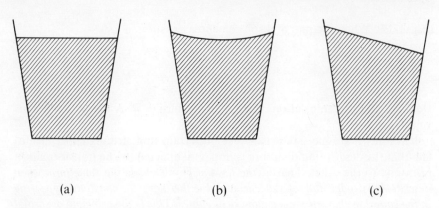

(a)   (b)   (c)

Figure 1-13.   Newton's water bucket accelerometer.

the bucket (Fig. 1-13b). This deformation results from the extra force associated with rotation, in this case called "centrifugal." By observing the shape of the water surface we can ascertain whether the water is rotating with respect to inertial frames. We note that accelerated translational motion of the bucket, for example to the right, would not curve the water surface but would tip it (Fig. 1-13c), as if the gravitational field of the earth had changed direction. With uniform rectilinear motion of the bucket and water, however, the surface retains its original shape and position; there is no way of detecting the motion by observations within the bucket.

The Galileo Transformation

Let us now examine the covariance of Newtonian mechanics with respect to uniform translation of the reference frame. Frame R is the "fixed" or "laboratory" frame. In it we have a system of standard measuring rods and clocks, permitting us to locate an event in space and time. We choose an origin O for a Cartesian coordinate system, and a set of axes Ox, Oy, Oz. The space coordinates (x, y, z) and the time t of an event can then be determined. Frame R' is the "moving" frame. In it we have an identical system of standard measuring rods and clocks. We pick an origin O' and axes $O'x', O'y', O'z'$ (Fig. 1-14). The same event specified by (x, y, z) and t in R is specified in R' by (x', y', z') and t'.

The Galileo transformation, which was accepted without question until the beginning of the twentieth century, states first that the coordinates in R and R' are related by the law of vector addition in a common space (Fig. 1-14)

$$\mathbf{r} = \overrightarrow{OO'} + \mathbf{r}' \qquad (1\text{-}7)$$

where, of course,

$$\mathbf{r} = \mathbf{i}x + \mathbf{j}y + \mathbf{k}z \qquad \mathbf{r}' = \mathbf{i}'x' + \mathbf{j}'y' + \mathbf{k}'z' \tag{1-15}$$

and second that

$$t = t' \tag{1-16}$$

These assumptions are simply the common sense view that relative motion does not affect geometry or timekeeping; measuring rods are not distorted nor are clocks slowed down or speeded up. In particular, two events that occur at the same time in one frame are also simultaneous in every other frame.

We specialize now to uniform translation of R' with respect to R. With $\mathbf{u}$ representing the velocity of R' with respect to R,

$$\overrightarrow{OO'} = \mathbf{u}t + \boldsymbol{\alpha} \tag{1-17}$$

where $\boldsymbol{\alpha}$ is the value of $\overrightarrow{OO'}$ at $t = t' = 0$. Let the event in question be the location of a particle at the point P. Differentiating Eq. (1-7), now written

$$\mathbf{r} = \mathbf{u}t + \boldsymbol{\alpha} + \mathbf{r}' \tag{1-18}$$

with respect to time, we find for the velocity of the particle

$$\frac{d\mathbf{r}}{dt} = \mathbf{u} + \frac{d\mathbf{r}'}{dt}$$

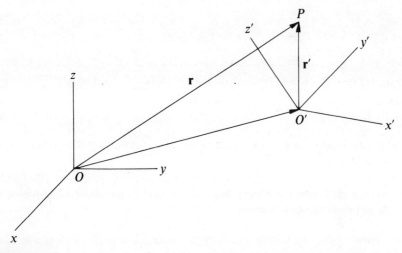

Figure 1-14.

which becomes, by Eq. (1-16),

$$\frac{d\mathbf{r}}{dt} = \mathbf{u} + \frac{d\mathbf{r}'}{dt'} \tag{1-19}$$

Since the basis vectors do not change orientation with time, they are constants in the differentiation of Eqs. (1-15). Thus,

$$\frac{d\mathbf{r}}{dt} = \mathbf{i}\frac{dx}{dt} + \mathbf{j}\frac{dy}{dt} + \mathbf{k}\frac{dz}{dt} \qquad \frac{d\mathbf{r}'}{dt'} = \mathbf{i}'\frac{dx'}{dt'} + \mathbf{j}'\frac{dy'}{dt'} + \mathbf{k}'\frac{dz'}{dt'} \tag{1-20}$$

The right-hand members are, respectively, the velocity of the particle with respect to R, $\mathbf{v}$, and the velocity of the particle with respect to R', $\mathbf{v}'$. Equation (1-19) becomes

$$\mathbf{v} = \mathbf{u} + \mathbf{v}' \tag{1-21}$$

This equation is just the Galilean law of addition of velocities, which states that the velocity of P with respect to R is the vector sum of its velocity with respect to R' and the velocity of R' with respect to R.

Differentiating Eq. (1-21), we find for the acceleration of the particle

$$\frac{d\mathbf{v}}{dt} = 0 + \frac{d\mathbf{v}'}{dt} = \frac{d\mathbf{v}'}{dt'} \tag{1-22}$$

Again,

$$\frac{d\mathbf{v}}{dt} = \mathbf{i}\frac{d^2x}{dt^2} + \mathbf{j}\frac{d^2y}{dt^2} + \mathbf{k}\frac{d^2z}{dt^2} \qquad \frac{d\mathbf{v}'}{dt'} = \mathbf{i}'\frac{d^2x'}{dt'^2} + \mathbf{j}'\frac{d^2y'}{dt'^2} + \mathbf{k}'\frac{d^2z'}{dt'^2} \tag{1-23}$$

The right-hand members are, respectively, the acceleration of P with respect to R, $\mathbf{a}$, and the acceleration of P with respect to R', $\mathbf{a}'$. Equation (1-22) becomes

$$\mathbf{a} = \mathbf{a}' \tag{1-24}$$

The acceleration of a particle is invariant under uniform translation of the reference frame.

Covariance of Newtonian Mechanics with Respect to the Galileo Transformation Between Inertial Frames

In Newtonian mechanics the inertial mass of a particle is a constant independent of the particle's velocity relative to the reference frame. This can be shown to be a necessary consequence of the law of conservation of momentum and the law of addition of velocities

$$\mathbf{v} = \mathbf{u} + \mathbf{v}' \qquad (1\text{-}21)$$

that we have just derived from the Galileo transformation for uniform translation. (The proof is given in Section 3-3.)

It follows that the inertial mass is an invariant with respect to transformations between inertial frames:

$$m = m' \qquad (1\text{-}25)$$

This result can be seen by analysis of the procedure used for measuring inertial mass. It consists essentially in comparing the velocity change with that of a standard body of unit inertial mass experiencing the same momentum change. The two bodies might be on a frictionless horizontal table (a cushion of gaseous carbon dioxide or compressed air provides essentially frictionless support), held together by a string against the force of a spring that is pushing them apart (Fig. 1-15). Initially, they are at rest; the momentum is zero. The string is burned away, and the bodies recoil from one another. The conservation of momentum reads

$$0 = mv_x + {}^s m \, {}^s v_x \qquad (1\text{-}26)$$

where the left superscript s refers to the standard body, giving

$$\frac{m}{{}^s m} = \left| \frac{{}^s v_x}{v_x} \right|$$

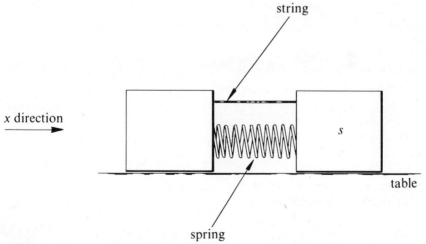

string

x direction
⟶

s

table

spring

Figure 1-15.   Lecture table measurement of inertial mass.

Measurement of the recoil velocities gives the inertial mass of the unknown in terms of that of the standard body. Let the x axis be along the direction of recoil of the standard body. Now let us describe the interaction in R' moving with speed u in the x direction. Before the bodies spring apart, they are moving in R' as one with velocity $-u$ in the x' direction. The momentum is

$$-(m' + {}^sm')u$$

Afterward, the momentum is

$$m'v_{x'} + {}^sm'{}^sv_{x'}$$

Conservation of momentum requires

$$-(m' + {}^sm')u = m'v_{x'} + {}^sm'{}^sv_{x'}$$

or

$$0 = m'(u + v_{x'}) + {}^sm(u + {}^sv_{x'}) \tag{1-27}$$

But the Galileo law of velocity addition, Eq. (1-21), gives

$$v_x = u + v_{x'} \qquad {}^sv_x = u + {}^sv_{x'}$$

so that Eq. (1-27) becomes

$$0 = m'v_x + {}^sm'{}^sv_x$$

Comparing with Eq. (1-26), we see that

$$\frac{m'}{{}^sm'} = \frac{m}{{}^sm}$$

A 2.0-kg body in R is a 2.0-kg body in R'!

With the inertial mass independent of the particle's velocity in a particular frame, the rate of change of momentum is equal to inertial mass times acceleration:

$$\frac{d\mathbf{p}}{dt} = \frac{d}{dt}(m\mathbf{v}) = m\frac{d\mathbf{v}}{dt} = m\mathbf{a}$$

With $m = m'$ [Eq. (1-25)] and $\mathbf{a} = \mathbf{a}'$, the product $m\mathbf{a}$ is invariant. The non-relativistic equation of motion

$$\mathbf{F} = m\mathbf{a} \tag{1-28}$$

will hold in all inertial frames if and only if $\mathbf{F} = \mathbf{F}'$. If the interactions in nature are such as to make the force on a particle the same in all inertial frames, the Newtonian equation of motion is covariant with respect to the Galileo transformation between inertial frames. Such *Galilean covariance* means that the law of motion, the Galileo transformation, and the principle of special relativity are consistent.

Galilean Invariance of Instantaneous Action at a Distance Force

The Galilean covariance of the Newtonian law of motion depends on the particular kind of interaction described by $\mathbf{F}$ in $\mathbf{F} = m\mathbf{a}$.

The contact forces between macroscopic bodies, such as the friction and pressure between members of a machine, are certainly invariant with respect to uniform translation. These pushes and pulls are the resultant of forces between neighboring atoms. They are basically electromagnetic and in most cases are essentially electrostatic or magnetostatic.

The Newtonian law of gravitation, in which the force is inversely proportional to the square of the separation and along the line joining the particles (Fig. 1-8),

$$\mathbf{F}_1^{(2)} = m_1 \frac{Gm_2}{|\mathbf{r}_2 - \mathbf{r}_1|^3}(\mathbf{r}_2 - \mathbf{r}_1) \tag{1-8}$$

gives an invariant force, because in the Galileo transformation [Eqs. (1-7) and (1-16)],

$$\mathbf{r}_2 - \mathbf{r}_1 = \mathbf{r}_2' - \mathbf{r}_1'$$

The location of O' relative to O has no effect on the instantaneous relative position of two particles. The Coulomb law of interaction between electric charges at rest is of the same form[4]

$$\mathbf{F}_1^{(2)} = e_1 \cdot \frac{-e_2}{|\mathbf{r}_2 - \mathbf{r}_1|^3}(\mathbf{r}_2 - \mathbf{r}_1) \tag{1-29}$$

and has the same invariance. Clearly any force determined by the position of the particle relative to the positions of other particles at the same instant is invariant with respect to any motion of R' relative to R. The static interaction between magnetic moments is also of this character.

The Galileo transformation for uniform translation makes relative

---

[4] We use Gaussian units. The functional form is, of course, the same with any units.

velocity also invariant. By Eq. (1-21),

$$\mathbf{v}_2 - \mathbf{v}_1 = \mathbf{v}_2' - \mathbf{v}_1'$$

If the force acting on a particle is a function only of its instantaneous position and velocity relative to other particles, then

$$\mathbf{F} = \mathbf{F}'$$

as between frames in uniform relative translation.

Such forces constitute the overwhelming bulk of the interactions studied in engineering and astronomical mechanics. Their instantaneous character flies in the face of common sense, implying that a particle *here* knows without delay the position and velocity of a particle *there*. But so successful was this classical Newtonian mechanics of instantaneous forces that many physicists (e.g., Helmholtz [6]) were encouraged to believe that a complete description of nature might be possible in terms only of particles obeying the Newtonian equation of motion and interacting with one another according to simple force laws depending only on instantaneous relative position (and possibly relative velocity). Newton himself had not been so naïve. He regarded the inverse square law of gravitation, Eq. (1-8), as an empirical description: "Hitherto I have not been able to discover the cause of those properties of gravity from phenomena, and I frame no hypotheses" ([1], General Scholium). Privately, he speculated on ethereal pressure gradients as the cause of gravity (letter to Boyle), but he wisely refrained from publication. He shared his contemporaries' aversion to instantaneous action at a distance: "... that one body may act upon another at a distance through a *vacuum*, without the mediation of anything else ... is to me so great an absurdity that I believe no man who has in philosophical matters a competent faculty of thinking can ever fall into it" (letter to Bentley, February 25, 1692). But the success of Newton's description of gravitation aroused false hopes. As electromagnetic phenomena were unraveled in the course of the next two centuries, it became at last evident that *this* interaction is not instantaneous but is actually transmitted with a finite velocity—that of light in vacuum. The force acting on a charged particle is determined by the electromagnetic field at the particle, which is in turn determined by the motion of the field sources at various earlier times. The reality of the electromagnetic field makes it impossible to write the force $\mathbf{F}$ in the single-time form required by action at a distance.

Breakdown of Galilean Invariance in the Electromagnetic Interaction

A simple example of the impossibility of fitting electromagnetic phenomena into a scheme requiring Galilean invariance of force is provided by the case of two charged particles moving in the same direction (Fig. 1-16). These might be electrons in an evacuated tube. In addition to the Coulomb repulsion between them, there is also a magnetic attraction. Electron 1 moving to the right corresponds to a positive current to the left, producing a magnetic field curling around the current in the sense given by the right-hand rule. So there is at electron 2 a magnetic field coming out of the plane of the paper, of magnitude proportional to the velocity of electron 1 with respect to the laboratory v_1. This field exerts a side thrust on electron 2 of magnitude proportional to the field strength and the velocity of electron 2 with respect to the laboratory v_2. The force is straight upward. Similarly, there is an equal and opposite magnetic force on electron 1 directed straight downward. The interaction force is the resultant of this magnetic attraction and the Coulomb repulsion. The former is not determined by the relative positions and velocities of the electrons; it depends on their velocities relative to the laboratory—in fact, on the product $v_1 v_2$. If the two electrons have the same velocity, $v_1 = v_2$, then their relative velocity is zero. In this case, if R' is moving with the electrons there is certainly no magnetic attraction in it. But in the laboratory frame R there is a very real magnetic attraction. Thus,

$$\mathbf{F}' \neq \mathbf{F}$$

This dependence of the force on the velocity of the reference frame with respect to the charges appears to contradict the principle of special relativity. This is not necessarily the case. Our example shows only that there is an

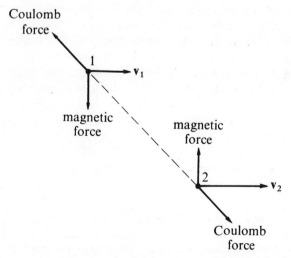

Figure 1-16.   Forces between two electrons moving in the same direction.

inconsistency between the experimentally confirmed laws of electromagnetism, the Newtonian equation of motion

$$\mathbf{F} = m\mathbf{a} \qquad (m = \text{const}) \qquad (1\text{-}28)$$

and the principle of special relativity using the Galileo transformation

$$t = t' \qquad (1\text{-}16)$$

$$\mathbf{r} = \mathbf{u}t + \boldsymbol{\alpha} + \mathbf{r}' \qquad (1\text{-}18)$$

The Lorentz Transformation

One conceivable way out of the dilemma is to give up the principle of special relativity. If this proved possible, we could say that there is indeed a privileged inertial frame, one in which we are supposed to calculate the electric and magnetic field. This frame was called the "luminiferous ether" or "world ether." It could also be identified with the absolute space in which Newton visualized matter as moving.

Many experiments have been carried out in efforts to detect this privileged frame. In these experiments one has tried to discover some effect of the velocity of the earth relative to other bodies on experiments carried out on the earth; they have all given negative results. All experiments agree with the original relativity principle of Galileo in the sense that there is no way of distinguishing whether the laboratory is in uniform motion or at rest with respect to the supposed privileged frame.

The solution to this problem was provided by Einstein in 1905 [2]. He saved the principle of special relativity by giving up the Galileo transformation. The new transformation equations for $\mathbf{r}$ and t, called the *Lorentz transformation*,[5] are such as to give the same experimental answers in all inertial frames even though force, inertial mass, acceleration, electric and magnetic field strength, and almost everything else are different. Einstein derived the new transformation by making a deep critique of the previously unanalyzed concepts of position and time, and developing a more realistic

---

[5] Einstein derived these "Lorentz transformation" formulas directly from fundamental considerations about clocks, measuring rods, and the equivalence of different inertial frames. Earlier, Larmor ([7], pp. 173–177) and then Lorentz [8] had used these equations in formally transforming the Maxwell field equations from the ether to a frame attached to a moving system of charged particles. Both regarded them as describing the effect on a material system of motion through the ether. In two papers essentially simultaneous with Einstein's, Poincaré [9] pointed out that this transformation, which he called the Lorentz transformation, reconciles the principle of special relativity and electrodynamics. His designation became established usage.

method of handling them. The changes do violence to our common sense ideas of space and time, but common sense is based on very limited experience.

We study the Lorentz transformation in detail in the next chapter. For the present, we merely state it. As in our presentation of the Galileo transformation [Eqs. (1-16) and (1-18)], **u** is the constant velocity of R' with respect to R. We select origins O' and O that coincide at an instant and measure both t' and t from that instant. For an event occurring at place **r** (components x, y, z) and epoch t in R, and at place **r**$'$ (components x', y', z') and epoch t' in R',

$$t' = \gamma \left[t - \frac{(\mathbf{r} \cdot \mathbf{u})}{c^2} \right] \tag{1-30a}$$

$$\mathbf{r}' = \mathbf{r} + (\gamma - 1)\frac{(\mathbf{r} \cdot \mathbf{u})}{u^2}\mathbf{u} - \gamma \mathbf{u} t \tag{1-30b}$$

where

$$\gamma = \frac{1}{\sqrt{(1 - u^2/c^2)}}$$

Equation (1-30b) relates the vectors **r**$'$ and **r** as plotted from their respective components in R' and R in an abstract three-dimensional position space. If in R the x axis is taken along the direction of motion of R' with respect to R, and in R' the x' axis is taken opposite to the direction of motion of R with respect to R', and the respective transverse axes are oriented so as to coincide at $t = t' = 0$, the Eqs. (1-30) reduce to the more familiar form

$$t' = \gamma \left(t - \frac{u}{c^2}x \right)$$

$$x' = \gamma(x - ut)$$

$$y' = y \tag{1-31}$$

$$z' = z$$

If the velocity of light c increases without limit, $c = \infty$, then $\gamma \to 1$, and Eqs. (1-30) reduce for finite r to

$$t' = t$$

$$\mathbf{r}' = \mathbf{r} - \mathbf{u}t \tag{1-32}$$

which is just the Galileo transformation, Eqs. (1-16) and (1-18). Thus the Galileo transformation is a limiting form of the Lorentz transformation and is expected to be a good approximation when the speed u is small compared with the speed of light. The speed of light—3×10^{10} cm sec^{-1}—is enormous on the scale of human size and reaction time, so it is no wonder that the Galileo transformation appears self-evident. For most terrestrial problems, the velocities involved are very small compared with c and there is no occasion to transform to a frame with large u. The Galileo transformation is then an excellent approximation for $u \ll c$. Anyone who tried to use relativistic transformation formulas in a study of, for instance, automobiles would be very foolish indeed.

The change in kinematics from the Galileo to the Lorentz transformation entails a corresponding change in dynamics. The Newtonian momentum–velocity relation

$$\mathbf{p} = m\mathbf{v} \qquad (m = \text{const})$$

which implies

$$\mathbf{F} = m\mathbf{a} \qquad (m = \text{const}) \tag{1-28}$$

is replaced by

$$\mathbf{p} = \frac{\mu}{\sqrt{1 - v^2/c^2}}\mathbf{v} \qquad (\mu = \text{const}) \tag{1-33}$$

The new equation includes the prerelativistic one as its low velocity limit. Experiment has confirmed the new relation [Eq. (1-33)], to very high precision. Classical electrodynamics is unaltered, finding in special relativity a particularly elegant formulation.

1-5. EARTHBOUND REFERENCE FRAME AND INERTIAL FRAMES

The effects of gravitation and of acceleration of the reference frame cannot be ignored, for our terrestrial laboratories are in a gravitational field and are spinning and orbiting with respect to celestial bodies. To compare our calculations for inertial frames with experiment, we must find out how to evaluate and, if necessary, correct for these effects. As the gravitational field near the edge of the earth is weak and the velocities involved are small (on the order of $10^{-4}c$), Newtonian mechanics and gravitation theory provide an adequate approximation for this purpose.

Inertial Forces

No longer limiting ourselves to uniform translation, we consider now the general motion of one rigid space R' with respect to another, R. We choose Cartesian axes in the two spaces, with origins at O' and O respectively.

The reader unfamiliar with the Galilean kinematics of accelerated reference frames should now study Appendix B. It is shown there, on the basis of the Galileo transformation,

$$t = t' \qquad (1\text{-}16)$$

$$\mathbf{r} = \mathbf{r}' + \overrightarrow{OO'} \qquad (1\text{-}7)$$

that the transformation equations for particle velocity and acceleration are

$$\mathbf{v} = \mathbf{u} + \mathbf{v}' + \boldsymbol{\omega} \times \mathbf{r}' \qquad (1\text{-}34)$$

$$\mathbf{a} = \mathbf{a}' + 2\boldsymbol{\omega} \times \mathbf{v}' + \boldsymbol{\omega} \times (\boldsymbol{\omega} \times \mathbf{r}') + \frac{d\boldsymbol{\omega}}{dt} \times \mathbf{r}' + \frac{d\mathbf{u}}{dt} \qquad (1\text{-}35)$$

Here $\mathbf{u}$ is the velocity of O' with respect to R, and $\boldsymbol{\omega}$ is the angular velocity of R' with respect to R.

If R' is in uniform translational motion with respect to R, $d\mathbf{u}/dt = 0$ and $\omega \equiv 0$. The equations then reduce to the familiar ones expressing the vectorial composition of velocities and the invariance of acceleration

$$\mathbf{v} = \mathbf{u} + \mathbf{v}' \qquad (1\text{-}21)$$

$$\mathbf{a} = \mathbf{a}' \qquad (1\text{-}24)$$

Let us denote the extra terms in the expression for $\mathbf{a}$ by $\mathbf{A}$:

$$\mathbf{A} = 2\boldsymbol{\omega} \times \mathbf{v}' + \boldsymbol{\omega} \times (\boldsymbol{\omega} \times \mathbf{r}') + \frac{d\boldsymbol{\omega}}{dt} \times \mathbf{r}' + \frac{d\mathbf{u}}{dt} \qquad (1\text{-}36a)$$

$$\mathbf{a} = \mathbf{a}' + \mathbf{A} \qquad (1\text{-}36b)$$

Here, $\mathbf{A}$ describes the effect of nonuniform motion of R' relative to R.
Suppose that R is the fixed star frame. In it,

$$\mathbf{F} = m\mathbf{a} \qquad (1\text{-}28)$$

where $\mathbf{F}$ is the force on particle P due to other particles. We call it the "real" force. Suppose that it depends only on the instantaneous position of P

relative to other particles, as with inverse square laws of gravitation and electrostatics [Eqs. (1-8) and (1-29)]. Then $\mathbf{F}$ is invariant, for, by Eq. (1-7),

$$\mathbf{r}_2 - \mathbf{r}_1 = \mathbf{r}_2' - \mathbf{r}_1'$$

no matter what the motion of R'. The inertial mass m must also be invariant in this nonrelativistic limit, for m is an intrinsic property of the particle. We have just seen how $\mathbf{a}$ transforms [Eq. (1-36b)]. The law of motion [Eq. (1-28)] becomes

$$\mathbf{F}' = m'(\mathbf{a}' + \mathbf{A})$$

where $\mathbf{F}' = \mathbf{F}$ is the invariant "real" force and $m' = m$ is the constant inertial mass. Rearranging, we have

$$\underset{\substack{\text{"real"}\\\text{forces}}}{\mathbf{F}'} \quad - \quad \underset{\substack{\text{"inertial"}\\\text{forces}}}{m'\mathbf{A}} \quad = m'\mathbf{a}' \qquad (1\text{-}37)$$

The law of motion in the accelerated frame R' has the standard form of force equals mass times acceleration, except that added to the invariant "real" force $\mathbf{F}'$ is the so-called *inertial force* $-m'\mathbf{A}$. The term $\mathbf{F}'$ contains physical forces like gravitation or electric attraction or repulsion. The second, inertial force, term is determined by the choice of coordinate system and has nothing to do with interactions with other particles. It results from our choice of some particular rigid body from which to survey and of some particular origin O' in that body.

In the impossibility of shielding, inertial force resembles gravitation. Another point of resemblance is the proportionality to inertial mass m' ($= m$). It is a very precisely established fact [14] that in a given gravitational field all bodies have the same acceleration. This means that the gravitational force on a body is proportional to its inertial mass.

We shall see later on how this parallelism led Einstein to a fusion of gravitation and inertia in the principle of equivalence. From the point of view of Newtonian physics, the two properties of matter are quite distinct. The gravitational field has directly identified sources—other particles—and depends only on relative position. The inertial force, on the other hand, depends on $\mathbf{r}'$ and $\mathbf{v}'$ in just the way given by Eq. (1-36a). There are checks and cross checks that one can make, permitting a practical judgment as to how much of the acceleration $\mathbf{a}'$ of P is due to other bodies and how much to the acceleration of the coordinate system. Note, however, that a translational acceleration $d\mathbf{u}/dt$ of the reference frame has the same effect as a gravitational field of distant bodies. (This is the fundamental ambiguity discussed in Section 1-2 that prevents us from positively identifying an inertial frame in the old-fashioned sense of the term.)

Because they are not directly due to other particles, the inertial forces have also been called fictitious. But their effect on motion with respect to an accelerated frame is every bit as real as that of particle–particle interactions. Centrifuges work!

Weightlessness in a Freely Falling Laboratory

We are accustomed to think of weightlessness in terms of a passive space-craft, where even the earth's gravitational field is transformed away by the free fall of the vehicle. Actually the earth is a spacecraft, and its free fall in the gravitational field arising from the sun, moon, and other outside bodies cancels out the effect of that field, leaving only the gravitational field of the earth itself (just as in the spacecraft there is a residual field due to the material of the vehicle).

To elucidate these statements we let R be the fixed star frame and R' the frame attached to the earth's crust. Assume, for the time being, that the earth is not spinning, only orbiting. Let $\mathbf{u}$ be the velocity of the earth with respect to the fixed stars. The gravitational field strength (i.e., the force per unit gravitational charge) can be written as the sum of an earth contribution $\mathbf{g}_{earth}$ and an extraterrestrial contribution $\mathbf{g}_{ext}$. In accordance with the Newtonian inverse square law, it depends only on distance from the sources and is the same in R and R'. In our system of units, gravitational charge equals inertial mass, and the force on a particle is

$$m\mathbf{g}_{earth} + m\mathbf{g}_{ext}$$

If there is in addition a nongravitational real force $\mathbf{F}_{non}$, the equation of motion of the particle in R is

$$\mathbf{F}_{non} + m\mathbf{g}_{earth} + m\mathbf{g}_{ext} = m\mathbf{a} \tag{1-38a}$$

In R' we have [by Eqs. (1-37) and (1-36a)],

$$\mathbf{F}_{non} + m\mathbf{g}_{earth} + m\mathbf{g}_{ext} - m\frac{d\mathbf{u}}{dt} = m\mathbf{a}' \tag{1-38b}$$

The acceleration $d\mathbf{u}/dt$ of the laboratory with respect to the fixed stars is the same as that of the center of mass of the earth. With M denoting the inertial mass of the earth and $\mathbf{G}_{ext}$ the external gravitational field averaged over the matter of the earth, the theorem of the motion of the center of mass in R gives

$$MG_{ext} = M\frac{d\mathbf{u}}{dt}$$

or

$$\frac{d\mathbf{u}}{dt} = G_{ext}$$

To the extent that the external gravitational field is homogeneous over the earth,

$$G_{ext} \approx g_{ext}$$

and hence

$$\frac{d\mathbf{u}}{dt} \approx g_{ext}$$

Equation (1-38b) can then be written

$$F_{non} + mg_{earth} \approx m\mathbf{a}' \qquad (1\text{-}39a)$$

To this very good approximation[6] the particle as observed in the laboratory feels only local forces, including earth gravity, and is weightless as regards gravitational fields of extraterrestrial origin. Since the earth and the particle are in essentially the same outside gravitational field and the earth is in free fall in it, this field has no effect on the motion of the particle with respect to the earth.

The gravitational force from outside the laboratory can be completely eliminated by detaching the laboratory from the earth's crust and letting it be in free fall. It is then a spaceship outside the atmosphere. Call its frame R'' and its velocity with respect to R, $\mathbf{u}'$. Its acceleration with respect to R is

$$\frac{d\mathbf{u}'}{dt} = g_{ext} + g_{earth}$$

---

[6] It is an approximation. Tides result from the inhomogeneity of the gravitational fields of the moon and sun.

where we have assumed that the change of gravitational field between the ground and the new location of the spaceship is negligible. The equation of motion of the particle in R'' is

$$\mathbf{F}_{\text{non}} + m\mathbf{g}_{\text{earth}} + m\mathbf{g}_{\text{ext}} - m\frac{d\mathbf{u}'}{dt} = m\mathbf{a}'' \qquad (1\text{-}38c)$$

or

$$\mathbf{F}_{\text{non}} \approx m\mathbf{a}'' \qquad (1\text{-}39b)$$

(The gravitational force on the particle from the material of the spaceship is assumed to be negligible.) A body released from rest in the cabin remains at rest; it does not fall with respect to R''. R'' is thus inertial in the modern sense of the word.

Relation Between Earth Frame and Inertial Frames

Comparing the equations of motion in the earth frame R' [Eq. (1-39a)] and the inertial frame R'' [Eq. (1-39b)], we find that the only difference is the presence of the earth gravity term. By including

$$m\mathbf{g}_{\text{earth}}$$

among the forces, we can proceed as if R' were inertial.

The earth's rotation has so far been ignored. If R' is now spinning uniformly as well as orbiting, the left-hand side of Eq. (1-38b) has, by Eq. (1-36a), the additional Coriolis and centrifugal terms

$$-m2(\boldsymbol{\omega} \times \mathbf{v}') - m\boldsymbol{\omega} \times (\boldsymbol{\omega} \times \mathbf{r}')$$

Denote (Fig. 1-17) the vector from the center of mass of the earth C to O' by $\boldsymbol{\rho}$. The velocities in R of C and O' are related by [Appendix B, Eq. (B-1)]

$$\mathbf{u} = \mathbf{v}_C + \boldsymbol{\omega} \times \boldsymbol{\rho}$$

so that

$$\frac{d\mathbf{u}}{dt} = \frac{d\mathbf{v}_C}{dt} + \boldsymbol{\omega} \times \frac{d\boldsymbol{\rho}}{dt}$$

Since

$$\frac{d\boldsymbol{\rho}}{dt} = \mathbf{u} - \mathbf{v}_C = \boldsymbol{\omega} \times \boldsymbol{\rho}$$

it follows that

$$\frac{d\mathbf{u}}{dt} = \frac{d\mathbf{v}_C}{dt} + \boldsymbol{\omega} \times (\boldsymbol{\omega} \times \boldsymbol{\rho})$$

As before, the motion of the center of mass gives

$$\frac{d\mathbf{v}_C}{dt} = \mathbf{G}_{ext} \approx \mathbf{g}_{ext}$$

so that

$$-m\frac{d\mathbf{u}}{dt} \approx -m\mathbf{g}_{ext} - m\boldsymbol{\omega} \times (\boldsymbol{\omega} \times \boldsymbol{\rho})$$

The new equation of motion in R' reads

$$\mathbf{F}_{non} + m\mathbf{g}_{earth} - m\boldsymbol{\omega} \times (\boldsymbol{\omega} \times \boldsymbol{\rho}) - m2(\boldsymbol{\omega} \times \mathbf{v}') - m\boldsymbol{\omega} \times (\boldsymbol{\omega} \times \mathbf{r}') \approx m\mathbf{a}'$$

or

$$\mathbf{F}_{non} + m\mathbf{g}_{earth} - m\boldsymbol{\omega} \times [\boldsymbol{\omega} \times (\boldsymbol{\rho} + \mathbf{r}')] - m2(\boldsymbol{\omega} \times \mathbf{v}') \approx m\mathbf{a}' \qquad (1\text{-}40)$$

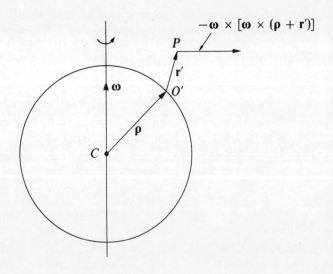

Figure 1-17.

By comparison with Eq. (1-39a), we see that the only effect of the earth's spinning is to replace the earth gravity field strength $\mathbf{g}_{earth}$ by

$$\mathbf{g}_{eff} = \mathbf{g}_{earth} - \boldsymbol{\omega} \times [\boldsymbol{\omega} \times (\boldsymbol{\rho} + \mathbf{r}')] - 2(\boldsymbol{\omega} \times \mathbf{v}') \qquad (1\text{-}41)$$

The second term on the right-hand side is the familiar centrifugal correction, pointing away from the axis of rotation (Fig. 1-17). It is strongest at the equator, where it amounts to -3.4 cm sec^{-2}, or about $1/3\%$ of 980 cm sec^{-2}. The third term is the Coriolis one, proportional to the particle's velocity and at right angles to it and the earth's axis. Its greatest possible value (extreme relativistic particle moving east or west) is

$$2\omega c = 2 \times 7.3 \times 10^{-5} \times 3 \times 10^{10} = 4.4 \times 10^{6} \text{ cm sec}^{-2}$$

$$= 4.5 \times 10^{3} \times 980 \text{ cm sec}^{-2}$$

This figure is still negligible compared even with the accelerations imparted to extreme relativistic charged particles by stray magnetic fields. A 20-GeV electron from the Stanford Linear Accelerator experiences in a transverse field of 1 G an acceleration of 1.3×10^{13} cm sec^{-2}. [A discussion of the application of Eq. (1-41) to projectile motion relative to the earth and of the detection of "absolute rotation" is given in Appendix C.]

We conclude that by including $m\mathbf{g}_{eff}$ among the forces acting on the particle we can pretend that the laboratory frame is inertial. There remains a question as to the correct form of the correction term when the particle has laboratory speed comparable with that of light. Our derivation of Eq. (1-40) is then no longer valid, for it uses the nonrelativistic equation of motion

$$\mathbf{F} = m\mathbf{a} \qquad (1\text{-}28)$$

A rigorous argument based on the principle of equivalence will be given in Chapter 6. The result (G. Ascoli, unpublished) is that a freely falling particle of velocity $\mathbf{v}'$ in the earth frame undergoes an acceleration

$$\mathbf{g}_{eff}\left[1 - \left(\frac{v'}{c}\right)^{2}\right] \qquad (1\text{-}42)$$

When other forces are present, this acceleration is only a tiny correction, and it can simply be added, if necessary, to the acceleration calculated as if the frame were inertial.

PROBLEMS

1-1. Prove that the Kronecker delta, δ_{jk}, defined by

$$\delta_{jk} = 1 \quad \text{if} \quad j = k$$

$$= 0 \quad \text{if} \quad j \neq k$$

is a tensor of rank 2.

Prove that the Levi–Civita epsilon, ε_{jkl}, defined by

$$\varepsilon_{jkl} = 0 \quad \text{if any two indices are equal}$$

$$= +1 \quad \text{if } j, k, \text{ and } l \text{ are an even permutation of 1, 2, and 3}$$

$$= -1 \quad \text{if } j, k, \text{ and } l \text{ are an odd permutation of 1, 2, and 3}$$

is a pseudotensor of rank 3.

1-2. A mass m is fastened by a horizontal spring (spring constant k) to a point of support that moves back and forth horizontally in simple harmonic motion of frequency v, amplitude A. The mass is on a frictionless horizontal table. It has zero velocity perpendicular to the line of motion of the point of support.

Set up and solve the equation of motion in a reference frame attached to the point of support.

Transform the solution to the laboratory reference frame (assumed inertial). This illustrates the principle of the seismograph.

1-3. In transforming from the earth crust frame to the fixed star frame, one needs to add to the forces on the particle the gravitational attraction of the moon and sun. Estimate the magnitude of these terms and compare with the attraction of the earth.

1-4. Consider an idealized wheel of radius a, all of whose mass m is in the rim. It is turning about its axis with angular velocity $\dot{\gamma}$. The axis is fixed in the earth's crust, making angle β with the earth's angular velocity ω. Determine the moment (torque) with respect to the center of the wheel of the Coriolis force on an element of the rim. Integrating it around the wheel, show that the resultant torque tends to tip the axis of the wheel into parallelism with ω and has magnitude

$$ma^2\omega\dot{\gamma}\sin\beta$$

This torque is the basis of Foucault's gyrocompass.

1-5. A ship of mass 50,000 tons is moving east on the equator at 30 knots. If she puts about and moves west at the same rate, what is the increase in her apparent weight? (1 knot = 51.5 cm sec^{-1}; 1 ton = 1000 kg.)

1-6. A tidal current is running south in a channel 20 miles wide at latitude 60° N with velocity 1.0 ft sec^{-1}. Show that the height of the water on the west coast exceeds that on the east coast by 0.50 in. (1 mile = 5280 ft, 1 in. = 1/12 ft, g = 32 ft sec^{-2}).

1-7. (Appendix C) A body is dropped from rest at height h above the surface of the earth. Neglect air resistance and the change of the earth's field with height.

Calculate the Coriolis force as a function of time, assuming it has a negligible effect on the motion.

Calculate the net displacement of the point of impact resulting from the calculated Coriolis force. Evaluate for $h = 30 \, \text{m}$, $\lambda = 36° \, \text{S}$ (Montevideo).

Give an alternative derivation of the net displacement by determining the motion in a nonrotating frame. You may assume that the earth's surface is spherical.

Chapter 2

The Lorentz Transformation and the Kinematics of a Particle

We enter now into the theory of special (or restricted) relativity. The name "relativity" is justified historically by the fact that the basic concern of the theory is with how the description of phenomena changes when one shifts from one reference frame to another. Einstein's elementary considerations on this problem have turned out to be enormously fruitful in physical consequences—such as, for example, the mass–energy relation—but the original, somewhat negative, name has meanwhile become established by usage.

In this first presentation we follow Einstein's classic paper [2].

2-1. THE POSTULATES OF SPECIAL RELATIVITY

Einstein's first postulate is that of the equivalence of all inertial reference frames. Our equations must be covariant with respect to a transformation between inertial frames—that is, the two sides of an equation must transform in the same way, so as to remain equal. This is the principle of special relativity of Galileo and Newton. As reaffirmed by Einstein, "The laws by which the states of physical systems undergo change are not affected whether these changes of state be referred to the one or the other of two systems of coordinates in uniform translational motion."

Although not explicitly stated by Einstein, the postulate includes a restriction to inertial frames. In a gravity-free laboratory that is moving at constant velocity relative to another gravity-free laboratory, all experiments proceed in the same way and give the same results. The laws have the same form in two such reference frames. Thus there is no such thing as absolute rest; there is no physical basis for preferring one inertial frame to another for describing nature.

The postulate leaves open the question as to just what the transformation equations are that relate quantities determined in the two frames. The laws of classical electrodynamics, as expressed in the Maxwell field equations, are not consistent with the postulate if they are used with the Galileo transformation of position and epoch

$$t' = t \tag{1-16}$$

$$\mathbf{r}' = \mathbf{r} - \mathbf{u}t - \boldsymbol{\alpha} \tag{1-18}$$

and the Newtonian assumption of constant inertial mass

$$\frac{p}{v} = m = \text{const} \tag{2-1}$$

Thus we saw in the preceding chapter that the magnetic force between charged particles involves not only their position and velocity relative to one another but also their velocities relative to the laboratory. Einstein's approach to the contradiction was to question Eqs. (1-16), (1-18), and (2-1) while retaining special relativity and the Maxwell field equations. In so doing, he cleaved to the experimental facts. Many very delicate experiments had been carried out with the object of determining the difference between inertial frames with regard to electromagnetic effects. All had had negative results. One of the most famous is the Michelson–Morley experiment (1887), which was a comparison of the to-and-fro travel time of light along two equal perpendicular paths. It failed to show any effect of orientation of the apparatus with respect to the earth. Experiments on the aberration of star light and on the speed of light in moving media showed that one could not explain away the null result by assuming that the ether is dragged along by the earth. The only interpretation possible for the whole body of experiments was that electromagnetic phenomena conform to the principle of special relativity. [There are, of course, detectable effects of the earth's rotation (Appendix C). The experiments we are now discussing are concerned not with acceleration but with velocity: Does it make any difference which way the earth is moving, and how fast?]

The conceptual difficulties come with Einstein's second postulate: "Light is always propagated in empty space with a definite velocity c which is independent of the state of motion of the emitting body." It is this brutal assertion, which also agrees exactly with all the experiments, that gives rise to the peculiar features of the theory, making it fascinating and "paradoxical." Offhand, the second postulate seems reasonable if one thinks of light as a wave phenomenon. Wave velocity is determined by the properties of the medium; a wave spreads on a pond with a velocity independent of the motion of whatever produced it. But Einstein is saying that one and the same wave spreads uniformly with the same speed in every inertial reference frame. Even if one reference frame is moving with a velocity of 2.997×10^{10} cm sec^{-1} with respect to the other, the light spreads spherically in each one

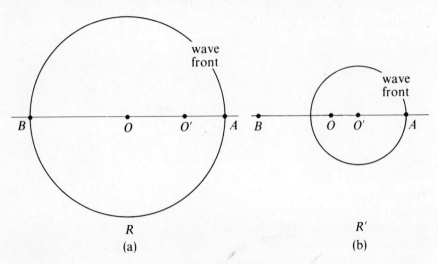

Figure 2-1.   (a) The situation in R when the wave front touches A. (b) The situation in R' when the wave front touches A.

with speed $c = 2.998 \times 10^{10}\,\mathrm{cm\,sec^{-1}}$. Each frame carries, so to speak, its own pond (ether), and they all have the same properties.[1]

We shall see that the two postulates taken together require that the position and time determination for events are intertwined in a special way.

An example will show this (Figs. 2-1a and b). Suppose that a light source flashes at the point O in R. A wave front is produced which in R is centered on O and expands spherically with velocity c. Let A and B be fixed, diametrically opposite points, equidistant from O. Figure 2-1a shows the wave front when it reaches A and B. Consider the same events from another reference frame R', which is moving with constant velocity u with respect to R in the BOA direction (to the right). Let O' be the point in R' coinciding with O when the flash at O occurs. In R', the light is emitted by a moving source, of speed u toward the left. According to the two postulates, the wave front in R' is centered on O' and expands spherically with velocity c. It meets A before it meets B, because both A and B are moving to the left with speed u. Figure 2-1b shows the situation in R' when the wave front gets to A. Evidently its arrival at B will occur later. The two events, arrival of the wave front at A and arrival of the wave front at B, are simultaneous in R and not simultaneous in R'. We thus have an example, without formulas, of the relativity of simultaneity. Two events can be simultaneous in one

---

[1] Maxwell's equations for the electromagnetic field in vacuum imply that light spreads isotropically with speed c. The second postulate could be replaced by the less general statement that Maxwell's equations are valid with the same constant c in every inertial frame.

inertial frame but occur at different times in another. It is not true in general that

$$t' = t \qquad (1\text{-}16)$$

This conclusion violates common sense. But common sense is based on experience with objects of low velocity, and one cannot expect it to apply in the newly discovered domain of high velocities.

The progress of experimental technique has made possible ever more precise tests of the postulates. One powerful tool has been the "maser." (Maser is an acronym for Microwave Amplification by Stimulated Emission of Radiation. A gas of molecules, which may be in the form of a beam, interacts with the electromagnetic field of a resonant cavity. If the frequency of the field overlaps that of a radiative molecular transition, stimulated emission occurs, and under appropriate conditions a large coherent amplitude builds up. The amplified frequency is very sharply defined. Depending on which has the smaller spectral width, the cavity or the molecule, it is determined either by the dimensions of the cavity or by the molecular energy level spacing. In the former case, a measurement of the amplified frequency provides a very sensitive measure of cavity length; in the latter it provides a very sensitive measure of molecular frequency, or clock rate.) A maser version [19] of the Michelson–Morley experiment uses two identical cavities mounted at right angles to one another, of the type with frequencies ($\sim 3 \times 10^{14}$ Hz) determined principally by the length. The beat frequency between the two cavities is recorded as a function of orientation with respect to the earth; a 90° rotation gives an effect of about 2.7×10^5 Hz, which is ascribed to magnetostriction in the earth's field. The important point is that the magnitude of this effect is independent of time of day, to better than 3×10^3 Hz. Thus, the experiment has a null result to within $(3 \times 10^3)/(3 \times 10^{14}) = 10^{-11}$, whereas a nonrelativistic ether drift theory predicts an effect equal to $(v/c)^2$, v being the velocity of the earth with respect to the ether. The orbital velocity of the earth around the sun is 30 km sec$^{-1} = 10^{-4}c$. In another experiment from the same laboratory [20], identical ammonia masers of frequency $\sim 2.4 \times 10^{10}$ Hz, determined by the molecular level spacing, were set up with the molecular beams going in opposite directions, east and west. The beat frequency between the two outputs was measured as the apparatus was turned 180° about the vertical. The change (1.08 ± 0.02 Hz) was again independent of time of day, and also of season. The constancy to 8×10^{-13} compares with an ether drift prediction of $8 \times 10^{-6}(v/c)$. Again, any effect is less than one thousandth of that predicted with $v = 30$ km sec^{-1}.

The Mössbauer effect has made possible an even more precise ether drift experiment, carried out in Dicke's laboratory [21]. (In the Mössbauer effect, an excited nucleus in a crystal lattice emits a photon with the entire lattice taking up the recoil. The energy of the photon is therefore very nearly the whole energy loss of the nucleus, and if the nuclear excited state is long lived, the energy of the photon is sharply defined. Such is the case with the 14.4-keV x ray from the excited state of iron 57 resulting from the decay of cobalt 57 embedded in an iron foil. One observes the resonant absorption of this photon by ground-state iron 57 nuclei in a similarly prepared iron foil. Since the photon frequency spectrum is very narrow, the attenuation is a very sensitive measure of any frequency shift of emitter relative to absorber.) A cobalt 57 in iron source is attached to a centrifuge near the rim. An iron absorber foil is placed on the same radius near the axis. The detector is on the axis. The transmission through the absorber is measured as a function of the angular position of the source. As the source spins around at $0.16 \, \text{km sec}^{-1}$, its velocity with respect to the fixed stars changes by something like the same amount (the exact number depending on the orientation of the centrifuge). No difference in transmission was observed among the four quadrants. There was, as expected, a transverse Doppler effect (Appendix F) that made the counting rate depend on the spin velocity, but no change occurred in the rate for different velocities of the source nucleus with respect to the stars. The upper limit set by this experiment corresponds to a possible ether drift of about $0.008 \, \text{km sec}^{-1} = 3 \times 10^{-4} \times 30 \, \text{km sec}^{-1}$.

Although historically the basis for the theory of special relativity is the absence of ether drift effects, we should not think that the theory rests only on the negative results of searches for minute effects. It makes many striking predictions that have been directly verified. Examples are the dependence of particle momentum on velocity, confirmed to high precision in the operation of electron and proton accelerators; the slowing down of moving clocks, confirmed by experiments on transition probabilities of moving unstable particles; and the mass–energy equivalence, confirmed by observations of creation and annihilation of particles. At the present time, the positive experimental evidence for the theory of special relativity is overwhelming.

An extremely precise test of the second postulate has recently been made at CERN (*Centre Européen de Recherches Nucléaires*, Geneva [22]). The experiment was a direct measurement of the velocity of very high energy photons from the decay in flight of π^0 mesons moving with a speed practically equal to that of light. The CERN accelerator is a huge evacuated ring in which protons are accelerated in short bunches up to an energy of about 3×10^{10} eV. At an appropriate time in the cycle, they are made to hit a small solid target in the ring; bursts of secondary particles lasting a few times 10^{-9} sec are produced. Among the secondaries are many π^0 mesons, each decaying with a mean life in its rest frame of about 10^{-16} sec into two photons. Even at these tremendous energies, this decay process is practically instantaneous in the laboratory frame. The time difference between the pulse of protons hitting the target A (Fig. 2-2) and the detection of a photon down the line at B was measured, as well as the time difference between the protons hitting the target A and the detection of another unrelated photon further down the path at C. Comparing the two time distributions, the authors obtained a very precise value of the time for photons to travel the distance BC (about 21 m). The photons detected were of more than 6×10^9 eV energy; therefore, the π^0's must have been $>6 \times 10^9$ eV, and their velocity at the time of emitting the photons must consequently have been [Eq. (3-20)]

$$\beta \geq 0.99975 \qquad (2\text{-}2a)$$

The measured velocity of the photons from this moving source was

$$c' = 2.9979(\pm 0.0005) \times 10^{10} \text{ cm sec}^{-1} \qquad (2\text{-}2b)$$

which is to be compared with the velocity of low energy photons ("light") from an essentially stationary source,

$$c = 2.997925(\pm 0.000003) \times 10^{10} \text{ cm sec}^{-1} \qquad (2\text{-}2c)$$

There is beautiful agreement within the estimated errors.

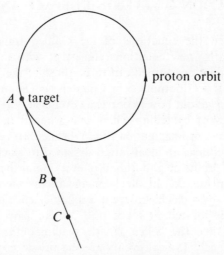

Figure 2-2.　　The CERN experiment to measure the speed of photons from a rapidly moving source. Neutral pions made in the target disintegrate in flight near A. The resulting photons are detected at B or C.

If we take a nonrelativistic point of view and say that

$$c' = c + \varkappa v \tag{2-3}$$

where v is the velocity of the moving source, from the experimental values [Eqs. 2-2)] we must conclude that

$$\varkappa = (0 \pm 1.3) \times 10^{-4}$$

If there is a departure from the second postulate, the coefficient dc'/dv is essentially zero within the errors.

One therefore must accept that in every inertial frame light spreads with the same speed c regardless of the motion of the source. Every inertial frame can, in effect, be regarded as its own ether; thus, there exists an infinity of ethers. It is meaningless to speak of a velocity with respect to *the* ether.

A third fundamental assumption of the theory, so compelling that it was not called a postulate by Einstein, is that the kinematics and dynamics of Galileo and Newton—essentially Eqs. (1-16), (1-18), and (2-1)—are valid in the low velocity limit. By low velocity, we mean $u \ll c$ and $v \ll c$. This is an assumption in the logical sense, but the experimental evidence for classical mechanics in its domain is so strong that no alternative is admissible. We shall see that this assumption is a powerful tool in deriving relativistic formulas, its role being analogous to that of Bohr's correspondence principle in quantum mechanics.

2-2. THE LOCATION OF EVENTS IN SPACE AND TIME

Einstein replaced the equations of the Galileo transformation [Eqs. (1-16) and (1-18)], by new equations consistent with the postulates just stated. The new equations turned out to be those of the Lorentz transformation [Eq. (1-31); see footnote 5 in Chapter 1). The argument involved basic considerations about how to correlate events at different places.

We locate an event by telling where and when it happens. The rigid reference frame is in our imagination (Fig. 1-1) filled with a lattice of monuments. This is, of course, an idealization designed to elucidate the concept "reference frame." In the neighborhood of every point there is an identified monument—a "benchmark" in the language of the Geological Survey—with a nameplate and a clock. An event is a coincidence of something with a monument. It is located in space (the "where") by the name of the monument and in time (the "when") by the reading of the local clock at the instant of coincidence. Typical events are the presence of a particle or a wave amplitude.

To correlate different events, we must introduce a metric and synchronize the clocks.

In an inertial frame, light travels in straight lines and the geometry of space is Euclidean; it is possible to survey the benchmarks using light rays and triangulation. We construct, in this way, a table of Cartesian coordinates of the monuments, based on a fundamental triple of mutually perpendicular rods of unit length. This table gives us the x, y, z coordinates of any event. There is no change here from prerelativity physics.

We need a recipe for synchronizing the host of clocks, which are, of course, of identical construction. Einstein's procedure is essentially that of the practicing radio engineer, who uses time signals from the Bureau of Standards radio transmitter at Washington (WWV). The engineer must allow for the travel time of the radio waves to his receiver, which he computes from the distance and the velocity of propagation. When the noon signal from Washington reaches Urbana, Illinois, the Urbana clock should read noon plus 3.7×10^{-3} sec, because Urbana has been surveyed to be 1.11×10^8 cm from Washington and $c = 3.0 \times 10^{10}$ cm sec^{-1}. But he has used here a *nominal* value for the speed of the radio wave, and that is not scientific! To measure the speed we must measure the time taken by the wave to go from the one place to the other. How can we do this before we have synchronized the clocks at the two places? The question is rhetorical: we cannot. Einstein's trick is to retransmit the signal without delay and determine the time on the original Washington clock when it returns there. The Urbana clock is said to be synchronized with the Washington clock when its recorded reading is midway between the two readings of the Washington clock. This procedure can be carried out in ignorance of the speed of the signal. It can be applied anywhere and allows us to synchronize any two clocks no matter how far apart.

Let t_A be the time read on the clock at A when the light signal leaves A, and t_B the time read on the clock at B when the signal reaches B (Fig. 2-3). A mirror or transponder at B sends the signal back to A without delay. When it gets to A, the clock there reads t'_A. The clocks at B and A are synchronized if, and only if,

$$t_B = \frac{t_A + t'_A}{2} \tag{2-4}$$

If t_B does not satisfy Eq. (2-4), we simply move the hands of the clock at B until it does.

Figure 2-3.

Note that Eq. (2-4) implies that

$$t_B - t_A = t'_A - t_B$$

With this rule for synchronization, it takes light the same "elapsed time" to travel from A to B as from B to A. By "elapsed time" we simply mean the difference in reading of duly synchronized clocks at the two places. Ether drift effects are ruled out by definition, because the line AB can have arbitrary orientation with respect to the velocity of R relative to the supposed ether, and it can, in particular, be parallel to it.

By measuring the distance between B and A, we can use the readings t_A and t'_A to determine the speed of light:

$$c = 2 \cdot \frac{\overline{AB}}{t'_A - t_A}$$

The assumed isotropy of inertial frames implies that c has the same value for all orientations of AB. This fact has been confirmed experimentally. Eq. (2-4) can now be written

$$t_B = t_A + \frac{\overline{AB}}{c} \tag{2-5}$$

EXERCISE *Check of the consistency of the synchronization procedure.*
Suppose the clock at B is synchronized with that at A and the clock at C is synchronized with that at A. Prove that clocks B and C are synchronized with one another.

SOLUTION Let a light signal leave A at t_A, reach B at t_B, be reflected from B to C, reach C at t_C, be reflected from C to A, and reach A at t''_A (Fig. 2-4).

$$t''_A - t_A = \frac{\overline{AB} + \overline{BC} + \overline{CA}}{c}$$

Since B and A are synchronized,

$$t_B - t_A = \frac{\overline{AB}}{c}$$

Since C and A are synchronized,

$$t''_A - t_C = \frac{\overline{CA}}{c}$$

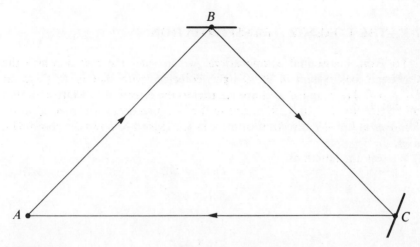

Figure 2-4.

Substituting for AB/c and CA/c in the first equation, it simplifies to

$$t_C - t_B = \frac{\overline{BC}}{c}$$

showing that the clocks at B and C are synchronized.

Suppose events associated with light flashes occur at A and B. If the clocks at A and B are synchronized and if the events occur "at the same time," each by its own clock, the light flashes get to the midpoint M of AB (Fig. 2-5) at the same instant. Conversely, a light flash from M arrives at A and B at the same time as read on the local clocks if they have been synchronized.

In general, events occurring at the same time as read on their local synchronized clocks are called *simultaneous*. Simultaneity so defined is relative to the reference frame, for it refers to a particular array of benchmarks and clocks.

The procedure has now been specified for determining in the frame R the spatial and temporal coordinates of an event, x, y, z, and t. We use the same standard equipment and procedures in every inertial frame. In R', the same event has the coordinates x', y', z', and t'. Let us consider now the relations between these quantities.

Figure 2-5.

2-3. THE LORENTZ TRANSFORMATION

To avoid unessential complications, we consider the case in which the Cartesian axis system in R', $O'x'y'z'$, coincides with that in R, $Oxyz$, at $t = t' = 0$. The x and x' axes are parallel to the direction of relative motion of the two frames—in R, R' moves in the $+x$ direction with speed u; in R', R moves in the $-x'$ direction with speed u. Figure 2-6 shows the coordinate axes.

We seek the functions

$$t' = t'(t, x, y, z)$$

$$x' = x'(t, x, y, z)$$

$$y' = y'(t, x, y, z)$$

$$z' = z'(t, x, y, z)$$

(2-6)

relating the temporal and spatial coordinates of an event in the two coordinate systems. They are to be the simplest forms satisfying the three postulates.

The third postulate requires that the form Eq. (2-6) reduce to the Galileo transformation

$$t' = t$$

$$x' = x - ut$$

$$y' = y$$

$$z' = z$$

(1-32a)

when $(c/u) = \infty$.

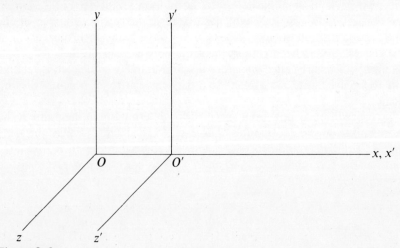

Figure 2-6.

The first postulate (equivalence of inertial frames) brings some drastic simplifications in the possible form of Eqs. (2-6). A uniform rectilinear motion in R must go over into a uniform rectilinear motion in R', otherwise one of the two reference frames would not be inertial. Comparing neighboring successive positions of a particle, we note that uniform rectilinear motion means that dx, dy, dz, and dt are in constant proportions:

$$dx:dy:dz:dt = v_x^0:v_y^0:v_z^0:1$$

There must be a similar constancy of the ratios in R'

$$dx':dy':dz':dt' = v_{x'}^0:v_{y'}^0:v_{z'}^0:1$$

for arbitrary values of v^0. Since

$$dx' = \frac{\partial x'}{\partial x}\,dx + \frac{\partial x'}{\partial y}\,dy + \frac{\partial x'}{\partial z}\,dz + \frac{\partial x'}{\partial t}\,dt$$

$$dy' = \frac{\partial y'}{\partial x}\,dx + \frac{\partial y'}{\partial y}\,dy + \frac{\partial y'}{\partial z}\,dz + \frac{\partial y'}{\partial t}\,dt$$

$$dz' = \frac{\partial z'}{\partial x}\,dx + \frac{\partial z'}{\partial y}\,dy + \frac{\partial z'}{\partial z}\,dz + \frac{\partial z'}{\partial t}\,dt$$

$$dt' = \frac{\partial t'}{\partial x}\,dx + \frac{\partial t'}{\partial y}\,dy + \frac{\partial t'}{\partial z}\,dz + \frac{\partial t'}{\partial t}\,dt$$

it is necessary as well as sufficient that all the partial derivatives be constant. The transformation [Eq. (2-6)] must be linear. With our choice of origins in space and time, there are no constant terms; the transformation is homogeneous. Letting the index values $0, 1, 2, 3$ stand, respectively, for t, x, y, z, we have

$$t' = a_{00}t + a_{01}x + a_{02}y + a_{03}z$$

$$x' = a_{10}t + a_{11}x + a_{12}y + a_{13}z$$

$$y' = a_{20}t + a_{21}x + a_{22}y + a_{23}z \tag{2-7}$$

$$z' = a_{30}t + a_{31}x + a_{32}y + a_{33}z$$

The elements a_{jk} of the transformation matrix depend only on the relative velocity u.

We first show that on the basis of the postulate of special relativity and the homogeneity and isotropy of inertial frames, the form Eq. (2-7) simplifies to Eq. (2-16) with only two undetermined coefficients, a_{00} and a_{01}. The second postulate is then used to determine a_{00} [Eq. (2-19)] and a_{01} [Eq. (2-20)].

The coincidence of the corresponding axes at $t = 0$ makes zero the six off-diagonal elements in the space–space $(1, 2, 3)$ part of the matrix. For example, a point of the y axis

$$x = 0 \qquad z = 0$$

must be on the y' axis

$$x' = 0 \qquad z' = 0$$

at $t = 0$. Hence, $a_{12} = a_{32} = 0$.

Forming differences between two events, number 1 and number 2, we get

$$\Delta t = t_2 - t_1 \qquad \Delta x = x_2 - x_1 \qquad \Delta y = y_2 - y_1 \qquad \Delta z = z_2 - z_1$$

$$(2\text{-}8)$$

$$\Delta t' = t'_2 - t'_1 \qquad \Delta x' = x'_2 - x'_1 \qquad \Delta y' = y'_2 - y'_1 \qquad \Delta z' = z'_2 - z'_1$$

Equation (2-7) gives

$$\Delta t' = a_{00} \, \Delta t + a_{01} \, \Delta x + a_{02} \, \Delta y + a_{03} \, \Delta z$$

$$\Delta x' = a_{10} \, \Delta t + a_{11} \, \Delta x$$

$$\Delta y' = a_{20} \, \Delta t \qquad\qquad + a_{22} \, \Delta y$$

$$\Delta z' = a_{30} \, \Delta t \qquad\qquad\qquad\qquad + a_{33} \, \Delta z$$

$$(2\text{-}9)$$

Let the two events be successive locations of a particle. If the particle is in R moving in the $+x$ direction with speed u,

$$\Delta x = u \, \Delta t \qquad \Delta y = 0 \qquad \Delta z = 0$$

it must be at rest in R' (there is still some common sense in relativity!)

$$\Delta x' = 0 \qquad \Delta y' = 0 \qquad \Delta x' = 0$$

It follows from Eq. (2-9) that

$$a_{10} + u a_{11} = 0 \qquad a_{20} = 0 \qquad a_{30} = 0$$

If the particle is at rest in R,

$$\Delta x = 0 \qquad \Delta y = 0 \qquad \Delta z = 0$$

it must in R' be moving in the $-x'$ direction with speed u

$$\Delta x' = -u \, \Delta t' = 0 \qquad \Delta y' = 0 \qquad \Delta z' = 0$$

It follows from Eq. (2-9) that

$$a_{10} + u a_{00} = 0 \qquad a_{20} = 0 \qquad a_{30} = 0$$

We see that

$$a_{11} = a_{00}$$

$$a_{10} = -u a_{00}$$

The transformation matrix has now become

$$\begin{pmatrix} a_{00} & a_{01} & a_{02} & a_{03} \\ -u a_{00} & a_{00} & 0 & 0 \\ 0 & 0 & a_{22} & 0 \\ 0 & 0 & 0 & a_{33} \end{pmatrix} \tag{2-10}$$

Arguments involving homogeneity and isotropy permit further simplification. Consider two events that in R are simultaneous ($\Delta t = 0$) and have the same x coordinate ($\Delta x = 0$) (Fig. 2-7). In R' they have the same x' co-

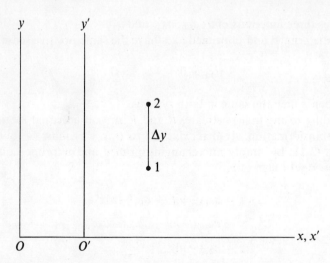

Figure 2-7.

ordinate ($\Delta x' = 0$). We demand that they also be simultaneous in $R'(\Delta t' = 0)$. Otherwise, the time of an event would depend on its distance from the x axis, and the transformation would single out a particular line in the frame as axis. This would run counter to the homogeneity of the space of an inertial frame. The transformation involves a direction (that of relative motion) but not a line.

To have $\Delta t' = 0$ with the events in either the (x, y) or the (x, z) plane requires $a_{02} = a_{03} = 0$.

The isotropy requirement prevents the transformation from introducing an azimuthal twist. For any event,

$$y:z = y':z'$$

entailing

$$a_{22} = a_{33}$$

The transformation now reads

$$t' = a_{00}t + a_{01}x$$
$$x' = -ua_{00}t + a_{00}x$$
$$y' = a_{22}y \qquad (2\text{-}11)$$
$$z' = a_{22}z$$

It involves three functions of u: a_{00}, a_{01}, and a_{22}.

Since the primed and unprimed axes have the same positive sense,

$$a_{00} > 0 \qquad a_{22} > 0$$

Time order is then the same in both systems.

According to the first postulate, R and R' are on an equal footing. The inverse transformation, from (t', x', y', z') to (t, x, y, z) must be obtainable from Eq. (2-11) by simply interchanging primed and unprimed quantities and reversing the sign of u.

$$t = a_{00}(-u)t' + a_{01}(-u)x'$$
$$x = ua_{00}(-u)t' + a_{00}(-u)x'$$
$$y = a_{22}(-u)y' \qquad (2\text{-}12)$$
$$z = a_{22}(-u)z'$$

One can also invert the transformation by the brute force procedure of solving Eq. (2-11) for the unprimed quantities. This gives

$$t = \frac{1}{a_{00}(u) + ua_{01}(u)}t' + \frac{-a_{01}(u)}{a_{00}(u)[a_{00}(u) + ua_{01}(u)]}x'$$

$$x = \frac{u}{a_{00}(u) + ua_{01}(u)}t' + \frac{1}{a_{00}(u) + ua_{01}(u)}x'$$

$$y = \frac{1}{a_{22}(u)}y'$$

$$z = \frac{1}{a_{22}(u)}z'$$

(2-13)

Matching coefficients, we get

$$a_{00}(-u) = \frac{1}{a_{00}(u) + ua_{01}(u)}$$

$$a_{01}(-u) = -\frac{a_{01}(u)}{a_{00}(u)}a_{00}(-u)$$

$$a_{22}(-u) = \frac{1}{a_{22}(u)}$$

(2-14)

The third of Eqs. (2-14) tells us that

$$a_{22}(-u)a_{22}(u) = 1$$

Now a_{22} is the factor by which transverse dimensions change. It must be the same whether R' is going to the left or to the right; that is, a_{22} is an even function of u. Therefore,

$$a_{22}^2(u) = 1$$

Since $a_{22} > 0$, it follows that

$$a_{22} = 1$$

(2-15)

The second of Eqs. (2-14) tells us that if a_{00} is an even function of u, a_{01} is an odd one. Now

$$a_{00} = \frac{\partial t'}{\partial t}$$

is the factor by which the time scale is changed, and it also must be the same whether R' is going to the right or to the left. It is an even function of u, and thus a_{01} is odd.

The transformation is thus simplified to

$$t' = a_{00}t + a_{01}x$$
$$x' = a_{00}(x - ut)$$
$$y' = y$$
$$z' = z$$

(2-16)

with

$$a_{00} > 0 \qquad a_{00}(-u) = a_{00}(u)$$
$$a_{01}(-u) = -a_{01}(u)$$

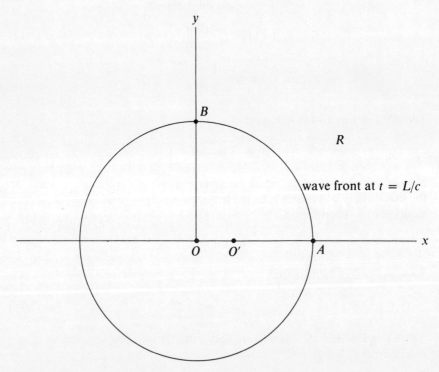

Figure 2-8a.   The situation in R when the light pulse reaches A and B. Note the position of O'.

This general form, involving only the equivalence of inertial frames (first postulate) and their homogeneity and isotropy, includes both the old (Galileo) transformation

$$a_{00} = 1 \qquad a_{01} = 0$$

and the new (Lorentz) transformation

$$a_{00} = \left(1 - \frac{u^2}{c^2}\right)^{-1/2} \qquad a_{01} = -\frac{u}{c^2}\left(1 - \frac{u^2}{c^2}\right)^{-1/2}$$

The old values result immediately if one assumes $t' = t$ (absolute simultaneity). The new dependence of a_{00} and a_{01} on u will be seen to follow from the second Einstein postulate.

We consider the thought experiment discussed in Section 2-1. As shown in Fig. 2-8a, a flash tube at O, fixed in R, emits a light pulse at $t = 0$. The pulse

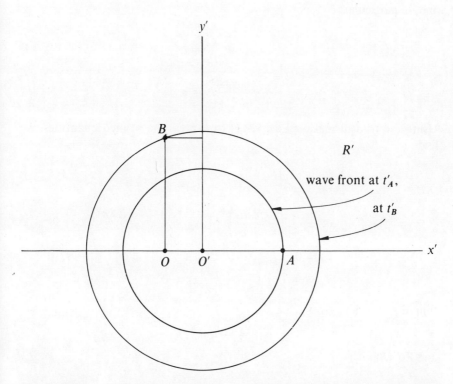

Figure 2-8b.   The situation in R' at two instants—when the light pulse reaches A (t'_A) and when it reaches B (t'_B). The position of O is shown at t'_B.

spreads spherically with speed c:

$$x^2 + y^2 + z^2 = c^2 t^2 \tag{2-17}$$

The event located in space and time by x, y, z, and t is the presence of the wave front. Let the subscripts A and B refer to arrival of the pulse at the points $A(L, 0, 0)$, $B(0, L, 0)$ on the sphere about O of radius L. According to the rule for synchronizing clocks [Eq. (2-5)],

$$t_A = 0 + \frac{L}{c} \qquad t_B = 0 + \frac{L}{c}$$

which is consistent with Eq. (2-17).

Consider the same phenomena as observed in R' (Fig. 2-8b). In R', points O, A, B, are moving in the $-x'$ direction with speed u. According to the second postulate, the pulse spreads spherically in R' with speed c despite the motion of the source

$$x'^2 + y'^2 + z'^2 = c^2 t'^2 \tag{2-18}$$

and, in particular,

$$\frac{x'_A}{t'_A} = c \tag{A}$$

$$\frac{(x_B'^2 + y_B'^2)^{1/2}}{t'_B} = c \tag{B}$$

Using the transformation Eqs. (2-16) to express the primed quantities, we find

$$t'_A = a_{00}\frac{L}{c} + a_{01}L \qquad t'_B = a_{00}\frac{L}{c}$$

$$x'_A = a_{00}\left(L - \frac{uL}{c}\right) \qquad x'_B = -a_{00}\frac{uL}{c}$$

$$y'_A = 0 \qquad\qquad y'_B = L$$

$$z'_A = 0 \qquad\qquad z'_B = 0$$

Equation (A) then gives

$$a_{01} = -\frac{u}{c^2}a_{00}$$

and Eq. (B) gives

$$a_{00} = \left(1 - \frac{u^2}{c^2}\right)^{-1/2}$$

so that

$$a_{01} = -\frac{u}{c^2}\left(1 - \frac{u^2}{c^2}\right)^{-1/2} \tag{2-20}$$

We have arrived at the Lorentz transformation. Note that a_{00} and a_{01} reduce to their Galilean values as $c \to \infty$, and that they have the required parity properties with respect to u. With the usual abbreviations

$$\beta = \frac{u}{c} \qquad \gamma = \left(1 - \frac{u^2}{c^2}\right)^{-1/2}$$

Equations (2-16) now read

$$t' = \gamma\left(t - \frac{\beta}{c}x\right)$$

$$x' = \gamma(x - \beta ct) \tag{1-31a}$$

$$y' = y$$

$$z' = z$$

They satisfy the third postulate, reducing to Eqs. (1-32a) of this section as $(u/c) \to 0$. They differ from the equations of the Galileo transformation in two respects. One is the presence of the term $-\beta x/c$ in the first equation, making the time of an event depend on its longitudinal position. The other is the presence of the factor γ in the first two equations, implying a change of clock rate and measuring rod length in the transformation. In the rest of this chapter, we shall consider the immediate consequences of the Lorentz transformation.

The sign in the parentheses of the second equation can be checked by considering the Galilean limit

$$x' = x - ut$$

It is useful to remember that the sign inside the parentheses of the first equation is the same as that in the second.

The equations look more symmetrical when ct is used instead of t:

$$\begin{array}{l} ct' = \gamma(ct - \beta x) \\ x' = \gamma(x - \beta ct) \\ y' = y \\ z' = z \end{array} \tag{1-31b}$$

The inverse transformation has primed and unprimed quantities interchanged, and the sign of the relative velocity is reversed.

$$ct = \gamma(ct' + \beta x')$$
$$x = \gamma(x' + \beta ct')$$
$$y = y'$$
$$z = z'$$
(1-31c)

The equations blow up for $u \geq c$, because the denominator is then imaginary or zero. The theory makes sense only if we limit our considerations to reference frames whose relative velocity is less than c

$$|u| < c$$

When we study relativistic dynamics we shall find that no body can be accelerated from a speed $<c$ to one $\geq c$. Thus no reference frame attached to a real body could have $|u| \geq c$. This property of the law of motion fits in nicely with the purely mathematical restriction on u in the Lorentz transformation.

A very important property of the Lorentz transformation is the invariance of the expression

$$\boxed{c^2t^2 - x^2 - y^2 - z^2 = c^2t'^2 - x'^2 - y'^2 - z'^2}$$
(2-21)

This relation holds for any event: (t, x, y, z) in R, (t', x', y', z') in R'; it is not limited to light signals, for which the invariant of Eq. (2-21) is zero. The identity [Eq. (2-21)] is immediately verified by squaring and adding Eqs. (1-31c). Equation (2-21) is the basic invariance property of the Lorentz transformation, just as

$$x^2 + y^2 + z^2 = x'^2 + y'^2 + z'^2$$

is the basic invariance property of the rotation transformation in space.

The Lorentz transformation Eq. (1-31b) can be derived from the invariance requirement [Eq. (2-21)] without considering any particular thought experiment. Starting with Eqs. (2-16),

$$t' = a_{00}t + a_{01}x$$
$$x' = a_{00}(x - ut)$$
$$y' = y$$
$$z' = z$$
(2-16)

we form

$$c^2t'^2 - x'^2 - y'^2 - z'^2 = c^2t^2a_{00}^2\left(1 - \frac{u^2}{c^2}\right) - x^2(a_{00}^2 - a_{01}^2c^2) - y^2 - z^2$$
$$+ 2xct\left(a_{00}a_{01}c + a_{00}^2\frac{u}{c}\right)$$

The identity Eq. (2-21) requires

$$a_{00}^2\left(1 - \frac{u^2}{c^2}\right) = 1$$

$$a_{00}^2 - a_{01}^2c^2 = 1$$

$$a_{01}c + a_{00}\frac{u}{c} = 0$$

Since $a_{00} > 0$, the first equation gives

$$a_{00} = \left(1 - \frac{u^2}{c^2}\right)^{-1/2} = \gamma \tag{2-19}$$

and the third gives

$$a_{01} = \frac{u}{c^2}a_{00} = \frac{-u}{c^2}\left(1 - \frac{u^2}{c^2}\right)^{-1/2} = \frac{-\beta\gamma}{c} \tag{2-20}$$

The second is consistent with these.

The invariance relation [Eq. (2-21)] is thus a sufficient as well as necessary condition for the Lorentz transformation (1-31b). The latter can be defined as that linear transformation [Eq. (2-16)] which keeps invariant the form

$$c^2t^2 - x^2 - y^2 - z^2$$

(or any function thereof).

Since the Lorentz transformation is linear, it applies to any linear combination of the corresponding coordinates of events. Applied in particular to the difference between events 1 and 2, Eq. (2-8), it becomes

$$c\,\Delta t' = \gamma\,(c\,\Delta t - \beta\,\Delta x)$$
$$\Delta x' = \gamma\,(\Delta x - \beta c\,\Delta t)$$
$$\Delta y' = \Delta y$$
$$\Delta z' = \Delta z$$

$$\tag{2-22}$$

applying also to the infinitesimal differences

$$c\,dt' = \gamma\,(c\,dt - \beta\,dx)$$
$$dx' = \gamma\,(dx - \beta c\,dt)$$
$$dy' = dy$$
$$dz' = dz$$

(2-23)

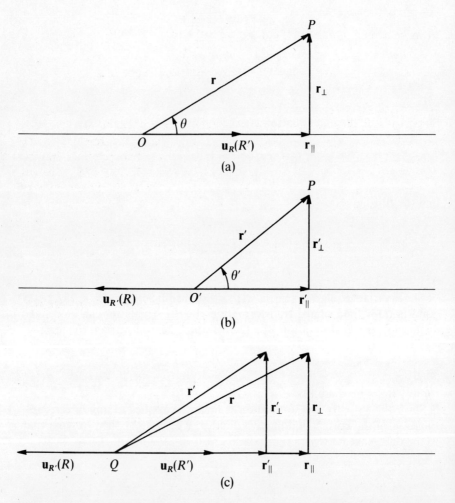

Figure 2-9.   (a) The vectors $\mathbf{r}$, $\mathbf{u}_R(R')$, $\mathbf{r}_\parallel$, $\mathbf{r}_\perp$ in R. (b) The corresponding vectors $\mathbf{r}'$, $\mathbf{u}_{R'}(R)$, $\mathbf{r}'_\parallel$, $\mathbf{r}'_\perp$ in R'. (c) Both sets of vectors plotted together in an abstract three-dimensional space.

The basic invariant [Eq. (2-21)] is in this case called the "infinitesimal interval squared" ds^2:

$$ds^2 = c^2\,dt^2 - dx^2 - dy^2 - dz^2 \qquad (2\text{-}24)$$

(A more explicit notation would use parentheses:

$$(ds)^2 = (c\,dt)^2 - (dx)^2 - (dy)^2 - (dz)^2$$

It is customary to leave them off.)

The orienting of the axes along the direction of relative motion has been just a matter of formal convenience. The only significant direction in the transformation is that of the relative motion of the two reference frames. Let $\mathbf{u}_R(R')$ denote the velocity of R' with respect to R (Fig. 2-9a). The coordinate x of an event at P is simply the component along the direction of $\mathbf{u}_R(R')$ of the radius vector $\mathbf{r}$ from O to P:

$$x = r_{\parallel} = r\cos\theta = \frac{\mathbf{r}\cdot\mathbf{u}_R(R')}{u_R(R')}$$

Here θ is the polar angle of the radius vector with respect to the direction of relative motion. The y and z coordinates are components in the transverse plane of the transverse part of r

$$\mathbf{r}_{\perp} = \mathbf{r} - \mathbf{r}_{\parallel} = \mathbf{r} - \frac{[\mathbf{r}\cdot\mathbf{u}_R(R')]\mathbf{u}_R(R')}{[u_R(R')]^2}$$

$$= \mathbf{u}_R(R') \times \frac{\mathbf{r}\times\mathbf{u}_R(R')}{[u_R(R')]^2}$$

In terms of azimuth φ measured from the y axis,

$$y = r_{\perp}\cos\varphi = r\sin\theta\cos\varphi$$
$$z = r_{\perp}\sin\varphi = r\sin\theta\sin\varphi$$

Corresponding relations hold in R' (Fig. 2-9b). With $\mathbf{u}_{R'}(R)$ denoting the velocity of R with respect to R',

$$x' = r'_{\|} = r' \cos \theta' = \frac{[\mathbf{r}' \cdot -\mathbf{u}_{R'}(R)]}{u_{R'}(R)}$$

$$\mathbf{r}'_{\perp} = \mathbf{r}' - \mathbf{r}'_{\|} = \mathbf{r}' - \frac{[\mathbf{r}' \cdot -\mathbf{u}_{R'}(R)][-\mathbf{u}_{R'}(R)]}{[u_{R'}(R)]^2}$$

$$= [-\mathbf{u}_{R'}(R)] \times \frac{\mathbf{r}' \cdot -\mathbf{u}_{R'}(R)}{[u_{R'}(R)]^2}$$

$$y' = r'_{\perp} \cos \varphi' = r' \sin \theta' \cos \varphi'$$

$$z' = r'_{\perp} \sin \varphi' = r' \sin \theta' \sin \varphi'$$

The magnitudes of the relative velocities are equal

$$u_{R'}(R) = u_R(R') = \beta c$$

The transformation Eqs. (1-31b) can be written

$$\boxed{\begin{aligned} ct' &= \gamma(ct - \beta r_{\|}) \\ r'_{\|} &= \gamma(r_{\|} - \beta ct) \\ \mathbf{r}'_{\perp} &= \mathbf{r}_{\perp} \end{aligned}}$$

(1-31d)

where the third equation means that $r'_{\perp} = r_{\perp}$ and the points of the plane of $\mathbf{r}'$ and $\mathbf{u}_{R'}(R)$ move in R in the plane of $\mathbf{r}$ and $\mathbf{u}_R(R')$. In this form, the Cartesian axes have disappeared from the transformation equations. The first two equations of Eqs. (1-31d) relate the times and the longitudinal distances from the origin. The third states that in the transverse plane distances and angles are invariant.

At the risk of some confusion, one can draw the vectors $\mathbf{r}$, $\mathbf{u}_R(R')$, $\mathbf{r}'$, and $\mathbf{u}_{R'}(R)$ from a common origin Q in an abstract three-dimensional space (Fig. 2-9c). The construction is made by plotting the components (x, y, z), $(\beta c, 0, 0)$, (x', y', z'), and $(-\beta c, 0, 0)$ in a common Cartesian coordinate system. The transformation equations can be written as vectorial identities in this hybrid space:

$$ct' = \gamma \left\{ ct - \frac{\mathbf{r} \cdot \mathbf{u}_R(R')}{c} \right\}$$

$$(1\text{-}31\text{e})$$

$$\mathbf{r}' = \mathbf{r} + (\gamma - 1)\frac{\mathbf{r} \cdot \mathbf{u}_R(R')}{[u_R(R')]^2}\mathbf{u}_R(R') - \gamma\frac{\mathbf{u}_R(R')}{c}ct$$

Finally, we give the Lorentz transformation in terms of spherical polar coordinates with the direction of relative motion as axis

$$ct' = \gamma(ct - \beta r \cos\theta)$$

$$r' \cos\theta' = \gamma(r \cos\theta - \beta ct)$$

$$r' \sin\theta' = r \sin\theta$$

$$(1\text{-}31\text{f})$$

$$\varphi' = \varphi$$

A *rotation* of axes in R leaves invariant the length squared of $\mathbf{r}$

$$x^2 + y^2 + z^2$$

and has no effect on the synchronization of clocks. It therefore preserves the numerical value of

$$c^2t^2 - x^2 - y^2 - z^2$$

and is consistent (trivially, because $u = 0$) with the three postulates. The same holds for a rotation of axes in R'. Rotations can thus be considered trivial special cases of the transformation under discussion. When we consider the composition of successive Lorentz transformations involving different directions of $\mathbf{u}$, we shall see that a rotation of axes is involved. It is then advantageous to include spatial rotations among the Lorentz transformations. Only then do the Lorentz transformations have the basic group property: The resultant of successive Lorentz transformations is a Lorentz transformation.

The transformation [Eq. 1-31b)] with a pair of corresponding axes (in this case x and x') in the direction of relative motion is called the *special Lorentz transformation* (SLT).

The *general Lorentz transformation*—that between an arbitrarily oriented coordinate system in R and an arbitrarily oriented coordinate system in R'—

can be realized in three steps:

(1) Rotate axes in R so that the new x axis is parallel to $\mathbf{u}_R(R')$. There are, of course, many such rotations, for one axis does not determine uniquely the orientation of a rigid frame. It is necessary to keep track of the one used. Knowing its matrix elements a_{jk}, we can calculate the new space coordinates by Eq. (A-3); t is invariant.

(2) Make a SLT to a parallel coordinate system in R' [Eqs. (1-31b)].

(3) Rotate axes in R' from these axes to the specified orientation. Again, Eq. (A-3) specifies the changes in space coordinates, and the time coordinate is unaffected.

If the handedness of the two coordinate systems happens to be different, we have to add at some stage another step, an *inversion*, which, like rotation, does not alter t or $c^2t^2 - x^2 - y^2 - z^2$.

The general Lorentz transformation is evidently characterized by linearity (the resultant of linear transformations is linear) and invariance of the quadratic form

$$c^2t^2 - x^2 - y^2 - z^2$$

A *pure Lorentz transformation* (PLT)—also called a *Lorentz transformation without rotation* or a "boost"—is defined as one in which the coordinate systems have the same handedness and the same orientation with respect to the direction of relative motion of the reference frames. That is, the x, y, and z axes make the same angles with $\mathbf{u}_R(R')$ as do, respectively, the x', y', and z' axes with $-\mathbf{u}_{R'}(R)$. The rotation (1) in R is then equal and opposite to the rotation (3) in R'. An example is shown in Fig. 2-10, where (1) is a rotation of 40° about the z axis and (3) is a rotation of $-40°$ about the z' axis. In this specific sense, the corresponding axes are "parallel." We shall see that simultaneous-in-R positions of the primed axes are not parallel to the corresponding unprimed axes in R, nor are simultaneous-in-R' positions of the unprimed axes parallel to the corresponding primed axes in R'. The axes do not coincide at any t or t', except in the special case of the SLT.

We have limited ourselves so far to homogeneous Lorentz transformations, for which the space origins coincide at $t = t' = 0$. In doing physics we never consider an isolated event *in situ* but are concerned with relations between events. We have above simply taken one of them as marking the origin of space and time coordinates. The quantities in the preceding equations, called t, x, y, z, are really Δt, Δx, Δy, Δz between the event in question and the origin event; the fundamental invariant of the Lorentz transformation is

$$\Delta s^2 = c^2 \, \Delta t^2 - \Delta x^2 - \Delta y^2 - \Delta z^2 \qquad (2\text{-}25)$$

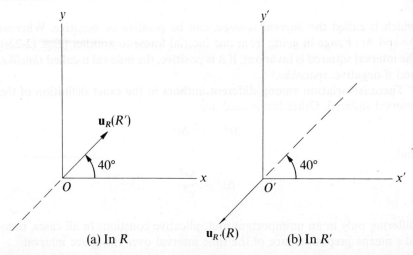

(a) In R 　$\mathbf{u}_R'(R)$ 　(b) In R'

Figure 2-10. An example of a pure Lorentz transformation (PLT). (a) In R, the axes are first rotated 40° about the z axis. A special Lorentz transformation (SLT) with velocity $\mathbf{u}_R(R')$ along the new x direction is then carried out. (b) In R', the primed axes are rotated $-40°$ about the z' axis.

independent of any choice of origins. The omission of the Δ's and restriction to homogeneous transformations is permissible so long as space and time in an inertial reference frame are homogeneous, as is assumed in special relativity.

2-4. INTERVALS

We now consider in detail the relation between different events. The separation between event 1 and event 2 has a spatial part

$$\Delta x = x_2 - x_1 \qquad \Delta y = y_2 - y_1 \qquad \Delta z = z_2 - z_1 \qquad (2-8)$$

whose magnitude Δr, called the distance or separation or space interval, satisfies

$$\Delta r^2 = \Delta x^2 + \Delta y^2 + \Delta z^2$$

and a temporal part, the time interval

$$\Delta t = t_2 - t_1 \qquad (2-8)$$

The expression formed from the difference of the squares

$$\Delta s^2 = c^2\,\Delta t^2 - \Delta r^2 \qquad (2-25)$$

which is called the *interval squared,* can be positive or negative. Whereas Δr and Δt change in going from one inertial frame to another [Eqs. (2-22)], the interval squared is invariant. If it is positive, the interval is called *timelike,* and if negative, *spacelike.*

There is variation among different authors in the exact definition of the interval squared. Other forms used are

$$\Delta r^2 - c^2 \Delta t^2$$

and

$$\Delta t^2 - \frac{\Delta r^2}{c^2}$$

differing only in an unimportant multiplicative constant. In all cases, time-like means preponderance of the time interval over the space interval

$$c^2 \Delta t > \Delta r^2 \tag{2-26a}$$

and spacelike means the opposite

$$\Delta r^2 > c^2 \Delta t^2 \tag{2-26b}$$

The essential feature, common to all definitions, is that the positive quantities Δr^2 and $c^2 \Delta t^2$ enter with opposite sign.

The invariance of the interval squared has important implications for the order in which events happen. If the interval is spacelike, there is no absolute time order of the events. If in R event 1 precedes event 2 $(\Delta t > 0)$, a frame R' can be found in which the events are simultaneous $(\Delta t' = 0)$, and there are a host of frames R'' in which the time sequence is reversed $(\Delta t'' < 0)$. Figure 2-11 shows the positions P_1 and P_2 of the events in R. Let R' move in the

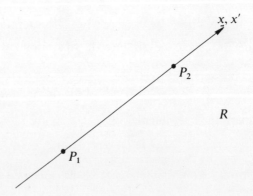

Figure 2-11.

direction $P_1 P_2$ with speed $u_R(R') = \beta c$. Choose the x and x' axes along $P_1 P_2$. According to Eq. (2-22), simultaneity is achieved for

$$\beta = \frac{c\,\Delta t}{\Delta x} = \frac{c\,\Delta t}{\Delta r} < 1$$

The spacelike character of the interval makes $\beta < 1$, and therefore physically realizable. To reverse the time order, let β exceed $c\,\Delta t/\Delta r$. It can do so and still be less than one. Conversely, if reference frames exist in which the events are simultaneous or reversed in time order, the interval between the events is spacelike. Such events cannot be related to one another physically, for in that case they would have to have a definite temporal order valid in all reference frames.

For a spacelike interval, the spatial separation of the events is greater than the distance traveled by light during their temporal separation. A field traveling with a speed not exceeding c could not get from one event to the other. If an interaction were to be discovered that propagated faster than light, it could in principle couple events having a spacelike interval. We would then have to give up either special relativity or the view that physically related events have an invariant time order. (This view is called "causality" in quantum field theory.) It would not do to keep the relativistic formulas but use a larger value of c, for several very precise experiments have established that the c of the relativistic formulas is indeed numerically equal to the characteristic velocity of the electromagnetic interaction.

Events with a spacelike interval have an invariant order along the line joining their positions. In Eq. (2-22), $\Delta x > \beta c\,\Delta t$, and the sign of $\Delta x'$ is that of Δx.

If an interval is timelike, the events may be causally related. In all inertial frames the time order is the same. In effect, again taking the x and x' axes along $P_1 P_2$, we have $\Delta x = \Delta r$ and

$$c\,\Delta t > \beta\,\Delta x \qquad (\beta < 1)$$

By Eq. (2-22), $c\,\Delta t'$ has the same sign as $c\,\Delta t$. In this case, the events can be made coincident in space by an appropriate Lorentz transformation, with

$$\beta = \frac{\Delta x}{c\,\Delta t} = \frac{\Delta r}{c\,\Delta t} < 1$$

and even interchanged as regards position on the line joining them.

We can represent the situation graphically, as in Fig. 2-12. The time separation between events is plotted vertically and the space separation is plotted horizontally. Of course, there are three space dimensions, so we cannot picture them all. Event 1 is at the origin. The lines $\Delta r = \pm c\,\Delta t$ are generators of the "light cone." If event 2 is anywhere in the upper nappe of the light cone, $c\,\Delta t > \Delta r$, the interval is timelike, and event 2 occurs after event 1 in every inertial frame. So this half of the cone can be called the "absolute future." If event 2 is in the lower nappe, the interval is again timelike, and event 2 is in the "absolute past." The region outside the cone is called "elsewhere." If event 2 is in this region,

$$\Delta r > c\,\Delta t$$

and the interval is spacelike. The events can have any time relation, depending on the reference frame used. Their spatial separation has the same sign in all frames, always exceeding a certain minimum value.

For any pair of events, there is evidently a reference frame in which the interval squared takes on the simplest form. If the interval between the events is timelike, the favored frame is that in which the events occur at the same place, the "rest frame" of the interval. In this reference frame the time interval has the smallest possible absolute value, because

$$\Delta s^2 = c^2\,\Delta t^2 - \Delta r^2 = c^2\,\Delta t'^2 - \Delta r'^2$$

If the interval is spacelike, the favored frame is that in which the events occur at the same time, the "equal-time" frame. In this reference frame, the space interval has the smallest possible absolute value, since its square equals the

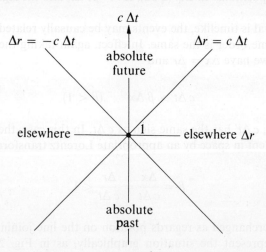

Figure 2-12.   Trace of the light cone, in time versus space plot.

invariant

$$-\Delta s^2 = \Delta r^2 - c^2 \, \Delta t^2$$

Events marking successive positions of a material particle (speed $v < c$) have a timelike interval, because

$$\Delta r \leq v_{max} \, \Delta t < c \, \Delta t$$

They must lie inside the light cone. Precisely stated, if the light cone is drawn for an event in the history of the particle, all other events in its history are inside the cone, neatly divided between past and future nappes. Events in the history of a photon or neutrino ($v = c$) are on the cone, not inside it.

EXERCISE A beam of negative pions passes through a liquid hydrogen target. An array of spark chambers and scintillation counters detects charged particles downstream. One of the pions interacts with a target proton at $x_1 = -50.0$ cm, $y_1 = 50.0$ cm, $z_1 = 20.0$ cm at $t_1 = 0.00$ sec (lab system coordinates). In this event—event 1—a K^0 meson is produced, which, being electrically neutral, moves with constant velocity until event 2 occurs. Event 2, breakup of the K^0 into two charged pions ($K^0 \to \pi^+ \pi^-$), occurs at $x_2 = 88.0$ cm, $y_2 = 48.0$ cm, $z_2 = 25.0$ cm at $t_2 = 5.31 \times 10^{-9}$ sec (Fig. 2-13). Compute the distance between the

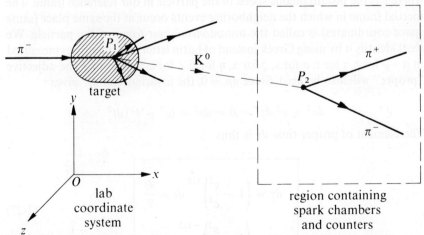

Figure 2-13. Production of a K^0 meson at P_1, its disintegration in flight at P_2.

events, the elapsed time, the velocity of the K^0, and the interval squared. Compute the velocity of the rest frame of the events and the lifetime of the K^0 in that frame.

SOLUTION   The distance is $(138.0^2 + 2.0^2 + 5.0^2)^{1/2} = 138.1$ cm. The elapsed time is 5.31×10^{-9} sec. The velocity of the K^0 is the distance divided by the difference in readings of the synchronized clocks, or 2.55×10^{10} cm sec^{-1}.

The interval squared is $(3.00 \times 10^{10} \times 5.31 \times 10^{-9})^2 - 138.1^2$ $= (159.3 - 138.1)(159.3 + 138.1) = 63.1 \times 10^2$ cm$^2 = 79.5^2$ cm^2.

The velocity of the rest frame, $\beta c = \Delta r/\Delta t$, is equal to the velocity of the K^0 meson, 2.55×10^{10} cm sec^{-1}.

The lifetime of the K^0 in its rest frame is found from the invariant interval squared, $\Delta s^2 = c^2\,\Delta t'^2 - 0$; thus $\Delta t' = \Delta s/c = 2.65 \times 10^{-9}$ sec. This lifetime is read on one and the same clock, moving with the K^0 from birth to death. It is one-half the lifetime in the laboratory, found by subtracting the readings of clocks at P_1 and P_2.

2-5. PROPER TIME AND PROPER LENGTH; TIME DILATION AND LENGTH CONTRACTION

When the successive events in the history of a particle are only infinitesimally different, the interval squared [Eq. (2-24)] is

$$ds^2 = c^2\,dt^2 - dr^2 = c^2\,dt^2 - v^2\,dt^2$$

where v is the instantaneous speed of the particle in our reference frame. The inertial frame in which the neighboring events occur at the same place (same space coordinates) is called the *instantaneous rest frame* of the particle. We shall identify it by using Greek instead of Latin letters for quantities measured in it—ρ for r, τ for t, σ for s, ξ for x, η for y, ζ for z, and so on. The adjective "proper" will also be used. Since $d\rho = 0$, the invariance of ds^2 gives

$$d\sigma^2 = c^2\,d\tau^2 - 0 = ds^2 = (c^2 - v^2)\,dt^2$$

The element of proper time $d\tau$ is thus

$$d\tau = \left(1 - \frac{v^2}{c^2}\right)^{1/2} dt = \frac{dt}{\gamma}$$

$$\gamma = \left(1 - \frac{v^2}{c^2}\right)^{-1/2}$$

(2-27)

It has the important feature that the readings τ and $\tau + d\tau$ are made on one and the same monument clock. In R, on the other hand, t and $t + dt$ are read on different, but synchronized, clocks. If the particle moves with constant velocity, Eq. (2-27) integrates to

$$\Delta\tau = \frac{\Delta t}{\gamma} \qquad\qquad (2\text{-}28)$$

A down-to-earth example of proper time and laboratory time is provided by a moving automobile. For events happening near the automobile, proper time is shown by the clock on the dashboard, laboratory time by the clocks on banks and churches along the road.

The relation Eq. (2-27) between laboratory time and proper time is the famous Einstein time dilation, or slowing down of moving clocks. The time difference read on a clock that is on the spot at both events (it is thus a proper time) is less than the elapsed time in any other inertial frame by the factor $(1 - v^2/c^2)^{1/2}$, v being the velocity of the moving clock. The exercise of the preceding section illustrates the effect—the clock moving with the K^0 meson makes half as many oscillations during the particle's life as the number of oscillations corresponding to the difference in readings of synchronized laboratory clocks at place of birth and place of death.

We shall discuss later (Section 2-7) the experimental tests of this prediction. At this time, we remark only that it has been brilliantly confirmed. The prediction follows logically from the rule for synchronization and the postulates. It must apply to all kinds of clocks, whether they use cesium atoms or grains of sand or even biological processes.

If the particle accelerates, the instantaneous rest frame is a different one at every instant. The factor γ^{-1} is a function of the time. One can still add the infinitesimal time intervals between successive events, obtaining

$$\tau_2 - \tau_1 = \int_{\tau_1}^{\tau_2} d\tau = \int_{\tau_1}^{\tau_2} dt \left[1 - \frac{v^2(t)}{c^2} \right]^{1/2} \leq (t_2 - t_1) \qquad (2\text{-}29)$$

But adding $d\tau$'s read on a host of different imagined clocks (each in an instantaneous rest frame) is different from cumulating the ticks of one accelerated clock. The accelerated clock is affected by the acceleration, to an extent depending on its specific structure; a pendulum clock is grossly disturbed; an iron-57 nucleus is essentially impervious. The effects of acceleration can be corrected out, like the effects of temperature and stray magnetic fields. After these corrections have been made, Eq. (2-29) applies, relating the elapsed time $(\tau_2 - \tau_1)$ read on an arbitrarily moving clock and the times t read on the monument clocks in some inertial reference frame.

Similarly, the proper length λ of a rod is the distance between the ends in the inertial reference frame in which the rod is instantaneously at rest. It is independent of the orientation of the rod (isotropy of inertial frame). To define length in a frame R in which the rod is moving, we consider two events. Event 1 is determination of (t, x, y, z) for one end of the rod; event 2 is determination of (t, x, y, z) for the other. When the two events are simultaneous in R ($t_1 = t_2$), the distance between them is called the length l of the rod:

$$l = [(x_2 - x_1)^2 + (y_2 - y_1)^2 + (z_2 - z_1)^2]^{1/2} = \Delta r \qquad \Delta t = 0$$

In the rest frame (Rho), these events are not simultaneous. Since the ends of the rod are at rest in Rho, the coordinate measurements can be made there at any time, and we can use the same events 1 and 2 to determine λ:

$$\lambda = [(\xi_2 - \xi_1)^2 + (\eta_2 - \eta_1)^2 + (\zeta_2 - \zeta_1)^2]^{1/2} = \Delta\rho$$

The relation between $\Delta\rho$ and Δr depends on the orientation of the rod with respect to the direction of motion. The transformation Eqs. (2-22)

$$c(\tau_2 - \tau_1) = \gamma[c(t_2 - t_1) - \beta(x_2 - x_1)] = -\gamma\beta(x_2 - x_1)$$

$$\xi_2 - \xi_1 = \gamma[(x_2 - x_1) - \beta c(t_2 - t_1)] = \gamma(x_2 - x_1)$$

$$\eta_2 - \eta_1 = y_2 - y_1$$

$$\zeta_2 - \zeta_1 = z_2 - z_1$$

give for the length $l_\perp$, if the rod is perpendicular to the direction of motion (Fig. 2-14a)

$$\lambda = l_\perp \tag{2-30a}$$

If it is parallel to it (Fig. 2-14b), the second equation gives for the length $l_\parallel$

$$\lambda = \gamma l_\parallel \tag{2-30b}$$

or

$$l_\parallel = \left(1 - \frac{v^2}{c^2}\right)^{1/2} \lambda$$

This equation is the famous Lorentz–Fitzgerald contraction hypothesized independently by Fitzgerald and Lorentz in the 1890's to explain in the framework of the stationary ether theory the null result of the Michelson–Morley experiment. A rod parallel to the direction of its motion is shortened by the factor $\gamma = (1 - v^2/c^2)^{-1/2}$. A rod perpendicular to its direction of motion is unchanged in length. From the relativistic point of view the contraction is not an absolute flattening of bodies resulting from their motion through the ether. It is, rather, a symmetrical relation between rods that are in motion with respect to one another. In the rest frame of each one the other is shortened by the factor γ. A meter stick—that is, a stick 1.00 m long in its rest frame—flying longitudinally through the laboratory with

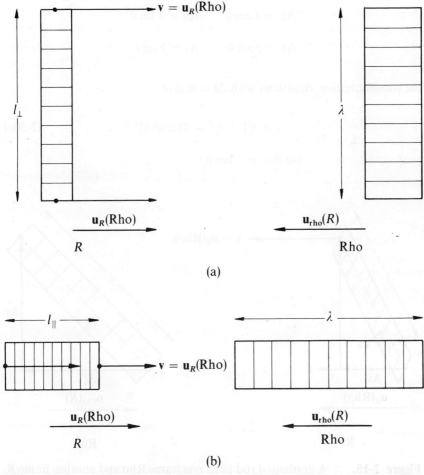

Figure 2-14. A graduated rod (meter stick) shown in its rest frameRho and another frame R. In (a), the rod is perpendicular to the direction of relative motion; in (b), it is parallel.

speed $\beta = 4/5$ has a length in the laboratory (distance between simultaneous positions of its ends) of 0.60 m. The laboratory meter stick on which this distance is read has in the flying frame a length (distance between simultaneous positions of its ends) of 0.60 m. The contraction results from the fact that simultaneous events in the rest frame of one rod are not simultaneous in the rest frame of the other.

If the rod's orientation is neither longitudinal nor transverse to the direction of relative motion, both the length and the inclination are different in the rod's rest frame and the laboratory frame (Fig. 2-15). Since

$$\Delta \xi = \lambda \cos \theta \qquad \Delta \eta = \lambda \sin \theta$$

$$\Delta x = l \cos h \qquad \Delta y = l \sin h$$

the transformation equations with $\Delta t = 0$ give

$$\lambda = l[1 + (\gamma^2 - 1) \cos^2 h]^{1/2} \qquad (2\text{-}30c)$$

$$\tan \theta = \gamma^{-1} \tan h$$

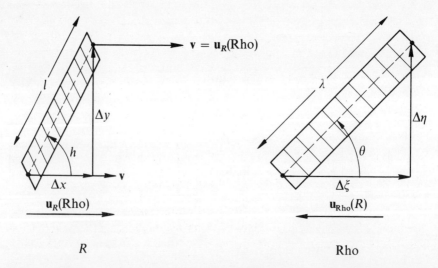

Figure 2-15.   A graduated rod in its rest frame Rho and another frame R. It is inclined to the direction of relative motion. Note the change in inclination (h, θ) as well as length (l, λ). The graduations, which are perpendicular to the edge in Rho, are not perpendicular to it in R.

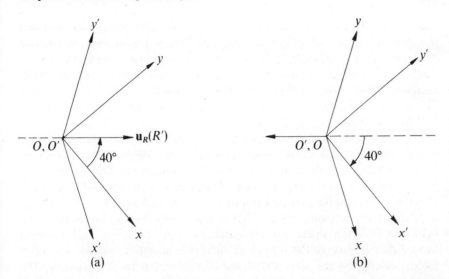

Figure 2-16. The coordinate axes at $t = t' = 0$ in the pure Lorentz transformation of Fig. 2-10. (a) Shown in R. (b) Shown in R'. The drawing is for $\gamma = 3$.

The effect is simply compression in the direction of relative motion by the factor γ, with no change of transverse dimensions. The first formula includes Eqs. (2-30b and a) as the special cases $h = 0$, $h = \pi/2$. The second formula states that a tilted rod is tipped so as to be more transverse to the direction of relative motion. In the nonrelativistic region ($\gamma \approx 1$), $\lambda \approx l$ and $\theta \approx h$; lengths and angles are invariant. In the extreme relativistic region ($\gamma \gg 1$), on the other hand, the rod is much shorter and essentially normal to the direction of relative motion, no matter what its orientation in the rest frame. Figure 2-15 is drawn for an intermediate case, $\gamma = 2$. The increase of inclination $\delta = h - \theta$ is by the addition law of tangents given by

$$\tan \delta = \tan \theta \, \frac{\gamma - 1}{1 + \gamma \tan^2 \theta} \tag{2-30d}$$

We must guard against giving an absolute significance to length and direction. A length must be thought of as the distance in a particular reference frame between events simultaneous in that reference frame. An angle must be thought of as the angle between two lines in the reference frame, thus a relation between vectors. The angle θ is the angle in Rho between the axis of the rod and the vector $-\mathbf{u}_{Rho}(R)$ (the line opposite the motion of a point fixed in R); the angle h is the angle in R between the axis of the rod and $\mathbf{u}_R(Rho)$ (the line of motion of a point fixed in Rho). The inequality of θ and h is as much of an affront to pre-Einsteinian habit as is the length contraction and the slowing down of clocks. A weathervane pointing northeast on a house flying east with speed such that $\gamma = 2.4$, points north northeast in the ground frame! In the pure Lorentz transformation shown in Fig. 2-10, the basis vectors in R' are spread apart in R, giving the situation at $t = t' = 0$ shown in Fig. 2-16. Only in the case of the special Lorentz transformation ($\theta = 0$, $\pi/2$) do the corresponding axes coincide at $t = t' = 0$.

The size and shape of a body depend on the reference frame, the deformation from the shape in the rest frame consisting in a longitudinal contraction by the factor γ with no change of transverse dimensions. We see from Eq. (2-30c), with $\lambda = $ const, that a sphere is squashed to an ellipsoid (Fig. 2-17). It needs to be emphasized that nothing is being said here about the visual appearance of a moving object. That is determined by the directions of the light rays from the object arriving simultaneously at the eye. Such photons leave different parts of the surface of the object at different times. We have, rather, calculated the size and shape of the region in R containing the monuments that coincide with the points of the body at a particular clock time in the laboratory frame.

If different parts of the rod have different velocity, there is no instantaneous rest frame for the whole rod. The relations Eq. (2-30) then apply only in the small, to an infinitesimal element of the rod or body:

$$d\lambda = dl[1 + (\gamma^2 - 1)\cos^2 h]^{1/2}$$

$$\tan \theta = \frac{\tan h}{\gamma}$$

(2-31)

with γ corresponding to the velocity of the element.

2-6. MINKOWSKI'S HYPERBOLIC GRAPH

Minkowski [23] devised an elegant geometrical representation of the special Lorentz transformation

$$ct' = \gamma(ct - \beta x)$$

$$x' = \gamma(x - \beta ct)$$

(1-31b)

using oblique axes in the plane. A point in the plane represents an event. The coordinates of the event in different reference frames—(ct, x), (ct', x'), and so on—are the parallel projections (Fig. 2-18) of the point on the corresponding axes in the plane (we have chosen one pair, for the laboratory reference frame, to be orthogonal). A continuous sequence of

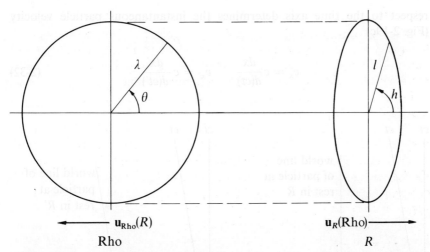

Figure 2-17.   A spherical body in its rest frame (Rho) is a flattened spheroid in the laboratory frame (R).

events marking the motion of a particle corresponds to a curve in the plane, called "world line" (Minkowski used the name "world" for the manifold of events, four-dimensional spacetime). If the world line is parallel to the ct axis (Fig. 2-19a), the particle is at rest in R. If it is parallel to the ct' axis (Fig. 2-19b), it is at rest in R'. In general, the inclination of the world line with

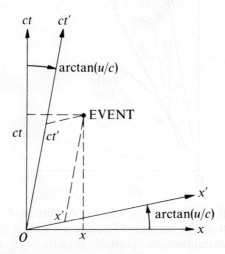

Figure 2-18.

respect to the time axis determines the instantaneous particle velocity (Fig. 2-19c)

$$v_x = c\,\frac{dx}{d(ct)} \qquad v_{x'} = c\,\frac{dx'}{d(ct')} \tag{2-32}$$

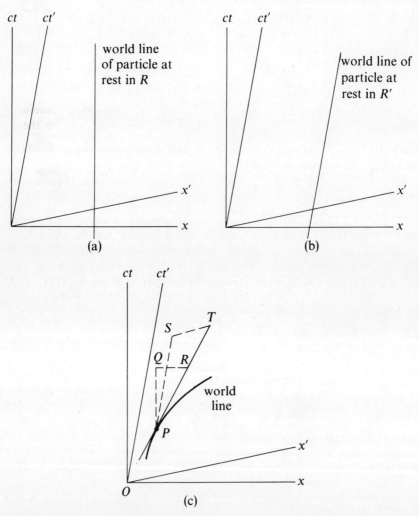

Figure 2-19.   World line of particle. (a) At rest in R. (b) At rest in R'. (c) The inclination of the tangent PT to the world line from the time axis determines the instantaneous particle velocity. $\overline{QR} \propto dx$, $\overline{PQ} \propto cdt$; $\overline{ST} \propto dx'$, $\overline{PS} \propto cdt'$, so $v_x = c(\overline{QR}/\overline{PQ})$ and $v'_x = c(\overline{ST}/\overline{PS})$.

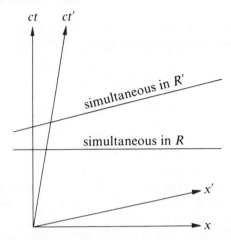

Figure 2-20.

A straight world line represents a particle in uniform motion. Curvature indicates acceleration.

We are, of course, describing longitudinal positions and velocities. The transverse components cannot be shown on this two-dimensional graph.

The origin O' in R' coincides with O at $t = t' = 0$. Its world line has the slope corresponding to $v_x = u$, so that its angle with the ct axis has u/c for its tangent. This world line traces out the ct' axis (the locus of events for which $x' = 0$; Fig. 2-18).

Simultaneous events are on straight lines parallel to the corresponding x or x' axis (Fig. 2-20). The x' axis is the locus of events occurring at $t' = 0$. We see from the first Eq. (1-31) that it makes an angle arctan (u/c) with the x axis (Fig. 2-18). Thus, the x' axis makes the same angle with the x axis that the ct' axis makes with the ct axis. Figure 2-21 shows the positive half-axes for several values of u/c. Note the symmetry about the 45° line, $ct = x$. The positive time axes are all in the upper quadrant bounded by the world lines of photons emitted at $t = 0$ from O in the positive and negative x directions ($x = \pm ct$). They are in the absolute future part of the light cone. The positive space axes are all in the right-hand quadrant outside the light cone, in the region called "elsewhere."

The relativity of simultaneity is immediately evident from the obliquity of the axes (Fig. 2-20). There is only one reference frame, if any, in which two given events are simultaneous—that for which the x' axis is parallel to the straight line joining the world points. In all other reference frames the events occur at different times.

We need to find the units of length along the axes. Draw the hyperbolas (Fig. 2-22)

$$c^2 t^2 - x^2 = \pm 1 \tag{2-33a}$$

They have as asymptotes the generators $x = \pm ct$ of the light cone. The invariance property of the transformation

$$c^2t^2 - x^2 = c^2t'^2 - x'^2$$

means that the events constituting the hyperbolas have for equation in the primed coordinates

$$c^2t'^2 - x'^2 = \pm 1 \tag{2-33b}$$

The points $(x = 0, ct = 1)$ and $(x = 1, ct = 0)$ are labeled "1" in the figure. The points $(x' = 0, ct' = 1)$ and $(x' = 1, ct' = 0)$ are labeled "1'." They mark the scale units in the two coordinate systems. In using these oblique axes we have to remember that the coordinates refer to parallel, not orthogonal, projections, and that the unit of length is different for each pair of axes.

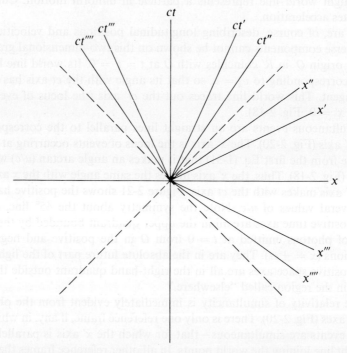

Figure 2-21.   Various pairs of axes in the spacetime plane. For each pair, the x' axis makes the angle $\arctan(u/c)$ with the x axis and the ct' axis makes the same angle with the ct axis. R' has $u = c/2$, R'' has $u = 3c/4$, R''' has $u = -c/2$, and R'''' has $u = -3c/4$. Only the positive halves of the axes are drawn.

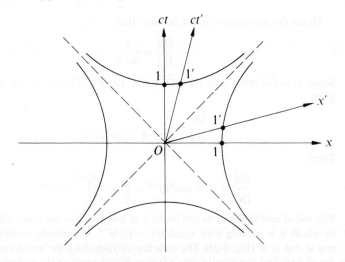

Figure 2-22. The hyperbolas $c^2t^2 - x^2 = c^2t'^2 - x'^2 = 1$ and $c^2t^2 - x^2 = c^2t'^2 - x'^2 = -1$ determine the scale of each axis.

Let us use the graph to depict the Fitzgerald–Lorentz contraction, Eq. (2-30b). In Fig. 2-23, we consider a rod at rest in R parallel to the x axis. It is of unit length with ends at $x = 0$ and $x = 1$. The world lines of the ends are shown, one being the ct axis and the other the parallel line AB. In R', the length of the rod is the difference of simultaneous (same t') x' coordinates, thus $\overline{OB}$. The unit of length in R' is $\overline{OC}$, so we see that the rod is shorter in the frame R', with respect to which it is moving with velocity $-u$.

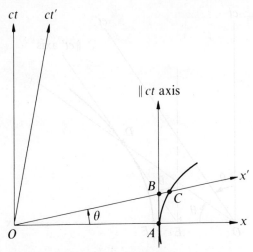

Figure 2-23. Rod at rest in R is shorter in R'.

To get the quantitative result, we note that

$$\frac{\overline{OB}}{\overline{OA}} = \frac{1}{\cos\theta}$$

Since $x_c = \overline{OC}\cos\theta$ and $ct_c = \overline{OC}\sin\theta$, the Eq. (2-33a) of the hyperbola gives

$$\frac{\overline{OC}}{\overline{OA}} = \frac{1}{(\cos^2\theta - \sin^2\theta)^{1/2}}$$

Thus,

$$\frac{\overline{OB}}{\overline{OC}} = (1 - \tan^2\theta)^{1/2} = \left(1 - \frac{u^2}{c^2}\right)^{1/2} = \gamma^{-1}$$

The rod of unit length in its rest frame is of length γ^{-1} in the frame with respect to which it is moving with speed $c(1 - 1/\gamma^2)^{1/2}$. Conversely, consider a unit rod at rest in R' (Fig. 2-24). The ends are moving along the world lines shown, the ct' axis and the parallel line ED (which is tangent to the hyperbola at D). In R', the length of $\overline{OF}$ is 1. In R, the length is the difference of simultaneous (same t) readings of x, thus $\overline{OE}$. Again $\overline{OE}$ is shorter than the unit length $\overline{OF}$. Quantitatively, the law of sines applied to ΔODE gives

$$\frac{\overline{OE}}{\overline{OD}} = \frac{\sin(\pi/2 - 2\theta)}{\sin(\pi/2 + \theta)} = \frac{\cos^2\theta - \sin^2\theta}{\cos\theta}$$

whereas Eq. (2-33a) gives

$$\frac{\overline{OF}}{\overline{OD}} = \frac{(\cos^2\theta - \sin^2\theta)^{1/2}}{1}$$

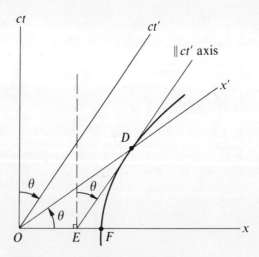

Figure 2-24.   Rod at rest in R' is shorter in R.

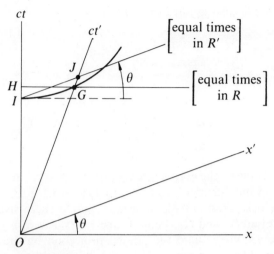

Figure 2-25.   Time dilation.

Therefore,

$$\frac{\overline{OE}}{\overline{OF}} = (1 - \tan^2 \theta)^{1/2} = \gamma^{-1}$$

The answer is the same, whether the rest frame of the rod is taken to be R or R'.

The graph shows very vividly that the contraction is a direct consequence of the breakdown of absolute simultaneity. The extension of the argument to apply to a longitudinal rod of any length or location is straightforward. We simply graph $(c\,\Delta t, \Delta x)$, $(c\,\Delta t', \Delta x')$ instead of (ct, x), (ct', x') and consider the hyperbola

$$c^2 \, \Delta t^2 - \Delta x^2 = -\lambda^2$$

EXERCISE   Use the hyperbolic graph to derive the Einstein dilation. Do it for two cases: clock at rest in R; clock at rest in R'.

SOLUTION   (See Fig. 2-25.) Consider a clock at rest at O. In one unit of time in R it moves along the ct axis from O to I. The events O and I might be successive ticks of the clock. The time of the second click in R' is found by parallel projection on the ct' axis. It is thus $\overline{OJ}$, with $\overline{OG}$ being unit time in R'. The equation of the hyperbola Eq. (2-33a) gives us

$$\overline{OG} = (\cos^2 \theta - \sin^2 \theta)^{-1/2} \qquad \overline{OI} = 1$$

and the law of sines applied to ΔOIJ gives

$$\frac{\overline{OJ}}{\overline{OI}} = \frac{\sin(\pi/2 + \theta)}{\sin(\pi/2 - 2\theta)} = \frac{\cos\theta}{\cos^2\theta - \sin^2\theta}$$

Thus,

$$\frac{\overline{OJ}}{\overline{OG}} = (1 - \tan^2\theta)^{-1/2} = \gamma$$

The time between the events is longer by the factor γ in the frame with respect to which the clock has the speed $c(1 - 1/\gamma^2)^{+1/2}$.

For a clock at rest in R', let it be at O', and let successive ticks correspond to events O and G. The unit time interval in R' is $\overline{OG}$. In R, the time of event G is found by parallel projection on the ct axis, with $\overline{OI}$ as unit of length. Again

$$\frac{\overline{OH}}{\overline{OG}} = \cos\theta$$

whereas the equation of the hyperbola gives

$$\frac{\overline{OI}}{\overline{OG}} = \frac{1}{(\cos^2\theta - \sin^2\theta)^{-1/2}}$$

so that

$$\frac{\overline{OH}}{\overline{OI}} = (1 - \tan^2\theta)^{-1/2} = \gamma$$

The time dilation depends only on the speed, not the direction, so it is immaterial whether the clock is moving to the right or to the left.

2-7. TESTS OF THE EINSTEIN DILATION; THE TWIN "PARADOX"

A long time, some 35 (earth) years, elapsed before the predicted slowing down of moving clocks could be tested experimentally. One test [24] used an atomic clock. Consider (Fig. 2-26) a beam of excited atoms moving in an evacuated tube with velocity v relative to the laboratory. A spectrograph is set up to measure the wavelength of the light from the atomic beam. When the angle of view χ is a right angle, classical theory predicts no Doppler shift, but special relativity does. In its rest frame the radiating atom emits

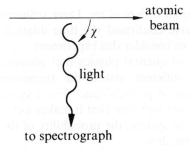

to spectrograph

Figure 2-26.   Transverse Doppler effect.

v_0 wave crests in one second. This number of crests is invariant—it is just a matter of counting—but the elapsed time interval is different in the laboratory, being $\gamma = (1 - v^2/c^2)^{-1/2}$ sec. Thus, the number of crests emitted per laboratory second is

$$v = \frac{v_0}{\gamma}$$

With $\chi = \pi/2$, there is no component of source motion along the line of sight, and v is equal to the frequency of the light entering the spectrograph. There is thus a relativistic transverse Doppler effect, a shift toward the red, of second order in v/c:

$$\lambda = \frac{c}{v} = \frac{\gamma c}{v_0} = \gamma \lambda_0 \tag{2-34}$$

With kinetic energies of the excited hydrogen atoms of the order of 10 keV, the expected wavelength shift of the H_β (4861 Å) line is only 0.05 Å. Nevertheless, by measuring the displacement from the unshifted line to the center of gravity of the lines from light emitted straight forward ($\chi \approx 0$) and straight back ($\chi \approx \pi$), Ives and Stilwell were able to confirm Eq. (2-34) to within 3%.[2]

Mössbauer technique has permitted a precise test [25] of the transverse Doppler effect formula [Eq. (2-34)] applied to the 0.86 Å (14.4 keV) x ray of iron 57. An arrangement with source on the axis, absorber in a radial hole in the rotor of an ultracentrifuge, and detector outside near the rim makes the line of sight transverse to the relative velocity. The formula is confirmed to within 1%.

Another test was a study of a certain transition rate in a rapidly moving system [26]. Here, because $\gamma \gg 1$, the effect was much larger, constituting

---

[2] The relativistic Doppler effect formula is derived in Appendix F.

a test of Eq. (2-28) to all orders of v/c. Later refinements of this "decay in flight" experiment have confirmed the time dilation formula to within a few percent [27]. Let us consider this experiment.

It is characteristic of quantal physics that phenomena are described as transitions between different states, the transitions being determined statistically by "transition probabilities." For a system in an initial state i there is a probability per unit time that it makes a transition to final state f. In the rest frame of the system, the probability of the transition occurring in the infinitesimal time $d\tau$ is

$$\varkappa(i \rightarrow f) \, d\tau = \varkappa \, d\tau \tag{2-35a}$$

We consider now the simple situation in which there are no alternative transitions, $i \rightarrow g$, $i \rightarrow h, \ldots$, to deplete state i. Let $w(\tau)$ be the probability of the system's surviving in state i for a time τ. Since $w(\tau + d\tau) = w(\tau) + dw$, w satisfies the differential equation

$$-dw = w \cdot \varkappa \, d\tau$$

In words, the probability that the system makes the transition in the time interval between τ and $\tau + d\tau$ equals the probability of surviving to τ times the probability of making the transition in $d\tau$. Integrating, we get the exponential law for the survival probability

$$w = \exp(-\varkappa\tau) \tag{2-35b}$$

(the same as that for the free path in the kinetic theory of gases). It follows that the mean life of the initial state is

$$\tau_0 = \langle\tau\rangle_{\text{av}} = \int_0^\infty d\tau \exp(-\varkappa\tau)\tau = \frac{1}{\varkappa}$$

In a series of experiments on an ensemble of identical systems in state i, the average time that elapses before the system goes to state f is $1/\varkappa$.

In the first experiment [26], i corresponded to an unstable particle called the muon (μ), and f corresponded to a three-particle state containing an electron (e), a beta neutrino (ν_β), and an (anti) mesic neutrino ($\bar{\nu}_\mu$).

$$\mu^+ \rightarrow e^+ + \bar{\nu}_\mu + \nu_\beta \tag{2-36}$$

There is no competing transition. Direct time measurements on the decay [Eq. (2-36)] of stopped muons show that there is indeed an exponential survival probability, with $\tau_0 = 2.20 \times 10^{-6}$ sec. The experiment on decay in flight used cosmic-ray muons. The cosmic protons getting to the earth interact with nuclei near the top of the atmosphere; in these interactions, pions (π-) and K mesons (K-) are produced. Many of these disintegrate to produce muons, which fly down through the air; many of the muons undergo the transition $i \rightarrow f$ while still in flight. By measuring the dependence of the muon flux density on altitude for various muon velocities, Rossi and his associates were able to determine the muon survival probability as a function of muon velocity relative to the earth.

Let us say that (Fig. 2-27) event A is the detection of the passing muon in a stationary counter, and event B is its transition to the state ($e + v + \bar{v}$). Special relativity predicts [Eq. (2-28)] that

$$t_B - t_A = \gamma(\tau_B - \tau_A)$$

Let us make an average of these time differences over an ensemble of muons descending with the same velocity. The average value (mean life) is

$$\langle t_B - t_A \rangle = \gamma \langle \tau_B - \tau_A \rangle \tag{2-37a}$$

giving the relation $t_0 = \gamma\tau_0$ between the mean life (t_0) in the laboratory frame and that (τ_0) in the muon rest frame. The mean life in the laboratory is always greater, and can be very much greater, because $1 \leq \gamma$. This formula applies to any transition of a moving system—such as, for example, the radiative transition of an excited atom to a state of lower energy.

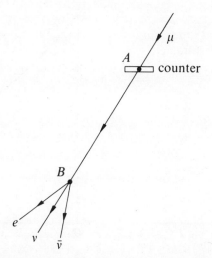

Figure 2-27. Cosmic ray experiment on decay in flight of muons.

The transition probability [Eq. (2-35a)] is invariant: A certain fraction of the muons decays regardless of what reference frame is used in locating the events. Thus,

$$k\,dt = \varkappa\,d\tau$$

and

$$k = \varkappa\,\frac{d\tau}{dt} = \frac{\varkappa}{\gamma} \tag{2-37b}$$

In the laboratory frame, as in the rest frame, there is an exponential survival probability:

$$w = \exp(-kt) \tag{2-35c}$$

$$t_0 = \frac{1}{k} \tag{2-35d}$$

In fact, w is the same number (Fig. 2-28). The only difference between the two exponentials is the stretching of the time scale; the time constant is greater in the laboratory frame.

For flight through vacuum we can convert directly to an exponential dependence of survival probability on distance. With v the velocity of the

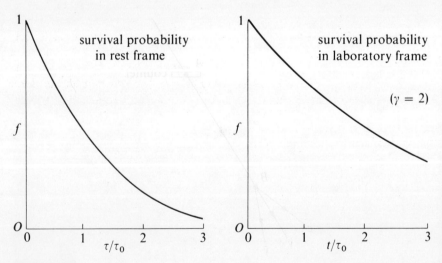

Figure 2-28.

system in state i, and x its coordinate along the line of flight,

$$dt = \frac{dx}{v}$$

The probability of the transition occurring in dx is, again, the same number as the probability of its occurring in the corresponding dt:

$$k\,dt = \frac{dt}{t_0} = \frac{dx}{vt_0} = \frac{dx}{v\gamma t_0} = \frac{1}{\beta\gamma c\tau_0}\,dx$$

This equation is of the form: a constant times dx, which means that the survival probability is an exponential function of x

$$w = \exp\left(\frac{-x}{\beta\gamma c\tau_0}\right) \tag{2-38a}$$

with mean path before decay

$$x_0(v) = \beta\gamma c\tau_0 \tag{2-38b}$$

The expression $\beta c\tau_0$ is the distance the system travels in laboratory time τ_0. The mean range before decay is increased over this nonrelativistic value by the factor γ. For the muon,

$$c\tau_0 = (3.00 \times 10^{10}\text{ cm sec}^{-1})(2.20 \times 10^{-6}\text{ sec}) = 0.660\text{ km}$$

The mean paths before decay for some values of v are given in Table 2-1. For cosmic-ray muons, β is in the range of values in the table. The muons are formed at heights on the order of 25 km. The mean path before decay is thus comparable to the distance to the ground. If, on the other hand, the mean life were independent of velocity, the mean path would be essentially 0.66 km

Table 2-1

β	$\beta\gamma$	$\gamma\beta c\tau_0$ (km)
0.9000	2.06	1.36
0.9900	7.02	4.63
0.9990	22.3	14.8
0.9999	70.7	46.7

for all these fast muons, and essentially none would survive even to mountain altitude. The experiment [26] showed, after the necessary correction was made for slowing down in the air, the predicted exponential attenuation due to decay in flight, with the mean path before decay given quantitatively by Eq. (2-38b).

More precise tests have since been made, using focused beams in vacuum of unstable particles from accelerators. An extremely precise confirmation of Eqs. (2-38) is a by-product of a recent experiment [27] comparing the decay probabilities of negative and positive pions. The number of pions surviving in an extremely well-focused monoenergetic beam of momentum 311 MeV/c was measured at ten positions along the beam for pions of each sign. The quality of the fit to the exponential law [Eq. (2-38a)] is indicated by comparison of the value of τ_0 calculated from these space attenuation data

$$\tau_0(\pi^+) = (26.02 \pm 0.04) \, \text{nsec}$$

with the value of τ_0 determined from direct time measurements on the decay of stopped pions

$$\tau_0(\pi^+) = (26.03 \pm 0.04) \, \text{nsec}$$

The slowing down of the transition rate of a moving system is an important fact to be borne in mind in designing experiments on the unstable particles made in high energy collisions. For example, separation of K^+ mesons from pions and protons of the same momentum requires long trains of magnets and electric deflectors. In a beam at the Argonne National Laboratory, the distance from the target in the accelerator to the bubble chamber is 104.6 m. For K^+ mesons of momentum 3.25 GeV/c ($\beta\gamma = 6.59$), the mean path before decay is $6.59 \times 3.00 \times 10^{10} \times 1.23 \times 10^{-8} = 2.43 \times 10^3$ cm, and the probability of survival to the bubble chamber is

$$\exp\left(\frac{-104.6}{24.3}\right) = 1.35 \times 10^{-2}$$

With typically 1.8×10^3 K^+ per accelerator pulse leaving the target in such directions as to enter the separator, the number reaching the bubble chamber is 24, assuming no losses at slit jaws and walls of the vacuum pipe. Without the Einstein dilation, the survival probability is (since $\gamma = 6.67$)

$$\exp\left(\frac{-104.6}{3.64}\right) = (1.35 \times 10^{-2})^{6.67} = 3.4 \times 10^{-13}$$

At most, one K^+ would reach the chamber every 1.6×10^9 accelerator pulses! The experiment would hardly be feasible.

In his first paper on relativity [2], Einstein pointed out a striking consequence of the time dilation that has recently received a good deal of publicity
under the name of the "twin paradox." It is not a logical paradox at all but
just a startling consequence of the theory. It is relevant to the synchronization
of distant clocks, showing that synchronization by transporting clocks to
the master clock is only possible at zero velocity of transportation.

We have seen [Eq. (2-29)] that the time measured on a clock moving with
a system is less than the time measured in any inertial frame, provided that
the moving clock is corrected for effects of acceleration. Apply this to the
following thought experiment. One young twin (A) stays on earth, assumed
inertial. The other (B) leaves him, shooting off with a velocity comparable
to that of light. Twin B has a clock in his spaceship reading the time τ. After
a long ride at constant velocity v, he reverses course and returns to earth
with the opposite contant velocity $-v$. Figure 2-29 shows the world lines
of A and B. The contributions to τ from the acceleration eras—startup (1),
reversal, and stop (2)—can be made negligible compared to those from the
uniform motion eras (to and fro). When the clocks are compared, B's reads
an earlier time than A's, Eq. (2-29) giving

$$\tau_2 - \tau_1 = \gamma^{-1}(t_2 - t_1) \tag{2-39}$$

This equation differs from the usual time dilation relation in having proper
times on both sides; t_2 and t_1 are read on the same A clock, τ_2 and τ_1 on the
same B clock. According to the relativity postulate, all processes in the
rocketship must have run off at the same rates in rocket times as the corresponding processes on earth in earth time. One such process is that of aging.

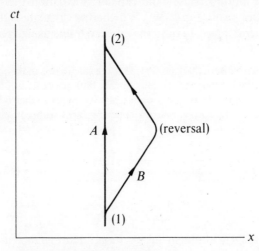

Figure 2-29.   World lines of twins.

If γ is large, the traveling twin returns still in the flush of youth, while his stay-at-home brother is lucky to be still alive.

The spacetime diagram, Fig. 2-29, shows clearly the asymmetry of the situation. The world line of A is straight; that of B is not. A reference frame attached to A is always inertial; one attached to B is not.

The apparent paradox results from switching the roles of the twins. In a frame attached to the traveling twin, the stay-at-home twin shoots off backward and later returns. Why should not he be less aged too? Because we have no right to interchange labels in this essentially asymmetrical situation.

It is wrong to confuse the principle of relativity with the belief that the behavior of the two clocks depends only on their motion relative to one another, as if there were nothing else in the universe. Their relative motion is equal and opposite, but their motion with respect to an outside reference frame is very different. As Laue remarks ([28], p. 36), "Inertial systems are observable realities; our thought experiment decides which clock [twin] remained at rest in the same system, which in different ones."

To describe the behavior of A's clock in B's frame requires knowledge of the transformation laws to an accelerated reference frame, that is, general relativity. According to the principle of equivalence (Chapter 6), things proceed in the accelerated B frame as if it were not accelerated but had instead a homogeneous gravitational field $\mathbf{g}$ equal and opposite to the acceleration of B with respect to its comoving inertial frame (Fig. 2-30). The clock of A is higher in this field than that of B, and therefore runs faster by an amount increasing with the distance between them (Chapter 6). This speeding up more than compensates its slowing down during the much longer uniform motion phase. The explicit calculation ([29], pp. 258–62) gives the invariant result [Eq. (2-39)]. Whether we calculate in the accelerated frame using general relativity or in the intertial frame using special relativity, A is older than B.

As shown by Born ([30], pp. 354–356), the calculation in the accelerated frame can be carried through in the low velocity limit (to order v^2/c^2) using only the principle of equivalence and special relativity. Suppose that B leaves A, moves with constant velocity v, reverses course, returns with constant velocity $-v$,

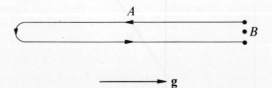

Figure 2-30.   Twin problem in B's frame. During the turnaround, the uniform gravitational field $\mathbf{g}$ prevails throughout the frame.

and stops near A. The accelerations at the beginning and end of the voyage are equal and opposite and also involve negligible distance between A and B; only the acceleration at reversal has a net effect. This acceleration is equivalent to an average gravitational field in B's frame of strength $2v/\Delta T$ and duration ΔT, where ΔT is the duration of the reversal era. As shown in Chapter 6 [Eq. (6-3)], A's clock runs faster than B's while the field is present by the factor

$$1 + \frac{\Delta\varphi}{c^2}$$

where $\Delta\varphi$ is the difference of gravitational potential between A and B. In the present case, this factor is

$$1 + \frac{(2v/\Delta T)(vT)}{c^2} = 1 + \frac{2v^2}{c^2}\frac{T}{\Delta T}$$

where T is the duration in B's frame of the outward voyage. The gain of A's clock over B's in the time ΔT is, therefore,

$$\frac{2v^2}{c^2}\frac{T}{\Delta T} \cdot \Delta T = \frac{2v^2}{c^2}T$$

During the field-free era, A's clock runs slow in the B (inertial) frame by the factor

$$\left(1 - \frac{v^2}{c^2}\right)^{-1/2} \approx 1 + \frac{1}{2}\frac{v^2}{c^2}$$

In the round-trip time $2T$ it loses altogether

$$\frac{1}{2}\frac{v^2}{c^2} \cdot 2T = \frac{v^2}{c^2}T$$

The net gain of A's clock over B's is, therefore,

$$\frac{2v^2}{c^2}T - \frac{v^2}{c^2}T = \frac{v^2}{c^2}T = \frac{1}{2}\frac{v^2}{c^2} \cdot 2T$$

which agrees to order $(v/c)^2$ with Eq. (2-39). One could alternatively reverse the argument and, assuming the result of the twin problem, Eq. (2-39), derive the gravitational red shift, Eq. (6-3).

Far from revealing a contradiction in special relativity, the twin problem illustrates the consistency of the slowing down of a moving clock (special relativity) and the red shift of a rising photon (principle of equivalence). Special relativity and the principle of equivalence are of one piece, and stand or fall together.

The experimental evidence for the special-relativistic time dilation is overwhelming. That for the gravitational red shift is also very strong (Chapter 6), though limited to weak gravitational fields. We must expect that when, some day, a spaceship returns to earth after many years of voyaging at high speed, its atomic clock will be found to have scored fewer oscillations than its twin clock left behind on earth. The crew may even look younger!

2-8. TRANSFORMATION OF VELOCITY

The motion of a particle is a continuous sequence of events (t, x, y, z), and is represented by a curve on the spacetime diagram. One particular way of parametrizing this curve is to give x, y, z as functions of t

$$x(t), y(t), z(t) \tag{2-40}$$

The velocity $\mathbf{v}$ has the components

$$v_x = \frac{dx}{dt} \qquad v_y = \frac{dy}{dt} \qquad v_z = \frac{dz}{dt}$$

found by differentiating the functions of Eq. (2-40). The transformation obtained by differentiating Eqs. (1-31c)

$$c\,dt = \gamma(c\,dt' + \beta\,dx')$$

$$dx = \gamma(dx' + \beta c\,dt')$$

$$dy = dy' \tag{2-41}$$

$$dz = dz'$$

determines the transformation of velocity components. (Note that β and γ refer to the constant velocity $\mathbf{u}$ of R' with respect to R.)

$$v_x = c\frac{dx}{c\,dt} = c\frac{\gamma(dx' + \beta c\,dt')}{\gamma(c\,dt' + \beta\,dx')} = \frac{(dx'/dt') + \beta c}{1 + (\beta\,dx'/c\,dt')} = \frac{v_{x'} + u}{1 + (uv_{x'}/c^2)}$$

$$\tag{2-42}$$

$$v_y = c\frac{dy}{c\,dt} = c\frac{dy'}{\gamma(c\,dt' + \beta\,dx')} = \gamma^{-1}\frac{dy'/dt'}{1 + (\beta\,dx'/c\,dt')} = \gamma^{-1}\frac{v_{y'}}{1 + (uv_{x'}/c^2)}$$

$$v_z = c\frac{dz}{c\,dt} = c\frac{dz'}{\gamma(c\,dt' + \beta\,dx')} = \gamma^{-1}\frac{dz'/dt'}{1 + (\beta\,dx'/c\,dt')} = \gamma^{-1}\frac{v_{z'}}{1 + (uv_{x'}/c^2)}$$

For the inverse transformation, we interchange primed and unprimed quantities and reverse the sign of u.

For the magnitude v and spherical polar angles (χ, ψ) of the velocity vector (Fig. 2-31),

$$v_x = v \cos \chi \qquad\qquad v_{x'} = v' \cos \chi'$$

$$v_y = v \sin \chi \cos \psi \qquad\qquad v_{y'} = v' \sin \chi' \cos \psi'$$

$$v_z = v \sin \chi \sin \psi \qquad\qquad v_{z'} = v' \sin \chi' \sin \psi'$$

we find, using Eqs. (2-42), that

$$v = (v_x^2 + v_y^2 + v_z^2)^{1/2} = \frac{[v'^2 + u^2 + 2uv' \cos \chi' - (uv' \sin \chi'/c)^2]^{1/2}}{1 + uv' \cos \chi'/c^2}$$

$$\tan \chi = \frac{(v_y^2 + v_z^2)^{1/2}}{v_x} = \gamma^{-1} \frac{v' \sin \chi'}{u + v' \cos \chi'} \tag{2-43}$$

$$\psi = \arctan \frac{v_z}{v_y} = \arctan \frac{v_{z'}}{v_{y'}} = \psi'$$

Naturally the transformation does not produce any azimuthal twisting when the polar axis is chosen parallel to the direction of relative motion. The speed and the inclination to this direction are, however, affected. Again, the inverse transformations are found by interchanging primed and unprimed quantities and reversing the sign of u.

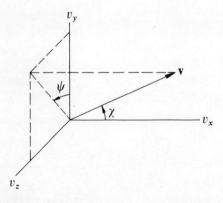

Figure 2-31.

Equations (2-42) and (2-43) reduce to the nonrelativistic limits when $c = \infty$:

$$v_x = v_{x'} + u$$

$$v_y = v_{y'}$$

$$v_z = v_{z'}$$

$$v = (v'^2 + u^2 + 2uv' \cos \chi')^{1/2}$$

$$\tan \chi = \frac{v' \sin \chi'}{u + v' \cos \chi'}$$

Plotting the vectors (v_x, v_y, v_z), $(v_{x'}, v_{y'}, v_{z'})$, $(u, 0, 0)$ in a common abstract velocity space, we see that the nonrelativistic limit corresponds to the Galilean law of vector addition of relative velocities

$$\mathbf{v} = \mathbf{v}' + \mathbf{u} \qquad (1\text{-}21)$$

Figure 2-32 shows the vector triangle. Here $\overrightarrow{RP}$ is the velocity of particle P in R; $\overrightarrow{RR'}$ is the velocity of R' in R; and $\overrightarrow{R'P}$ is the velocity of P in R'. The equations above are trigonometric identities in the Euclidean velocity space.

They do not hold relativistically. We shall see that a particle with speed $< c$ in one inertial reference frame has speed $< c$ in all others. And a particle with speed c in one reference frame has speed c in all others. The existence of the limiting speed c makes it impossible for the scale-free law of vector addition, Eq. (1-21), to hold.

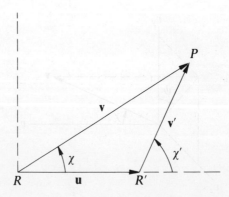

Figure 2-32.   Nonrelativistic velocity triangle, valid in the limit $c = \infty$.

From the first Eq. (2-43) we can derive

$$\left(1 - \frac{v^2}{c^2}\right)^{1/2} = \frac{(1 - u^2/c^2)^{1/2}(1 - v'^2/c^2)^{1/2}}{1 + uv' \cos \chi'/c^2} \qquad (2\text{-}44)$$

If $v' < c$ (and, of course, $u < c$), the denominator is always positive, and

$$\left(1 - \frac{v^2}{c^2}\right)^{1/2} > 0$$

that is,

$$v < c$$

If the speed is less than c in one inertial frame, it is less than c in every other inertial frame. It is impossible to make something go faster than light just by transforming to an appropriate frame. This is another indication of the internal consistency of special relativity. In nonrelativistic mechanics, on the other hand, there is no built-in speed scale corresponding to c, and any speed can be attained by suitable choice of **u** for the transformation.

For a particle moving with speed c, $v' = c$, we see from Eq. (2-44) that $v = c$; that is, a particle moving with the speed of light in one inertial frame moves with the speed of light in every inertial frame. The speed cannot be changed by transforming, although the direction can. Of course this result is just Einstein's second postulate expressed in the language of light quanta; light propagates with the same speed in every inertial frame. (Historically, the relativistic transformation formulas for velocity, momentum, and energy played a key role in the discovery of the quantal wave–particle relationship (Appendix F).)

The speed c is the only one having this invariance property. In fact, for every speed less than c there is a frame, the rest frame, in which the speed is zero. In Eq. (2-44), $u = v'$, $\chi' = \pi$ gives $v = 0$. The R frame is evidently moving along with the particle.

If a particle were to move faster than light, $v' > c$, one could transform with $u < c$ to a reference frame in which its motion is reversed. In effect, take the x' and x axes along the direction of motion of the particle. Eqs. (2-42) give

$$v_x = \frac{v' + u}{1 + uv'/c^2} = v$$

Negative values of u with $|u| < c$ exist such that the denominator is negative

$$1 + \frac{uv'}{c^2} < 0 \quad \text{if} \quad \frac{|u|}{c} > \frac{c}{v'}$$

whereas the numerator is always positive. Thus v is negative and v' is positive! Consider successive events in the history of the particle, occurring at P_1 and P_2 (Fig. 2-33). In R',

$$t_2' - t_1' = \frac{x_2' - x_1'}{v'} > 0$$

In R,

$$t_2 - t_1 = \frac{x_2 - x_1}{v} < 0$$

The time order is reversed. In R the creation of the particle would occur after its destruction, illustrating vividly the causality requirement already stated in Section 2-4: A sequence of physically related events must have an invariant time order, the same in all inertial frames. For our theory to satisfy this requirement, its dynamics (Chapter 3) must be such that a particle can never be accelerated to the velocity of light.

With $v = v' = c$, the second and third equations in Eq. (2-43) give the theory of optical aberration, the change in direction of a light source resulting from its motion

$$\tan \chi = \gamma^{-1} \frac{\sin \chi'}{\beta + \cos \chi'}$$

$$\psi = \psi'$$

(2-45)

These equations determine the angular distribution of photons and neutrinos emitted by rapidly moving sources, such as neutral pions ($\pi^0 \rightarrow \gamma + \gamma$),

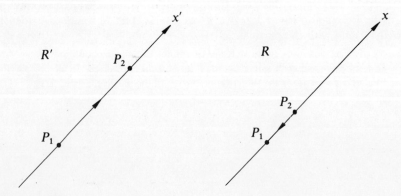

Figure 2-33.

charged pions $(\pi^+ \to \mu^+ + \nu_\mu)$, neutral Σ hyperons $(\Sigma^0 \to \Lambda + \gamma)$, and electrons in circular accelerators $(e^- \to e^- + \gamma)$.

For all particle speeds there is a collimating effect of the Lorentz transformation. In the second Eq. (2-43), which can be written

$$\tan \chi = (1 - \beta^2)^{1/2} \frac{\tan \chi'}{1 + (\beta c/v') \sec \chi'}$$

we see that as $\beta \to 1$, $\tan \chi \to 0$ for any χ'. The higher the speed of R' with respect to R, the more the particles get beamed forward in R. If $v' < c$, there is a range of transformation velocities βc

$$\beta c > v'$$

such that even particles moving straight back in $R'(\chi' = \pi)$ move forward in $R(\chi = 0)$. For $v' = c$, there is a small backward cone of opening angle arcos β that remains backward in R; all other rays are pulled into the forward hemisphere. In high energy particle reactions, the outgoing particles are emitted symmetrically fore and aft in the frame R' ("center of mass" frame) in which the total momentum is zero. This frame moves very fast with respect to the laboratory; therefore, the particles move forward in the laboratory, making but small angles with the direction of the incident particle beam. The effect can be seen in the photograph (Fig. 1-4) of the tracks in the laboratory frame resulting from the reaction

$$p + p \to p + p + \pi^+ + \pi^-$$

Here the velocity of the center-of-mass frame is $\beta = 0.89$. The measured angles χ of the outgoing tracks with respect to the incident proton direction are given in Table 2-2, along with the corresponding angles χ' computed in the center-of-mass frame.

Table 2-2

Track	χ'	χ
2	125.4°	48.0°
3	22.1°	4.9°
4	171.7°	23.6°
5	55.5°	14.6°

The collimation in the laboratory reference frame is apparent.

EXERCISE   A Σ^0 hyperon disintegrates into a Λ hyperon and a photon

$$\Sigma^0 \rightarrow \Lambda + \gamma$$

In the rest frame of the Σ^0, the velocities of the Λ and γ are, respectively, $6.70 \times 10^{-2}c$ and c; they are opposite. The velocity of the Σ^0 in the laboratory frame is $0.8c$. Calculate the speed and direction of the Λ and γ in the laboratory reference frame for three cases: (a) γ emitted straight forward in the Σ^0 rest frame (Fig. 2-34a); (b) γ emitted broadside in the Σ^0 rest frame (Fig. 2-34b); (c) γ emitted straight back in the Σ^0 rest frame (Fig. 2-34c).

SOLUTION   We merely substitute in Eqs. (2-43) and (2-45). The results are given in Table 2-3 and Fig. 2-34d–f, showing the velocities in the laboratory in the three cases.

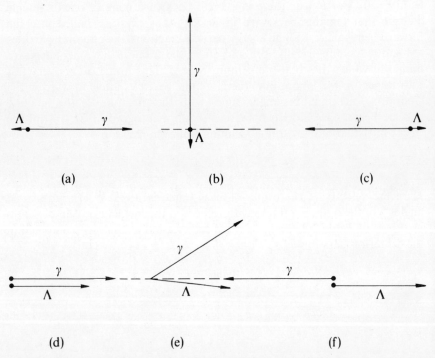

Figure 2-34.   Three samples of the velocity transformation applied to the transition $\Sigma^0 \rightarrow \Lambda + \gamma$. In (a), the photon is emitted forward in the rest frame of the Σ^0; the laboratory velocities are shown in (d). In (b), the photon is emitted transversally; the laboratory velocities are shown in (e). In (c), the photon is emitted backward; the laboratory velocities are shown in (f).

Table 2-3

Case	Photon Speed	Angle with forward direction	Λ Hyperon Speed	Angle with forward direction
(a)	c	0	$0.77c$	0
(b)	c	arctan $0.75 = 36.9°$	$0.80c$	arctan $0.05 = 2.9°$
(c)	c	π	$0.82c$	0

The azimuths are, of course, unaffected. Note the forward collimation. Note also the striking departure from nonrelativistic velocity addition for both the photon and the material particle. In case (a), $1.8c \neq c$, $0.733c \neq 0.77c$; in case (c), $-0.2c \neq -c$, $0.876c \neq 0.82c$.

2-9. SPACETIME

In the Lorentz transformation, space and time are inextricably intertwined. In considering intervals between events (Section 2-4), we saw that neither the spatial separation Δr nor the time difference Δt is any longer invariant; rather it is $c^2 \Delta t^2 - \Delta r^2$ that has the same numerical value in all inertial frames. It was Minkowski [23] who emphasized that the new transformation law ends the isolation between physical space and physical time, and forces us to treat them together as coordinates in a four-dimensional manifold of *events*. Minkowski called this manifold "world" (*Welt*), and a trace of his nomenclature remains in the designation "world line" for a particle track in this continuum. In current English usage, the name *spacetime* is preferred, and the prefix "four-" is used to label spacetime quantities (as in four-vector).

The points of spacetime are events. The time of occurrence and the position of an event are its four coordinates, projections of the event on a set of axes belonging to the reference frame. The Lorentz transformation is a linear transformation of coordinates in spacetime, giving the projections of the same event on a new set of axes belonging to another reference frame. This coordinate transformation in spacetime is defined by the requirement that it leave invariant the form Eq. (2-21) (which we now write with opposite sign)

$$x^2 + y^2 + z^2 - c^2t^2 \tag{2-46a}$$

It is analogous to a rotation of axes in three-dimensional space, which is a linear transformation defined by the requirement that it leave invariant the form

$$x^2 + y^2 + z^2 \tag{2-47a}$$

The substitution

$$x_1 = x$$

$$x_2 = y$$

$$x_3 = z \tag{2-48a}$$

$$x_4 = ict$$

makes the analogy exact, the invariants being

$$(x_1{}^2 + x_2{}^2 + x_3{}^2 + x_4{}^2) \quad \text{(Lorentz transformation)} \tag{2-46b}$$

and

$$(x_1{}^2 + x_2{}^2 + x_3{}^2) \quad \text{(rotation in space)} \tag{2-47b}$$

Just as in Section 1-3 we used the transformation behavior of quantities under rotation to classify them as scalars, vectors, and tensors in three-dimensional space, so now we shall use the transformation behavior under a shift from a coordinate system in one inertial reference frame to a coordinate system in another to classify physical quantities as four-scalars, four-vectors, and four-tensors in spacetime.

We denote an index running over the four spacetime coordinates by a Greek letter, one running over the three space coordinates by a Latin letter. Thus, x_μ runs over x_1, x_2, x_3, x_4; x_m runs over x_1, x_2, x_3. The subscript zero is reserved to indicate the real part of the time component,

$$x_0 = \frac{x_4}{i} = ct \tag{2-48b}$$

When we refer to the geometry of spacetime we shall always use the prefix "four-" or "world-." Omission of the prefix means that we are referring to three-dimensional space.

A four-scalar is a number whose value is unchanged in a Lorentz transformation, an invariant. An example is the expression (2-46a) or (2-46b).

A four-vector is a quantity whose four components transform in the same way as the components x_1, x_2, x_3, x_4 of an event.[3]

The special Lorentz transformation Eq. (1-31b) becomes, in the notation of Eqs. (2-48a),

$$x'_1 = \gamma(x_1 + i\beta x_4)$$
$$x'_2 = x_2$$
$$x'_3 = x_3 \tag{2-49a}$$
$$x'_4 = \gamma(-i\beta x_1 + x_4)$$

Notice that the new way of writing the transformation using i in the timelike component has opposite signs in the two equations coupling time and longitudinal position. The four components A_λ of a four-vector $\tilde{A}$ transform in this SLT according to

$$A'_1 = \gamma(A_1 + i\beta A_4)$$
$$A'_2 = A_2$$
$$A'_3 = A_3$$
$$A'_4 = \gamma(-i\beta A_1 + A_4) \tag{2-49b}$$

(The tilde or "twiddle" is the spacetime counterpart of the arrow, or boldface type, of space.)

The three space components are the components of a vector **A**

$$A_1 = A_x \qquad A_2 = A_y \qquad A_3 = A_z \tag{2-48c}$$

When we write plain A, sans twiddle, we mean the magnitude of this vector

$$A = |\mathbf{A}| = (A_1{}^2 + A_2{}^2 + A_3{}^2)^{1/2} \tag{2-48d}$$

The four-vector $\tilde{A}$ may be specified in terms of components

$$\tilde{A} = (A_\mu) = (A_1, A_2, A_3, A_4) = (\mathbf{A}, A_4) \tag{2-48e}$$

---

[3] We ignore the distinction between contra- and covariant components of a four-vector. Only in general relativity does it become necessary to introduce it.

The real, imaginary, or complex character of a physical quantity must be invariant under both rotation and special Lorentz transformation. The three components A_1, A_2, A_3 must have the same character. If they are real, then A_4 must be imaginary in order for A'_1, A'_2, A'_3 to be real also. If A_1, A_2, and A_3 are imaginary, A_4 must be real.

The *scalar product* of two four-vectors $\tilde{A}$ and $\tilde{B}$ is defined as

$$\tilde{A}\tilde{B} = A_\mu B_\mu = A_1 B_1 + A_2 B_2 + A_3 B_3 + A_4 B_4 \qquad (2\text{-}50a)$$

(summation convention as in Section 1-3: sum over index if repeated in a product).

By substituting Eq. (2-49b) into Eq. (2-50a) one verifies that the scalar product of two four-vectors is invariant under a SLT:

$$\tilde{A}\tilde{B} = A_\mu B_\mu = A'_\nu B'_\nu$$

If the scalar product of two four-vectors is zero, the four-vectors are said to be orthogonal.

The scalar product of a four-vector with itself is its "length squared" or "square"

$$\tilde{A}^2 = \tilde{A}\tilde{A} = A_\mu A_\mu = A_1{}^2 + A_2{}^2 + A_3{}^2 + A_4{}^2 \qquad (2\text{-}51a)$$

An example with which we are familiar is the interval squared. The four-interval between events 1 and 2

$$\tilde{x}_2 - \tilde{x}_1 = \Delta\tilde{x} \qquad (2\text{-}8)$$

has the square

$$\Delta\tilde{x}^2 = \Delta x^2 + \Delta y^2 + \Delta z^2 - c^2 \Delta t^2 = \Delta r^2 - c^2 \Delta t^2$$

which is (-1) times the interval squared [Eq. (2-25)]:

$$\Delta s^2 = c^2 \Delta t^2 - \Delta r^2$$

The appellations "timelike" and "spacelike" come from consideration of this basic four-interval. A four-vector is called spacelike if its square has the same sign as that of a spacelike four-interval, with our convention positive. If the square has the same sign as that of a timelike four-interval, the four-vector is timelike. If the square is zero, the four-vector is lightlike. In contrast with the situation in three-space, the square of a four-vector is not positive definite, and lengths are real or imaginary.

The algebra of four-vectors is exactly the same as that of vectors. The operation of addition, for example,

$$\tilde{A} + \tilde{B} = \tilde{C}$$

means that in all coordinate systems corresponding components add as numbers do,

$$A_\mu + B_\mu = C_\mu$$

The analogue of the 3×3 rotation matrix of elements a_{jk} formed from the direction cosines between new and old axes [Eq. (A-3)] is the 4×4 matrix of elements $a_{\mu v}$ describing the spacetime coordinate transformation

$$A'_\mu = a_{\mu v} A_v \qquad (2\text{-}52)$$

For the SLT of Eqs. (2-49), the matrix elements are

$$\mathbf{a} = \begin{pmatrix} \gamma & 0 & 0 & i\beta\gamma \\ 0 & 1 & 0 & 0 \\ 0 & 0 & 1 & 0 \\ -i\beta\gamma & 0 & 0 & \gamma \end{pmatrix} \qquad (2\text{-}49\text{c})$$

Like the 3×3 matrix of direction cosines, this 4×4 matrix is orthonormal

$$a_{\mu\lambda} a_{\mu\rho} = \delta_{\lambda\rho}$$
$$a_{\lambda\mu} a_{\rho\mu} = \delta_{\lambda\rho}$$

For example, $i\beta\gamma^2 - i\beta\gamma^2 = 0$, $\gamma^2 - \beta^2\gamma^2 = 1$, and so forth. It must be, to make the scalar product, Eq. (2-50a), invariant.

A four-tensor (of rank 2) is a quantity whose 16 components, labeled by two indices, transform like the products of components of four-vectors. The four-tensor transformation equation is

$$T'_{\mu v} = a_{\mu\lambda} a_{v\rho} T_{\lambda\rho} \qquad (2\text{-}53)$$

Four-tensors are useful in the relativistic theory of fields. As shown in Appendix D, the electric and magnetic field vectors $\mathscr{E}$ and $\mathscr{B}$ are related through the Maxwell equations in such a way that they are essentially components of one four-tensor of the electromagnetic field, $\tilde{\tilde{\mathscr{F}}}$.

$$\mathcal{F}_{\mu\nu} = \begin{pmatrix} 0 & \mathcal{B}_z & -\mathcal{B}_y & -i\mathcal{E}_x \\ -\mathcal{B}_z & 0 & \mathcal{B}_x & -i\mathcal{E}_y \\ \mathcal{B}_y & -\mathcal{B}_x & 0 & -i\mathcal{E}_z \\ i\mathcal{E}_x & i\mathcal{E}_y & i\mathcal{E}_z & 0 \end{pmatrix} \tag{2-54}$$

This antisymmetric four-tensor has six independent components, determining $\mathcal{B}$ and $\mathcal{E}$. The field strengths $\mathcal{B}$ and $\mathcal{E}$ are essentially projections of the field four-tensor relative to the coordinate system used. Their transformation law is Eq. (2-53), yielding for the SLT Eq. (2-49c),

$$\begin{aligned} \mathcal{E}_{x'} &= \mathcal{E}_x & \mathcal{B}_{x'} &= \mathcal{B}_x \\ \mathcal{E}_{y'} &= \gamma(\mathcal{E}_y - \beta\mathcal{B}_z) & \mathcal{B}_{y'} &= \gamma(\mathcal{B}_y + \beta\mathcal{E}_z) \\ \mathcal{E}_{z'} &= \gamma(\mathcal{E}_z + \beta\mathcal{B}_y) & \mathcal{B}_{z'} &= \gamma(\mathcal{B}_z - \beta\mathcal{E}_y) \end{aligned} \tag{2-55a}$$

or, in a form independent of the coordinate axes,

$$\begin{aligned} \mathcal{E}'_{\parallel} &= \mathcal{E}_{\parallel} & \mathcal{B}'_{\parallel} &= \mathcal{B}_{\parallel} \\ \mathcal{E}'_{\perp} &= \gamma\left(\mathcal{E} + \frac{\mathbf{u}}{c} \times \mathcal{B}\right)_{\perp} & \mathcal{B}'_{\perp} &= \gamma\left(\mathcal{B} - \frac{\mathbf{u}}{c} \times \mathcal{E}\right)_{\perp} \end{aligned} \tag{2-55b}$$

The subscripts $\parallel$ and $\perp$ refer to the direction of the velocity of R' with respect to R. Notice the intermingling of $\mathcal{B}$ and $\mathcal{E}$ in this beautiful realization of Faraday's ideal: One field describing all electric and magnetic interactions.

A four-vector can, of course, be regarded as a four-tensor of rank 1 (one index), and a four-scalar as a four-tensor of rank 0 (no index). Any tensorial expression in which all dummy indices are summed over is necessarily invariant because of the invariance of the scalar product.

In these considerations of the analytic geometry of spacetime there is no need to restrict ourselves to special Lorentz transformations. We have seen (Section 2-3) that the general homogeneous Lorentz transformation can be analyzed into two spatial rotations and a SLT, with possibly also an inversion of space axes. The four-tensorial classification of physical quantities is not affected by inclusion of rotations and inversions. The space part $\mathbf{A}$ of a four-vector $\tilde{A}$ transforms under rotation and inversion like any other vector; A_4 is unchanged, and the invariance of

$$A_\mu{}^2 = A^2 + A_4{}^2$$

is respected. Rotations, inversions, and SLT's can be regarded as special cases of the general Lorentz transformation.

In manipulating spacetime components, we can avoid explicit use of the unit imaginary i by specifying the timelike component of a four-vector by the "zero-" component instead of the "four-" component.

$$x_0 = \frac{x_4}{i} = ct \qquad (2\text{-}48\text{b})$$

$$A_0 = \frac{A_4}{i} \qquad (2\text{-}48\text{f})$$

This procedure has the advantage of reducing the risk of errors of sign. When we use this notation we shall write the time transformation first instead of last. The special Lorentz transformations [Eqs. (2-49a and b)] become

$$\begin{aligned}
x'_0 &= \gamma(x_0 - \beta x_1) \\
x'_1 &= \gamma(x_1 - \beta x_0) \\
x'_2 &= x_2 \\
x'_3 &= x_3
\end{aligned} \qquad (2\text{-}49\text{aa})$$

and

$$\begin{aligned}
A'_0 &= \gamma(A_0 - \beta A_1) \\
A'_1 &= \gamma(A_1 - \beta A_0) \\
A'_2 &= A_2 \\
A'_3 &= A_3
\end{aligned} \qquad (2\text{-}49\text{bb})$$

In this form, the sign inside the parentheses is the same in the time and longitudinal space equations; whether to use $+$ or $-$ is evident from the Galilean limit

$$x'_1 \sim x_1 - ux_0$$

The zero component of a four-vector has the same real, imaginary, or complex character as the three spacelike components.

In the scalar product of four-vectors, a minus sign appears before the zero–zero term.

$$\tilde{A}\tilde{B} = A_1B_1 + A_2B_2 + A_3B_3 - A_0B_0 = -A_0B_0 + \mathbf{A} \cdot \mathbf{B} \qquad (2\text{-}50\text{b})$$

$$\tilde{A}^2 = A_1^2 + A_2^2 + A_3^2 - A_0^2 = -A_0^2 + \mathbf{A} \cdot \mathbf{A} \qquad (2\text{-}51\text{b})$$

This sign change makes the $(0, 1, 2, 3)$ notation inconvenient when one is doing algebra with running indices. For such cases, we shall continue to use the $(1, 2, 3, 4)$ notation, with i appearing explicitly in the transformation equations.

When we use the $(0, 1, 2, 3)$ basis, curly brackets serve to identify the timelike and spacelike components of a four-vector

$$\tilde{A} = \{A_0, \mathbf{A}\}$$

in contrast with the ordinary parentheses used with the $(1, 2, 3, 4)$ basis.

$$\tilde{A} = (\mathbf{A}, A_4) \equiv (\mathbf{A}, iA_0)$$

2-10. SPACETIME DESCRIPTION OF PARTICLE MOTION: FOUR-VELOCITY, FOUR-ACCELERATION

Along the world line of a particle there are functional relationships between x_1, x_2, x_3, and x_4, because

$$x_1 = x = x(t) = x\left(\frac{x_4}{ic}\right)$$

$$x_2 = y = y(t) = y\left(\frac{x_4}{ic}\right) \qquad (2\text{-}56)$$

$$x_3 = z = z(t) = z\left(\frac{x_4}{ic}\right)$$

The independent variable $t = x_4/ic$ suffers from the inconvenience of not being a four-scalar. It is, rather, the timelike component of a four-vector. A convenient invariant parameter for labeling points on the world line is the proper time τ of the particle. It is related to x_1, x_2, x_3, x_4 by Eq. (2-27) of Section 2-5

$$d\tau = \frac{dt}{\gamma} = \frac{dx_4}{ic\gamma} \qquad (2\text{-}57)$$

with

$$\gamma = \left(1 - \frac{v^2}{c^2}\right)^{-1/2}$$

$$v^2 = \left(\frac{dx}{dt}\right)^2 + \left(\frac{dy}{dt}\right)^2 + \left(\frac{dz}{dt}\right)^2 = -c^2\left[\left(\frac{dx_1}{dx_4}\right)^2 + \left(\frac{dx_2}{dx_4}\right)^2 + \left(\frac{dx_3}{dx_4}\right)^2\right]$$

As discussed in Section 2-5, τ is the time shown on a clock moving with the particle, provided any effects of acceleration on the rate of the clock are corrected for. It equals to within a factor of ic the four-dimensional arc length

$$\int_1 \sqrt{dx_\mu^2} = \int_1 \sqrt{dr^2 - c^2 \, dt^2} = \int_1 \sqrt{-c^2 \, d\tau^2} = ic \int_1 d\tau = ic(\tau - \tau_1)$$

The development of the world line $\tilde{x}(\tau)$

$$\tilde{x}(\tau) \begin{cases} x_1 = x_1(\tau) \\ x_2 = x_2(\tau) \\ x_3 = x_3(\tau) \\ x_4 = x_4(\tau) \end{cases} \tag{2-58}$$

from event to event is determined by the derivatives

$$w_1 = \frac{dx_1}{d\tau}$$

$$w_2 = \frac{dx_2}{d\tau}$$

$$w_3 = \frac{dx_3}{d\tau} \tag{2-59a}$$

$$w_4 = \frac{dx_4}{d\tau}$$

These are the components of a four-vector, since $(d\tau)^{-1}$ is a four-scalar.

$$\tilde{w} = \frac{d\tilde{x}}{d\tau} \tag{2-59b}$$

It is called the *four-velocity*. Its spacelike components w_1, w_2, w_3—components of a vector **w**—are closely related to the corresponding components of the velocity **v**,

$$v_x = \frac{dx}{dt} = \frac{dx_1}{dt}$$

$$v_y = \frac{dy}{dt} = \frac{dx_2}{dt}$$

$$v_z = \frac{dz}{dt} = \frac{dx_3}{dt}$$

differing only in the replacement of t by τ as independent variable. In the low velocity limit $v \ll c$, they become equal, since there $\gamma \approx 1$ and [Eq. (2-57)] $d\tau \approx dt$. The exact relationship is given by Eq. (2-57):

$$w_1 = \frac{dx_1}{d\tau} = \gamma \frac{dx_1}{dt} = \gamma v_x = v_x \left(1 - \frac{v^2}{c^2}\right)^{-1/2}$$

$$w_2 = \frac{dx_2}{d\tau} = \gamma \frac{dx_2}{dt} = \gamma v_y = v_y \left(1 - \frac{v^2}{c^2}\right)^{-1/2}$$

$$w_3 = \frac{dx_3}{d\tau} = \gamma \frac{dx_3}{dt} = \gamma v_z = v_z \left(1 - \frac{v^2}{c^2}\right)^{-1/2}$$

or

$$\mathbf{w} = \gamma \mathbf{v} \tag{2-60a}$$

Note that $w(= |\mathbf{w}|)$ has no upper bound, because $\gamma = \infty$ as $v \to c$. The timelike component of $\tilde{w}$ is

$$w_4 = \frac{dx_4}{d\tau} = ic\gamma = ic\left(1 - \frac{v^2}{c^2}\right)^{-1/2} \tag{2-60b}$$

or

$$w_0 = \frac{w_4}{i} = c\gamma = c\left(1 - \frac{v^2}{c^2}\right)^{-1/2} \tag{2-60bb}$$

Evidently the timelike component is determined by the magnitude of $\mathbf{w}$, $w = (w_1{}^2 + w_2{}^2 + w_3{}^2)^{1/2}$, because, by Eq. (2-60a),

$$\left(1 - \frac{v^2}{c^2}\right)^{-1/2} = \left(1 + \frac{w^2}{c^2}\right)^{1/2} \tag{2-61a}$$

The constraint is expressed four-dimensionally by the remark that the square of the four-velocity is constant,

$$w_\mu{}^2 = \frac{dx_\mu{}^2}{d\tau^2} = \frac{-c^2 \, d\tau^2}{d\tau^2} = -c^2 \tag{2-61b}$$

in fact, it is a universal constant, the same everywhere on every world line. It is negative—the four-velocity of a material particle is a timelike four-vector. The nonindependence of the four components of $\tilde{w}$ results from the fact that the adjacent events determining $\tilde{w}$ are physically related through the particle's velocity.

The moving automobile of Section 2-5 provides vivid illustrations of velocity and four-velocity. The velocity v is the distance on the ground, determined from milestones along the road, divided by the time, determined from the difference of readings of synchronized clocks along the road. The spatial part of the four-velocity w is the distance on the ground, determined from the same milestones, divided by the elapsed time shown on the dashboard clock. It might seem that four-velocity is a hybrid quantity, the quotient of a length determined in one reference frame by a time interval determined in another. The time interval is, however, a proper time, thus belonging in a sense to all reference frames. The four-velocity is a four-vector with components belonging to a particular reference frame, just as the velocity components belong to a particular reference frame. Both are determined in practice from distance and time measurements in one reference frame—that of the laboratory. A "time of flight" or "Cerenkov pulse" determination of the speed v of a fast particle can just as well be regarded as a determination of w, because

$$w = v\left(1 - \frac{v^2}{c^2}\right)^{-1/2}$$

The synchronization of clocks by means of light flashes means that any laboratory measurement of v or w is essentially the timing on a laboratory clock of a race between the particle and a photon over the same laboratory distance in vacuum.[4]

The transformation behavior of the velocity components [Eq. (2-42)] is rather complicated, because these quantities are ratios of components of a four-vector $d\tilde{x}$. The components of $\tilde{w}$, on the other hand, transform as the components of a four-vector.

Four-velocity is not defined for photons or neutrinos, since they have no rest frame and thus no proper time.

One can proceed in the same way to define the *four-acceleration*, $\tilde{b}$, the derivative of the four-velocity with respect to proper time.

$$\tilde{b} = \frac{d\tilde{w}}{d\tau} = \frac{d^2\tilde{x}}{d\tau^2} \tag{2-62}$$

Its four-vector transformation law is in even greater contrast to that of the components of the acceleration, a complicated set of equations that can be

---

[4] A name is needed for the vector **w**. *Proper velocity* and *relativistic velocity* are sometimes used. They may be confusing—the velocity of a particle in its rest frame is zero, and, relativistically, velocity is velocity—but any name can be used if it is clearly defined.

found by differentiating Eqs. (2-42). If one needs to transform acceleration components, it is simpler to convert to four-acceleration, transform that, and convert back to acceleration in the new frame. To that end we now derive the expressions relating b_μ to the acceleration **a** and velocity **v**. Substituting from Eq. (2-57) and Eqs. (2-60), we get

$$b_1 = \frac{dw_1}{d\tau} = \frac{d[v_x(1 - v^2/c^2)^{-1/2}]}{dt(1 - v^2/c^2)^{1/2}} = a_x\left(1 - \frac{v^2}{c^2}\right)^{-1} + \frac{v_x}{c^2}(\mathbf{v} \cdot \mathbf{a})\left(1 - \frac{v^2}{c^2}\right)^{-2}$$

$$b_2 = \frac{dx_2}{d\tau} = \frac{d[v_y(1 - v^2/c^2)^{-1/2}]}{dt(1 - v^2/c^2)^{1/2}} = a_y\left(1 - \frac{v^2}{c^2}\right)^{-1} + \frac{v_y}{c^2}(\mathbf{v} \cdot \mathbf{a})\left(1 - \frac{v^2}{c^2}\right)^{-2}$$

$$b_3 = \frac{dw_3}{d\tau} = \frac{d[v_z(1 - v^2/c^2)^{-1/2}]}{dt(1 - v^2/c^2)^{1/2}} = a_z\left(1 - \frac{v^2}{c^2}\right)^{-1} + \frac{v_z}{c^2}(\mathbf{v} \cdot \mathbf{a})\left(1 - \frac{v^2}{c^2}\right)^{-2}$$

or

$$\mathbf{b} = \mathbf{a}\left(1 - \frac{v^2}{c^2}\right)^{-1} + \mathbf{v}\frac{(\mathbf{v} \cdot \mathbf{a})}{c^2}\left(1 - \frac{v^2}{c^2}\right)^{-2} \tag{2-63a}$$

For the timelike component,

$$b_4 = \frac{dw_4}{d\tau} = \frac{d[ic(1 - v^2/c^2)^{-1/2}]}{dt(1 - v^2/c^2)^{1/2}} = \frac{c}{i}\frac{(\mathbf{v} \cdot \mathbf{a})}{c^2}\left(1 - \frac{v^2}{c^2}\right)^{-2} \tag{2-63b}$$

or

$$b_0 = \frac{b_4}{i} = \frac{c(\mathbf{v} \cdot \mathbf{a})}{c^2}\left(1 - \frac{v^2}{c^2}\right)^{-2} \tag{2-63bb}$$

Again, the four components of $\tilde{b}$ are not independent. Combining the equations, we find

$$\mathbf{b} = \mathbf{a}\left(1 - \frac{v^2}{c^2}\right)^{-1} + \mathbf{v}\frac{b_4}{ic}$$
$$= \mathbf{a}\left(1 - \frac{v^2}{c^2}\right)^{-1} + \mathbf{v}\frac{b_0}{c} \tag{2-64}$$

The constancy of the four-velocity [Eq. (2-61b)] implies orthogonality of four-acceleration and four-velocity

$$\frac{d(w_\mu^2)}{d\tau} = 2w_\mu\frac{dw_\mu}{d\tau} = 2w_\mu b_\mu = 2\tilde{w}\tilde{b} = 0 \tag{2-65}$$

These equations hold in any inertial reference frame. Apply them now in the particle's rest frame. We now use italic letters with a superscript dagger (†) to label quantities in the rest frame. Since the particle is at rest in $R^\dagger$,

$$v^\dagger = w^\dagger = 0 \qquad w_4^\dagger = ic \qquad (2\text{-}66a)$$

Equations (2-63) give

$$\left.\begin{array}{l} b_1^\dagger = a_x^\dagger \\ b_2^\dagger = a_y^\dagger \\ b_3^\dagger = a_z^\dagger \end{array}\right\} \quad \mathbf{b}^\dagger = \mathbf{a}^\dagger \qquad (2\text{-}66b)$$

$$b_4^\dagger = 0 = b_0^\dagger$$

We could have inferred the vanishing of the timelike component in the rest frame from the orthogonality condition, Eq. (2-65). The square of the four-acceleration is in the rest frame

$$(b_\mu^\dagger)^2 = (a^\dagger)^2 \qquad (2\text{-}66c)$$

and this is invariant. Thus we have a restriction on the components of $\tilde{b}$ in any frame: The square of the four-acceleration equals the square of the acceleration in the instantaneous rest frame. It is positive; thus, the four-acceleration is spacelike.

Let us now find the transformation equations between acceleration components in the rest frame ($R^\dagger$) and the laboratory frame R. The inverse of the SLT Eq. (2-49bb) gives, with the x axis parallel to the instantaneous particle velocity,

$$b_0 = \gamma(b_0^\dagger + \beta b_1^\dagger)$$
$$b_1 = \gamma(b_1^\dagger + \beta b_0^\dagger)$$
$$b_2 = b_2^\dagger$$
$$b_3 = b_3^\dagger$$

so that

$$\begin{aligned} b_0 &= \gamma\beta b_1^\dagger = \beta b_1 \\ b_1 &= \gamma b_1^\dagger = \gamma a_x^\dagger \\ b_2 &= b_2^\dagger = a_y^\dagger \\ b_3 &= b_3^\dagger = a_z^\dagger \end{aligned} \qquad (2\text{-}67)$$

Since we are considering the rest frame, $\mathbf{u}_R(R^+) = \mathbf{v}$ and there is no ambiguity in the use of γ and β in the equations. Equation (2-64) in the laboratory frame gives

$$b_1 = a_x\gamma^2 + \frac{\beta c}{c}b_0 = a_x\gamma^2 + \beta^2 b_1$$

$$b_2 = a_y\gamma^2 \tag{2-68}$$

$$b_3 = a_z\gamma^2$$

Substituting for b_1, b_2, b_3 from Eq. (2-67), we obtain

$$a_x = a_x^\dagger\left(1 - \frac{v^2}{c^2}\right)^{3/2}$$

$$a_y = a_y^\dagger\left(1 - \frac{v^2}{c^2}\right) \tag{2-69a}$$

$$a_z = a_z^\dagger\left(1 - \frac{v^2}{c^2}\right)$$

or, because the x direction is parallel to the velocity,

$$a_\parallel = a_\parallel^\dagger\left(1 - \frac{v^2}{c^2}\right)^{3/2}$$

$$a_\perp = a_\perp^\dagger\left(1 - \frac{v^2}{c^2}\right) \tag{2-69b}$$

For finite acceleration in the rest frame, all components of acceleration in the laboratory frame approach zero as $v \to c$. This is, of course, consistent with the role of c as limiting velocity.

These simple transformation equations [Eqs. (2-69)] hold only if $R^\dagger$ is the instantaneous rest frame. They will be useful in dealing with accelerated motion, where one can write down a simple Newtonian equation of motion in the rest frame (third postulate) and then transform it to the laboratory frame to get the correct relativistic equation of motion.

2-11. UNIFORM LONGITUDINAL ACCELERATION

The simplest type of motion, after that with constant velocity, is uniform linear acceleration. Since acceleration is not invariant, one must specify the reference frame in which the acceleration is constant. The case we consider now is that in which the acceleration is parallel to the velocity and is constant in the instantaneous rest frame of the particle. The successive rest frames move faster and faster, in the direction of the initial velocity. This situation is realized for an ion in an electrostatic accelerator moving in the direction of a constant electric field. The magnitude of the acceleration in the rest frame is, by Newton's second law [Eq. (1-28)],

$$a^\dagger = \frac{e\mathscr{E}_\parallel^\dagger}{\mu}$$

{e is the charge, an invariant [see Eq. (D-2)]; μ is the proper mass, equal to the inertial mass in the rest frame}. By Eq. (2-55b),

$$a^\dagger = \frac{e\mathscr{E}_\parallel}{\mu}$$

Another example is a rocket with a burning rate controlled to keep constant the ratio of thrust to inertial mass as measured in the rocket frame.

With the x axis parallel to the velocity,

$$a_x^\dagger = a^\dagger = \text{const.}$$
$$a_y^\dagger = 0 \qquad\qquad\qquad (2\text{-}70)$$
$$a_z^\dagger = 0$$

Transforming to the laboratory by Eq. (2-69a), we get

$$a_x = a^\dagger\left(1 - \frac{v_x^2}{c^2}\right)^{3/2} = a^\dagger\gamma^{-3}$$
$$a_y = 0 \qquad\qquad\qquad (2\text{-}71)$$
$$a_z = 0$$

Note that the acceleration in the laboratory frame is not constant but diminishes as v_x increases. The first of Eqs. (2-71) reads

$$\frac{dv_x}{(1 - v_x^2/c^2)^{3/2}} = a^\dagger\, dt$$

integrating immediately to

$$\frac{v_x/c}{(1 - v_x^2/c^2)^{1/2}} = \frac{a^\dagger t}{c} \qquad\qquad (2\text{-}72a)$$

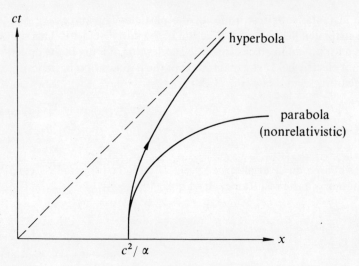

Figure 2-35.   World line of a particle undergoing constant acceleration in its rest frame. Also shown is the Galilean parabola that approximates the hyperbola in the small-velocity region.

the origin of time being put at the epoch when $v_x = 0$. Solving for v_x yields

$$v_x = c\frac{a^\dagger t/c}{[1 + (a^\dagger)^2 t^2/c^2]^{1/2}} \tag{2-72b}$$

As t increases, $v_x \to c$. In the nonrelativistic region, $a^\dagger t \ll c$; this formula gives the familiar expression for constant acceleration

$$v_x \approx a^\dagger t$$

as expected. Integrating Eq. (2-72b), we find

$$x = \frac{c^2}{a^\dagger}\left[1 + \frac{(a^\dagger)^2 t^2}{c^2}\right]^{1/2} \tag{2-73}$$

where we have placed the origin so that $x = c^2/a^\dagger$ at $t = 0$. Equation (2-73) can be written

$$x^2 - c^2 t^2 = \left(\frac{c^2}{a^\dagger}\right)^2$$

The world line (Fig. 2-35) is a hyperbola. The hyperbola osculates at $t = 0$ the familiar nonrelativistic parabola

$$x = \frac{c^2}{a^\dagger} + \frac{1}{2}a^\dagger t^2$$

which is also shown in the figure.

2-12. CONSTANT TRANSVERSE ACCELERATION

Another simple case occurs when the acceleration in the rest frame is constant and perpendicular to the line of flight of the laboratory in the rest frame (Fig. 2-36). The particle is set into motion in $R^\dagger$, and the new rest frame at a later time has, of course, a different velocity with respect to the laboratory. It will turn out that in the laboratory not the speed but only the direction is changed. This case is realized when a charged particle moves in a transverse magnetic field (Fig. 2-37). In effect, Eqs. (2-55b) show that in the rest frame there is an electric field $\mathscr{E}^\dagger$ transverse to both the velocity and the magnetic field and of magnitude $\beta\gamma\mathscr{B}$, where β and γ refer to the velocity of the particle and $\mathscr{B}$ is the laboratory magnetic field; the magnetic field in $R^\dagger$ has no effect on a particle instantaneously at rest. Newton's second law then gives for the magnitude of the acceleration in $R^\dagger$,

$$a^\dagger = \frac{e\beta\gamma\mathscr{B}}{\mu} \tag{2-74a}$$

Another example is provided by a charged particle moving at right angles to an electric field. Equations (2-55b) lead to a transverse electric field in the

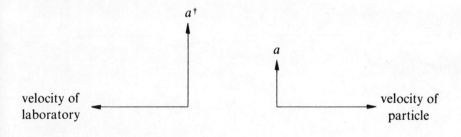

rest frame, $R^\dagger$ laboratory frame, R

Figure 2-36.   Transverse acceleration.

rest frame of magnitude $\gamma\mathscr{E}$, so that

$$a^\dagger = \frac{e\gamma\mathscr{E}}{\mu} \qquad (2\text{-}74b)$$

the same expression as above with $\mathscr{E}$ replacing $\beta\mathscr{B}$.

With the $x^\dagger$ axis in the direction of $-\mathbf{u}_{R^\dagger}(R)$ and the $y^\dagger$ axis in the direction of the acceleration,

$$a_x^\dagger = 0$$
$$a_y^\dagger = a^\dagger \qquad (2\text{-}75)$$
$$a_z^\dagger = 0$$

Transforming to the laboratory by Eqs. (2-69a), we get

$$a_x = 0$$
$$a_y = a^\dagger\left(1 - \frac{v_x^2}{c^2}\right) = a^\dagger\gamma^{-2} \qquad (2\text{-}76)$$
$$a_z = 0$$

In the laboratory the acceleration is also transverse, as drawn in Fig. 2-36. The motion is, therefore, a turning of the velocity vector without change of

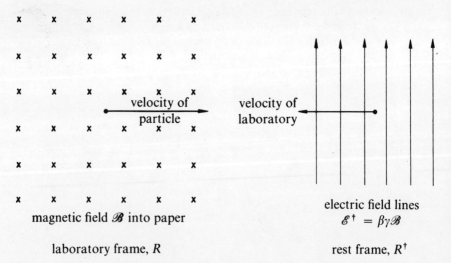

magnetic field $\mathscr{B}$ into paper

velocity of particle

velocity of laboratory

electric field lines
$\mathscr{E}^\dagger = \beta\gamma\mathscr{B}$

laboratory frame, R

rest frame, $R^\dagger$

Figure 2-37. Particle moving in transverse magnetic field.

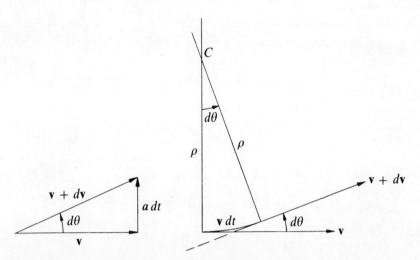

Figure 2-38. Turning of velocity vector by transverse acceleration, showing angular deflection $d\theta$ and radius of curvature ρ.

speed (Fig. 2-38), the angle in R being

$$d\theta = \frac{a_y dt}{v} = \frac{a^\dagger \gamma^{-2}}{\beta c} dt$$

The angular velocity with which the velocity vector turns is

$$\frac{d\theta}{dt} = \frac{a^\dagger}{\beta \gamma^2 c} = \begin{cases} \dfrac{e\mathcal{B}}{\mu \gamma c} & \text{(magnetic deflection)} \\[4mm] \dfrac{e\mathcal{E}}{\mu \beta \gamma c} & \text{(electric deflection)} \end{cases} \qquad (2\text{-}77)$$

The radius of curvature ρ is (Fig. 2-38),

$$\rho = \frac{v\,dt}{d\theta} = \frac{\beta^2 \gamma^2 c^2}{a^\dagger} = \begin{cases} \dfrac{\mu \beta \gamma c^2}{e\mathcal{B}} = \dfrac{\mu \gamma v c}{e\mathcal{B}} \\[4mm] \dfrac{\mu \beta^2 \gamma c^2}{e\mathcal{E}} = \dfrac{\mu \gamma v^2}{e\mathcal{E}} \end{cases} \qquad (2\text{-}78a)$$

In terms of the spatial part of the four-velocity $\tilde{w}$,

$$\mathbf{w} = \gamma\mathbf{v} \tag{2-60a}$$

these equations take the form

$$\rho = \frac{w^2}{a^\dagger} \tag{2-78b}$$

or

$$\frac{e}{c}\mathscr{B}\rho = \mu w \quad \text{(magnetic)}$$

$$e\mathscr{E}\rho = \mu w v \quad \text{(electric)} \tag{2-78c}$$

If the magnetic field is homogeneous, the acceleration remains transverse, and the motion is uniform circular at the "cyclotron" angular frequency given by Eq. (2-77) and the radius given by Eqs. (2-78). The electric case is more complicated. If, however, the center of curvature of the orbit coincides with the center from which the electric field lines are diverging, acceleration remains transverse and one again has a circular orbit at constant speed. Such is the case for an atomic electron in a circular Bohr orbit about the nucleus.

PROBLEMS

2-1. A thought clock [R. B. Leighton, *Principles of Modern Physics* (McGraw-Hill, New York, 1959)] is made from a flashtube, a mirror, a photocell, and some pulse circuits. Referring to the sketch, we note that the circuits in the black box trigger the flashtube with negligible delay whenever the photocell detects a light pulse. The time between successive pulses is the travel time of the light from the flashtube to the mirror to the photocell, a distance $2d$ in the clock frame. The clock reads time as the number of flashes.

This clock is observed from a reference frame in which the clock is moving at right angles to the light path with velocity u. Calculate the time between flashes in this frame, assuming only Einstein's second postulate and the invariance of transverse dimensions. Verify that the Lorentz transformation gives the same answer.

Now the clock is observed from a frame in which the clock is moving parallel to the light path with velocity u. Again calculate the time between flashes, this time assuming Einstein's second postulate and the Lorentz contraction. Verify that the Lorentz transformation gives the same answer.

This problem shows that the rate of a clock as observed from a moving frame depends only on the relative speed and is independent of the direction of motion with respect to the axis of the clock.

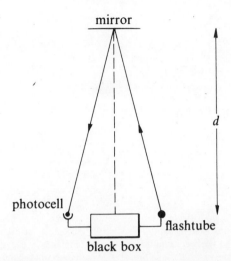

mirror

d

photocell

flashtube

black box

2-2. Prove the statement made at the beginning of Section 2-3 that the transformation Eq. (2-6) must be linear if it is to take uniform rectilinear motion in R into uniform rectilinear motion in R' and inversely.

2-3. In the derivation of the special Lorentz transformation Eq. (1-31a) from Eq. (2-16), we considered a thought experiment in which a light is flashed at O at $t = t' = 0$ (Fig. 2-8). Assume instead of Einstein's second postulate that the Galileo transformation is valid. Find the size and shape of the wave front in R'. In R', what is the velocity of light in the direction of x', of y', of $-x'$?

2-4. [From J. H. Smith, *Introduction to Special Relativity* (Benjamin, New York, 1965).] A rocketship of proper length 10 m is moving away from the earth at speed $4c/5$. A light signal is sent after it that arrives at the rocket's tail at time zero according to rocket clocks and earth clocks. Calculate in both frames the time at which the light signal reaches the head of the rocket. It is there reflected back by a mirror. Calculate in both frames the time at which the light signal again reaches the tail of the rocket. Compare the elapsed time in the two frames for the round trip tail–head–tail.

2-5. A laboratory L has an optical bench 2.0 m long, the ends of which are labeled A and B. (See sketch, top of next page.) At each end is a photocell connected to a clock so as to record the time of arrival of a light pulse. This laboratory is in uniform motion with respect to an identical laboratory M containing an identical optical bench with ends labeled C and D and having identical clocks. Its velocity is $\sqrt{99}\, c/10$ ($\gamma = 10$, $1 - \beta \approx 1/200$, $1 + \beta \approx 2$). The benches are parallel to the direction of relative motion and are essentially on the same line. At the instant that the midpoints of the benches coincide, a flashlamp at the midpoint of the bench in M is triggered. Designate by α, β, γ, and δ the arrival of the light flash at the respective bench ends A, B, C, and D. Choose suitable origins of position and time in both laboratories. Let the unit of time be the time needed for light to travel 1.0 m ($c = 1.0$).

What are the four clock readings for events α, β, γ, and δ? Which events are simultaneous in which laboratory?

Calculate the position and time of each event, in each reference frame. Calculate in each frame the spatial interval, temporal interval, and four-interval between α and β and between γ and δ.

Discuss the relation between your results, the Lorentz contraction and the Einstein dilation.

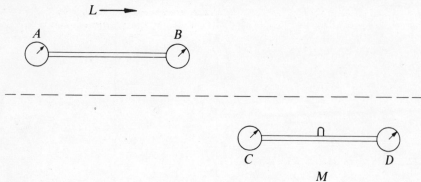

2-6. A rod of proper length 10 cm is moving parallel to itself, along line PQ, with a constant velocity three-fifths that of light ($c = 3 \times 10^{10}$ cm sec^{-1}). At each end of the rod is a lamp that can be flashed instantaneously on command. A camera with open shutter produces an image on a film. The line $P'Q'$ is the image of the line PQ in the film plane; $P'Q'$ is parallel to PQ. At time $t = 0$ in the laboratory reference frame, the midpoint of the rod is at S, directly above the lens.

At $t = 0$ in the laboratory frame, the lamps are flashed. What is the distance between their images on the film?

In another experiment, a light signal is emitted from the middle of the rod, at time $t = -2/3 \times 10^{-8}$ sec in the laboratory frame. When the signal reaches each end of the rod it causes the lamp there to flash immediately. What is the distance between the images of the two end lights on the film?

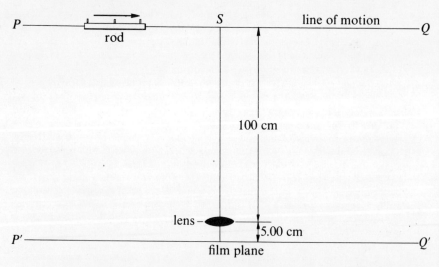

2-7. [E. F. Taylor and J. A. Wheeler, *Spacetime Physics* (Freeman, San Francisco, 1963.)] In a thought experiment, a man running with speed $\sqrt{3}\,c/2$ carries a horizontal pole that is 20 m long in his reference frame. On the ground is a barn 10 m long open at both ends. The runner runs through it. Compute the length of the pole in the ground reference frame (R). Compute the length of the barn in the runner's reference frame (R').

Call the ends of the pole P_A and P_B and the ends of the barn B_A and B_B. Let $t = 0$, $x = 0$, and $t' = 0$, $x' = 0$ when P_A meets B_A. Calculate in both frames the time and position of the following events: (1) P_A meets B_B; (2) P_B meets B_A; (3) P_B meets B_B. Discuss in both R and R' the layman's question: Does the pole fit in the barn?

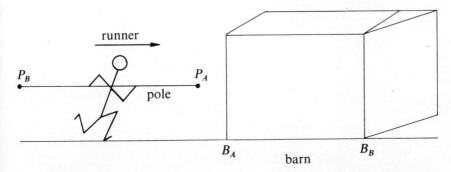

2-8. An event A occurs at $(ct, x, y, z) = (0, 0, 0, 0)$. Another event, B, occurs at $(3, 3, 4, 0)$ cm (the velocity of light is 3×10^{10} cm sec^{-1}). Find a frame R' in which B precedes A in time by 1×10^{-10} sec. Is there a frame in which B and A occur at the same place?

Suppose $ct_B = 6$ instead of 3 cm. Discuss the possibilities of bringing the events into spatial and temporal coincidence.

2-9. An interaction (event 1 at point A) gives rise to a proton of speed $0.5c$. The proton travels 100 m at constant velocity to point B where it interacts (event 2). In the rest frame of the proton $R^†$, what is

    (a) the spatial separation of the two events;

    (b) the distance between A and B;

    (c) the temporal separation of the two events;

    (d) the velocity of the laboratory?

What is the interval between the events? Consider all possible inertial frames, subject only to $|u| < c$; what is the range of values of the spatial separation? Of the temporal separation?

2-10. A particle at the origin executes simple harmonic motion along the y direction, with frequency 60 Hz and amplitude 1.0 cm. Determine the position and velocity as functions of time in a reference frame R' moving with velocity V in the x direction. Calculate the time and distance between successive maxima of y'.

2-11. (From University of Illinois Ph.D. Qualifying Examination.) Relative to the earth, the stars A, B, and C are moving in the direction E with a common velocity three-fifths that of light. At the same time in the earth frame, they are at the vertices of an equilateral triangle of side $L = 100$ lightyears. Determine the sides and orientation of the triangle ABC in the rest frame of the three stars. (See sketch, top of next page.)

The earth is in the direction S relative to the three stars at a distance $\gg 100$ lightyears. In what direction in the stars' rest frame is the light emitted that travels S in the earth frame? What frequency does the H_β line ($\nu_0 = 0.616 \times 10^{14}$ Hz) from a star have in the earth frame?

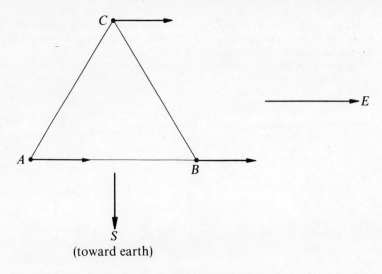

S
(toward earth)

2-12. We know that R' is moving with velocity $c/2$ with respect to R and that R'' is moving in the same direction with velocity $c/2$ with respect to R'. Compounding the SLT from R to R' with the SLT from R' to R'', find the transformation equations between R and R''. What is the velocity with respect to R of a point fixed in R''?

2-13. Show that the transformation equations for particle acceleration are

$$\mathbf{a}_\parallel = \left(1 + \frac{uv'_\parallel}{c^2}\right)^{-3}\left(1 - \frac{u^2}{c^2}\right)^{3/2}\mathbf{a}'_\parallel$$

$$\mathbf{a}_\perp = \left(1 + \frac{uv'_\parallel}{c^2}\right)^{-2}\left(1 - \frac{u^2}{c^2}\right)\mathbf{a}'_\perp - \frac{ua'_\parallel}{c^2}\left(1 + \frac{uv'_\parallel}{c^2}\right)^{-3}\left(1 - \frac{u^2}{c^2}\right)\mathbf{v}'_\perp$$

where $\mathbf{u}$ is the velocity of R' with respect to R, and the particle velocity and acceleration are resolved into parts parallel and perpendicular to $\mathbf{u}$:

$$\mathbf{v} = \mathbf{v}_\parallel + \mathbf{v}_\perp \qquad \mathbf{v}' = \mathbf{v}'_\parallel + \mathbf{v}'_\perp$$
$$\mathbf{a} = \mathbf{a}_\parallel + \mathbf{a}_\perp \qquad \mathbf{a}' = \mathbf{a}'_\parallel + \mathbf{a}'_\perp$$

Note that when R' is the rest frame ($v' = 0$), these equations reduce to Eqs. (2-69).

2-14. If a rocket has a constant acceleration in its rest frame of $g = 32$ ft sec^{-2}, how long does it take to get to a velocity of $(1/\sqrt{2})c$ with respect to the earth? Give the answer both in proper time and in earth time. Neglect the acceleration of the earth, and gravitational fields.

In tests on the ground it is found that the rate of consumption of fuel is proportional to the thrust, the constant of proportionality being 4.0×10^{-3} lb of fuel per second per pound of thrust. If the ratio of initial mass to residual mass after burnout is 10, for how long can acceleration equal to g be maintained? Discuss the possible use of a multistage system to get to relativistic speeds.

2-15. Ignoring the result of the preceding problem, we undertake a direct test of the twin problem. It starts on April 1, 1984, when the twins are 20 years old. One twin rides off in a spaceship;

the other remains on earth. The ship has a constant acceleration equal to $980\,\mathrm{cm\,sec^{-2}}$ in its rest frame, and can maintain it for 8.0 years of its proper time (!). It accelerates in a straight line for 2.0 years, decelerates at the same rate for 2.0 more years, turns around, accelerates for 2.0 years, decelerates for 2.0 years, and lands on earth. Neglect gravitational fields and the earth's acceleration. On his return, the traveling twin is 28 years old (proper time). What is the date on earth? How far away from the earth did the ship get?

Chapter 3

Dynamics of a Particle

In dynamics we consider the effect of interactions on motion. We shall see that in special relativity a central position is occupied by a certain four-vector, the four-momentum $\tilde{p}$. A natural generalization of Newton's momentum, or quantity of motion, the four-momentum involves momentum and energy together, in the same way that the localized event four-vector $\tilde{x}$ involves position and epoch together.

The theory of special relativity can be regarded as the culmination of classical electrodynamics. It is an adjustment of kinematics and dynamics to fit Maxwell's theory of the electromagnetic field. Einstein derived relativistic dynamics from the behavior of a charged particle in an electromagnetic field. The resulting formalism applies, so far as we can ascertain at present, to all interactions, and can be presented without reference to electrodynamics. But it is illuminating to follow initially the historical order, as we shall do.

3-1. THE LAW OF MOTION FOR A CHARGED PARTICLE

The argument has been anticipated in the treatment of special cases of constant acceleration in Sections 2-11 and 2-12. One assumes the correctness of Newtonian mechanics in the particle's rest frame, and uses the relativistic transformation equations to transform the equation of motion to any other inertial frame. The relevant equations from electrodynamics are the transformation equations of field strength and charge (Appendix D).

For a special Lorentz transformation (SLT) with x, x' axes in the direction of relative motion

$$
\begin{aligned}
\mathscr{E}_{x'} &= \mathscr{E}_x & \mathscr{B}_{x'} &= \mathscr{B}_x \\
\mathscr{E}_{y'} &= \gamma(\mathscr{E}_y - \beta\mathscr{B}_z) & \mathscr{B}_{y'} &= \gamma(\mathscr{B}_y + \beta\mathscr{E}_z) \\
\mathscr{E}_{z'} &= \gamma(\mathscr{E}_z + \beta\mathscr{B}_y) & \mathscr{B}_{z'} &= \gamma(\mathscr{B}_z - \beta\mathscr{E}_y)
\end{aligned}
\tag{2-55a}
$$

The electric charge of a system turns out to be invariant:

$$
e = e'
\tag{D-2}
$$

136

An electron has a charge of 4.80×10^{-10} Gaussian units of charge (or 1.60×10^{-19} C) in every inertial frame.

In any frame, the force on a charged particle is

$$\mathbf{F} = e\left(\mathscr{E} + \frac{1}{c}\mathbf{v} \times \mathscr{B}\right) \tag{D-1}$$

(This is, in fact, the definition of electric and magnetic field strength). In the rest frame $R^\dagger$, the velocity is zero; so

$$\mathbf{F}^\dagger = e^\dagger \mathscr{E}^\dagger = e\mathscr{E}^\dagger$$

Newton's law of motion in the form Eq. (1-28) holds in the rest frame

$$\mathbf{F}^\dagger = m^\dagger \mathbf{a}^\dagger \qquad (m^\dagger = \text{const})$$

We denote the inertial mass in the rest frame by μ, and call it the *proper mass* or *rest mass*. Thus,

$$e\mathscr{E}^\dagger = \mu \mathbf{a}^\dagger \tag{3-1}$$

is the equation of motion in the rest frame. We now make a special Lorentz transformation to the laboratory frame, with the x axis parallel to the particle velocity in the laboratory.

$$u = v = v_x$$

Equations (2-55a) enable us to express the left-hand side of Eq. (3-1) in terms of laboratory quantities

$$e\mathscr{E}_x^\dagger = e\mathscr{E}_x$$

$$e\mathscr{E}_y^\dagger = e\gamma(\mathscr{E}_y - \beta\mathscr{B}_z) = \gamma e\left(\mathscr{E}_y - \frac{v_x}{c}\mathscr{B}_z\right) \tag{3-2a}$$

$$e\mathscr{E}_z^\dagger = e\gamma(\mathscr{E}_z + \beta\mathscr{B}_y) = \gamma e\left(\mathscr{E}_z + \frac{v_x}{c}\mathscr{B}_y\right)$$

and Eqs. (2-69a) give for the right-hand side,

$$\mu a_x^\dagger = \mu a_x\left(1 - \frac{v^2}{c^2}\right)^{-3/2} = \mu\gamma^3 a_x$$

$$\mu a_y^\dagger = \mu a_y\left(1 - \frac{v^2}{c^2}\right)^{-1} = \mu\gamma^2 a_y$$

$$\mu a_z^\dagger = \mu a_z\left(1 - \frac{v^2}{c^2}\right)^{-1} = \mu\gamma^2 a_z$$

Equation (3-1) becomes

$$e\mathscr{E}_x = \mu\gamma^3 a_x$$

$$e\left(\mathscr{E}_y - \frac{v_x}{c}\mathscr{B}_z\right) = \mu\gamma a_y \tag{3-3a}$$

$$e\left(\mathscr{E}_z - \frac{v_x}{c}\mathscr{B}_y\right) = \mu\gamma a_z$$

These are the equations of motion of a charged particle in an arbitrary electromagnetic field. They determine **a**, the rate of change of velocity, in terms of $\mathscr{E}, \mathscr{B}$, and **v**:

$$a_x = \frac{e\mathscr{E}_x}{\mu\gamma^3}$$

$$a_y = \frac{e[\mathscr{E}_y - (v_x/c)\mathscr{B}_z]}{\mu\gamma} \tag{3-3b}$$

$$a_z = \frac{e[\mathscr{E}_z + (v_x/c)\mathscr{B}_y]}{\mu\gamma}$$

We are free to multiply the two sides of an equation by anything we please, so we cannot say that a particular factor belongs on one side or the other. In the form Eq. (3-3a), the left-hand side consists of the longitudinal and the two transverse components of the force [Eq. (D-1)]:

$$e\left(\mathscr{E} + \frac{\mathbf{v}}{c} \times \mathscr{B}\right)_x = (\mu\gamma^3)a_x$$

$$e\left(\mathscr{E} + \frac{\mathbf{v}}{c} \times \mathscr{B}\right)_y = (\mu\gamma)a_y \tag{3-4a}$$

$$e\left(\mathscr{E} + \frac{\mathbf{v}}{c} \times \mathscr{B}\right)_z = (\mu\gamma)a_z$$

If one is wedded to $\mathbf{F} = m\mathbf{a}$, one is then inclined to call $\mu\gamma^3$ the "longitudinal mass" and $\mu\gamma$ the "transverse mass." This was, in fact, done by Einstein in his first paper [2]. He was, however, quick to agree [31] with Planck that a more illuminating form of the right-hand side is as the time derivative of the components of a certain vector, $\mu\gamma\mathbf{v}$:

$$\mu\gamma^3 a_x = \frac{d}{dt}(\mu\gamma\mathbf{v})_x$$

$$\mu\gamma a_y = \frac{d}{dt}(\mu\gamma\mathbf{v})_y \qquad\qquad (3\text{-}4b)$$

$$\mu\gamma a_z = \frac{d}{dt}(\mu\gamma\mathbf{v})_z$$

Let us verify this identity. Recalling that

$$\mu\gamma\mathbf{v} = \frac{\mu\mathbf{v}}{\sqrt{1 - v^2/c^2}}$$

we see that its time derivative will take different forms depending on whether the acceleration is longitudinal or transverse. If it is longitudinal, the change involves both the change in $\mathbf{v}$ in the numerator and the change in $v^2 = \mathbf{v}\cdot\mathbf{v}$ in the denominator. If it is transverse, the speed is constant and only the numerator changes. The longitudinal component is

$$\frac{d}{dt}\left[\mu v_x\left(1 - \frac{v^2}{c^2}\right)^{-1/2}\right] = \mu a_x\left(1 - \frac{v^2}{c^2}\right)^{-1/2}$$

$$+ \mu v_x\left(-\frac{1}{2}\right)\left(1 - \frac{v^2}{c^2}\right)^{-3/2}\frac{-2(\mathbf{v}\cdot d\mathbf{v})}{c^2\, dt}$$

Since

$$\frac{(\mathbf{v}\cdot d\mathbf{v})}{dt} = \mathbf{v}\cdot\mathbf{a} = v_x a_x$$

this equation becomes

$$\mu a_x\left(1 - \frac{v^2}{c^2}\right)^{-1/2}\left[1 + \left(1 - \frac{v^2}{c^2}\right)^{-1}\frac{v_x^2}{c^2}\right] = \mu a_x\left(1 - \frac{v^2}{c^2}\right)^{-3/2}$$

and the first of Eqs. (3-4b) has been arrived at. For the y component,

$$\frac{d}{dt}\left[\mu v_y\left(1 - \frac{v^2}{c^2}\right)^{-1/2}\right] = \mu a_y\left(1 - \frac{v^2}{c^2}\right)^{-1/2}$$

$$+ \mu v_y\left(-\frac{1}{2}\right)\left(1 - \frac{v^2}{c^2}\right)^{-3/2}\cdot\frac{-2(\mathbf{v}\cdot d\mathbf{v})}{c^2\, dt}$$

Because $v_y = 0$, the second term on the right-hand side vanishes, and we are left with

$$\mu a_y \left(1 - \frac{v^2}{c^2}\right)^{-1/2}$$

for the second of Eqs. (3-4b). Similarly for the third.

The three equations of motion are seen to be one vector equation:

$$e\left(\mathscr{E} + \frac{\mathbf{v}}{c} \times \mathscr{B}\right) = \frac{d}{dt}(\mu\gamma\mathbf{v})$$

$$\gamma = \left(1 - \frac{v^2}{c^2}\right)^{-1/2}$$

(3-5)

It is evidently covariant with respect to rotation and inversion of the space axes (Appendix D). It is, of course, covariant with respect to special Lorentz transformations, because it was derived by use of the relativistic transformation formulas. It is thus covariant with respect to the general Lorentz transformation.

The equation of motion, Eq. (3-5), has been confirmed experimentally to very high precision. Until 1914, the "best" experimental data seemed to disagree with it (for an interesting discussion by Einstein, see [31], pp. 436–439). The more precise modern experiments have given beautiful agreement. The most sensitive test is provided by the large circular proton or electron accelerators, such as the Brookhaven alternating gradient synchrotron, in which the beam particles go round and round literally hundreds of thousands of times under conditions where the slightest departure from the expression would show itself immediately in the output of the machine.

The left-hand side of Eq. (3-5) is the force expression Eq. (D-1). The right-hand side is the time derivative of a quantity having all the properties we associate with momentum. It is a vector parallel to the velocity, vanishing for a particle at rest. Its magnitude increases with the speed. For a given speed its magnitude is proportional to μ, the inertial mass of the particle in its rest frame. At $v \ll c$, it reduces to the prerelativistic expression for the momentum, $\mu\mathbf{v}$, the difference being of second order in v/c:

$$\mu\gamma\mathbf{v} = \mu\mathbf{v}\left(1 + \frac{1}{2}\frac{v^2}{c^2} + \cdots\right)$$

It is, according to Eq. (3-5), a constant of the motion if the particle is free (zero field). We therefore take it to be the *momentum*. The relativistic equation relating momentum and velocity is thus

$$\mathbf{p} = \frac{\mu \mathbf{v}}{\sqrt{1 - v^2/c^2}} \tag{1-33}$$

The new feature of this relation is that the scalar coefficient of the velocity is no longer a constant independent of the speed. This coefficient, called the *inertial mass m*,

$$\mathbf{p} = m\mathbf{v} \tag{1-1}$$

is rather

$$m = \frac{\mu}{\sqrt{1 - v^2/c^2}} \tag{1-2}$$

This equation [Eq. (1-2)] gives the famous relativistic variation of mass with velocity. It is just another way of stating the momentum–velocity relation Eq. (1-33).

We shall save the Latin letter m for *inertial mass* and the Greek letter μ for *proper mass*. Among particle physicists the word "mass" used without qualification usually means proper mass. We shall always use a defining adjective.

The force is by definition the time rate of change of momentum:

$$\mathbf{F} = \frac{d\mathbf{p}}{dt} \tag{1-5}$$

The derivation of Eq. (3-5) is a proof of the consistency of this definition and the expression used in electrodynamics:

$$\mathbf{F} = e\left(\mathcal{E} + \frac{\mathbf{v}}{c} \times \mathcal{B}\right) \tag{D-1}$$

based on the assumption of the relativistic transformation properties of e, $\mathcal{E}$, and $\mathcal{B}$.

Equations (3-2a) are the transformation law of electromagnetic force:

$$F_x^\dagger = F_x$$
$$F_y^\dagger = \gamma F_y \tag{3-2b}$$
$$F_z^\dagger = \gamma F_z$$

or

$$\mathbf{F}_\| = \mathbf{F}_\|^\dagger$$
$$\mathbf{F}_\perp = \gamma^{-1} \mathbf{F}_\perp^\dagger \tag{3-2c}$$

Einstein argued that these same transformation equations must hold for any kind of force. For one could balance an electromagnetic force by a nonelectromagnetic force in the rest frame and the equilibrium should exist in all inertial frames. But this requires that the two forces transform in the same way.

Equation (3-5) follows from the transformation laws of electromagnetic force and acceleration. Its derivation thus generalizes to the derivation from

$$\mathbf{F}^\dagger = \mu \mathbf{a}^\dagger \tag{3-6}$$

of

$$\mathbf{F} = \frac{d}{dt}(\mu\gamma\mathbf{v}) \tag{3-7}$$

$$\gamma = \left(1 - \frac{v^2}{c^2}\right)^{-1/2}$$

This equation of motion extends beyond electrodynamics to apply to all interactions. The expression for the momentum:

$$\mathbf{p} = \mu\gamma\mathbf{v} \tag{1-33}$$

must be of general validity.

Since force increases momentum and the momentum can increase without limit while the speed remains less than c, *no material* ($\mu > 0$) *particle can ever be accelerated to speed c*. Thus, $|u| < c$ for every physical reference frame, and relativistic kinematics cannot become absurd.

The transformation law of force [Eq. (3-2c)] imposes a restriction on the form of admissible force laws. The Newtonian law of gravitation [Eq. (1-8)], like Coulomb's law [Eq. (1-29)], is certainly incompatible with it, as is seen by considering the case of two particles moving line abreast with the same velocity. In essence, the force must be derivable from a field theory that is relativistically covariant, like classical electrodynamics.

3-2. INTERACTION BETWEEN CHARGED PARTICLES MOVING WITH THE SAME VELOCITY

We saw in Section 1-4 from an example of two charged particles moving with the same velocity that there is an inconsistency between the old law of motion, $\mathbf{F} = m\mathbf{a}$ (m = const), the principle of special relativity using the Galileo transformation, and classical electrodynamics. We can now reexamine this example, using the new, instead of the old, law of motion and the Lorentz, instead of the Galileo, transformation. It will turn out that there is no longer any inconsistency between the description in the laboratory frame and that in the rest frame of the particles, even though in the former there is magnetic attraction plus electrostatic repulsion and in the latter electrostatic repulsion alone.

Nothing is lost by simplifying to the concrete case of two electrons traveling line abreast along the axis of a cathode ray tube (Fig. 3-1). (The general case is worked out in [32].) We calculate their spreading apart as they move to the screen. At $t = 0$, electron A has coordinates $(0, d/2, 0)$ and electron B has coordinates $(0, -d/2, 0)$. The screen is at $x = L$. The electrons are far enough apart so that their deviation is small: $y_A - y_B \approx d$ and $\mathbf{v}_A \approx \mathbf{v}_B \approx \mathbf{i}v$.

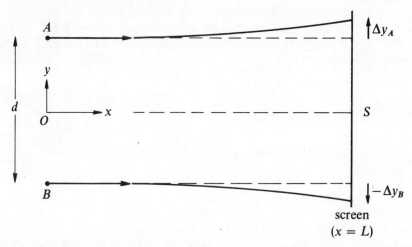

Figure 3-1.   Two electrons, A and B, travel line abreast with equal velocity parallel to the axis of a cathode ray tube. Their separation increases from its initial value of d to the value $d + \Delta y_A - \Delta y_B$ when they hit the screen.

In the rest frame $R^\dagger$, the equation of motion is

$$\mu_A \frac{d^2 y_A^\dagger}{dt^{\dagger 2}} = \frac{e_A e_B}{(y_A^\dagger - y_B^\dagger)^2} \approx \frac{e_A e_B}{d^{\dagger 2}} \qquad \mu_B \frac{d^2 y_B^\dagger}{dt^{\dagger 2}} = \frac{- e_A e_B}{(y_A^\dagger - y_B^\dagger)^2} \approx \frac{- e_A e_B}{d^2}$$

Each electron undergoes uniform acceleration, like a body in free fall, until it is hit by the screen moving to the left at time $T^\dagger$. The distance spread apart is $\Delta y_A^\dagger - \Delta y_B^\dagger = 2\Delta y_A^\dagger$, where

$$\Delta y_A^\dagger = \frac{1}{2} \frac{e^2}{\mu d^2} (T^\dagger)^2$$

To express $T^\dagger$ in terms of laboratory quantities, we note that $\gamma T^\dagger = T$ and $T = L/v$, so that $T^\dagger = L/v\gamma$. Alternatively, the distance OS is Lorentz contracted in the electrons' rest frame to L/γ. The tube is moving to the left with speed v in $R^\dagger$, so that $T^\dagger = L/v\gamma$. Thus,

$$\Delta y_A^\dagger = \frac{1}{2} \frac{e^2}{\mu d^2} \frac{L^2}{\gamma^2 v^2}$$

Note that the spreading apart is reduced below the nonrelativistic value by the factor γ^2.

Now we do the same problem in the laboratory frame R. We need to calculate the electric and magnetic fields at the position of one electron due to the charge on the other. They can be found by using Eqs. (2-55), substituting the known purely electrostatic field in $R^\dagger$. For the fields at the position of A due to B,

$$\mathscr{E}_x(A) = \mathscr{E}_x^\dagger(A) = 0$$

$$\mathscr{E}_y(A) = \gamma \mathscr{E}_y^\dagger(A) = \gamma \frac{e_B}{(y_A^\dagger - y_B^\dagger)^2} = \gamma \frac{e_B}{(y_A - y_B)^2}$$

$$\mathscr{E}_z(A) = \gamma \mathscr{E}_z^\dagger(A) = 0$$

$$\mathscr{B}_x(A) = 0$$

$$\mathscr{B}_y(A) = \gamma[- \beta \mathscr{E}_z^\dagger(A)] = 0$$

$$\mathscr{B}_z(A) = \gamma[+ \beta \mathscr{E}_y^\dagger(A)] = \beta \gamma \frac{e_B}{(y_A - y_B)^2}$$

While the relativistic transformation law of the field permits us to write down these expressions directly, they can also be derived from the Maxwell field equations without reference to relativity [33]. The law of motion, Eq. (3-3a), now gives

$$\mu_A \gamma \frac{d^2 y_A}{dt^2} = e_A \left[\frac{\gamma e_B}{(y_A - y_B)^2} - \beta \frac{\beta \gamma e_B}{(y_A - y_B)^2} \right] = \frac{e_A e_B}{(y_A - y_B)^2} \frac{1}{\gamma}$$

or

$$\frac{d^2 y_A}{dt^2} = \frac{e_A e_B}{\mu_A (y_A - y_B)^2} \frac{1}{\gamma^2}$$

Note how the magnetic attraction is β^2 times the Coulomb repulsion. A similar expression holds for the downward acceleration of electron B. Each electron is uniformly accelerated with acceleration of magnitude

$$\frac{e^2}{\mu d^2} \frac{1}{\gamma^2}$$

for time $T = L/v$. On meeting the screen,

$$\Delta y_A = \frac{1}{2} \frac{e^2}{\mu d^2} \frac{L^2}{\gamma^2 v^2}$$

which is the same answer we obtained in the rest frame.

We see that the relativistic reduction in the transverse spreading appears in the laboratory frame as the result of a magnetic attraction counteracting the Coulomb repulsion. In the rest frame, where there is no magnetic field, the same reduction appears as the result of the shortening of the time till the screen meets the electrons. In transforming between reference frames, we must use the appropriate transformation laws for all the quantities involved; otherwise, "paradoxes" arise. In the present example, contradiction occurs only if one erroneously assumes that force is invariant or that duration is invariant or that distance is invariant.

The $1/\gamma^2$ effect on the transverse spreading in a charged beam in vacuum is of practical importance for high intensity accelerator operation, where it makes the broadening of beams owing to electric repulsion ("space charge defocusing") much less of a problem than it would otherwise be [34]. In plasma physics, the same physical considerations explain the "pinch effect." A stationary space charge of the opposite sign cancels the Coulomb repulsion in the beam, and the magnetic attraction then squeezes the beam down to smaller diameter.

The preceding example illustrates an important fact: With the kinematics embodied in the Lorentz transformation, the dynamics embodied in Eqs. (1-5) and (1-33), and Maxwell's field equations for vacuum, we have a consistent classical theory of charged particles and electromagnetic fields.[1] This theory agrees with experiment everywhere, except in those cases where quantal effects are appreciable.

3-3. DYNAMICS BASED ON MOMENTUM CONSERVATION

It is possible to arrive at the relativistic momentum–velocity relation without reference to electrodynamics, basing the argument on the transformation law of velocity and the principle of conservation of momentum.

Define the momentum of a particle as a vector

$$\mathbf{p} = m(v)\mathbf{v} \tag{3-8}$$

Its direction is that of the velocity, and its magnitude is equal to some as yet undetermined function of the speed times the speed. The assumption of a scalar coefficient $m(v)$ corresponds to the isotropy of space. If space had different properties in different directions, the coefficient would no doubt differ for different directions of $\mathbf{v}$. We know how $\mathbf{v}$ transforms between inertial frames [Eq. (2-42)] and so can transform the above expression.

We require that for an isolated system of particles the vector sum of the momenta (3-8) be the same before and after a collision, and that this be the case in every inertial frame.

By applying this requirement to an extremely symmetrical (and artificial but nevertheless conceivable) situation, Lewis and Tolman [36] were able to find the form of $m(v)$.

Suppose that two highly trained experimentalists have practiced long and hard at throwing billiard balls so as to collide in flight. Experimenter I has ball 1; experimenter II has an identical ball, 2. The experimenters are in uniform motion with respect to one another and to an inertial frame. Let R be the inertial frame in which experimenter I is at rest and R' the one in which experimenter II is at rest. Let u be the velocity of R' with respect to R. Choose the x and x' axes along the direction of relative motion and the y and y' axes so that one is in the plane swept out by the other. Then the Lorentz transformation between the two frames gives for the velocity of a ball

$$v_x = \frac{v_{x'} + u}{1 + uv_{x'}/c^2}$$

$$v_y = (1 - u^2/c^2)^{1/2}\frac{v_{y'}}{1 + uv_{x'}/c^2} \tag{2-42}$$

---

[1] A classic presentation is given in [35], Chapters 1–9.

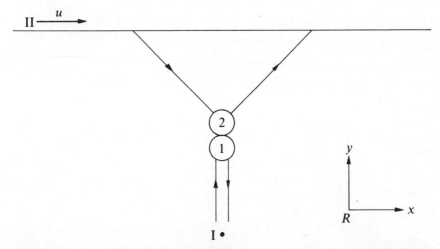

Figure 3-2.   Lewis–Tolman thought experiment. Paths of balls 1 and 2 and physicist II in the rest frame of physicist I.

Experimenters I and II have learned how to throw their balls so that they collide with the collision diameter perpendicular to the direction of relative motion. Figure 3-2 shows the paths as seen in R, Fig. 3-3 shows them in R'. By symmetry, the speed of ball 1 in R before impact equals the speed of ball 2 in R' before impact. Call it α_{in}. Using the left superscript 1 or 2

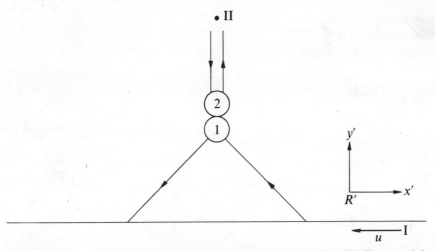

Figure 3-3.   Lewis–Tolman thought experiment. Paths of balls 1 and 2 and physicist I in the rest frame of physicist II.

to label the ball, and the right superscript "in" for initial, we have

$$^1v_x^{in} = 0 \qquad ^1v_y^{in} = \alpha_{in}$$
$$^2v_{x'}^{in} = 0 \qquad ^2v_{y'}^{in} = -\alpha_{in}$$

Since the collision diameter is in the y direction, the x components of the velocities are the same after the collision as before. The y components may be different. With "fin" for final,

$$^1v_x^{fin} = 0 \qquad ^1v_y^{fin} = -\alpha_{fin}$$
$$^2v_{x'}^{fin} = 0 \qquad ^2v_{y'}^{fin} = \alpha_{fin}$$

Again, symmetry requires the speed of ball 1 in R to be equal to the speed of ball 2 in R'. The signs in the equations have been chosen so that α_{in} and α_{fin} are positive.

We now use the velocity transformation law just quoted [Eq. (2-42)] to find the velocity components of ball 2 in R. Before the collision,

$$^2v_x^{in} = \frac{0 + u}{1 + 0} = u$$

$$^2v_y^{in} = \left(1 - \frac{u^2}{c^2}\right)^{1/2} \frac{(-\alpha_{in})}{1 + 0} = -\alpha_{in}\left(1 - \frac{u^2}{c^2}\right)^{1/2}$$

and the speed is

$$^2v^{in} = \left[u^2 + \alpha_{in}^2\left(1 - \frac{u^2}{c^2}\right)\right]^{1/2}$$

After the collision,

$$^2v_x^{fin} = u$$

$$^2v_y^{fin} = \alpha_{fin}\left(1 - \frac{u^2}{c^2}\right)^{1/2}$$

and the speed is

$$^2v^{fin} = \left[u^2 + \alpha_{fin}^2\left(1 - \frac{u^2}{c^2}\right)\right]^{1/2}$$

We now require that conservation of momentum hold in R. For the x component,

$$\cancel{m(\alpha_{in})\cdot 0} + m\left(\left[u^2 + \alpha_{in}^2\left(1 - \frac{u^2}{c^2}\right)\right]^{1/2}\right)\cdot u = \cancel{m(\alpha_{fin})\cdot 0}$$
$$+ m\left(\left[u^2 + \alpha_{fin}^2\left(1 - \frac{u^2}{c^2}\right)\right]^{1/2}\right)\cdot u$$

Hence,

$$\alpha_{in} = \alpha_{fin} = \alpha$$

In R, ball 1 bounces back with its speed unchanged, as does ball 2 in R'. For the y component of momentum,

$$m(\alpha)\cdot\alpha - m\left(\left[u^2 + \alpha^2\left(1 - \frac{u^2}{c^2}\right)\right]^{1/2}\right)\cdot\alpha\left(1 - \frac{u^2}{c^2}\right)^{1/2} \doteq -m(\alpha)\cdot\alpha$$
$$+ m\left(\left[u^2 + \alpha^2\left(1 - \frac{u^2}{c^2}\right)\right]^{1/2}\right)\alpha\left(1 - \frac{u^2}{c^2}\right)^{1/2}$$

so that

$$m(\alpha) = m\left(\left[u^2 + \alpha^2\left(1 - \frac{u^2}{c^2}\right)\right]^{1/2}\right)\left(1 - \frac{u^2}{c^2}\right)^{1/2} \tag{3-9}$$

This equation gives the functional form of the dependence of inertial mass on speed. It holds for any value of α and the limit $\alpha \to 0$ gives

$$m(0) = m(u)\cdot\left(1 - \frac{u^2}{c^2}\right)^{1/2}$$

or

$$m(u) = m(0)\left(1 - \frac{u^2}{c^2}\right)^{-1/2} \tag{3-10}$$

The dependence Eq. (3-10) satisfies Eq. (3-9) for any u and α. The inertial mass of the billiard ball depends on its speed as

$$m(v) = \frac{\mu}{\sqrt{1 - v^2/c^2}} \qquad \mu = m(0) \tag{1-2a}$$

and

$$\mathbf{p} = \frac{\mu}{\sqrt{1 - v^2/c^2}} \mathbf{v} \tag{1-33}$$

Although this result has been proved for one very complex physical system (a billiard ball), it must hold in general, even for a single particle. The dependence of momentum on velocity must be independent of structural details so long as they are not changed by the external forces.

Equation (1-2a) contains the Newtonian result of constant inertial mass as the limit $c = \infty$, for

$$m(v) = \mu\left(1 + \frac{1}{2}\frac{v^2}{c^2} + \cdots\right) \tag{1-2b}$$

It is nevertheless illuminating to obtain the nonrelativistic result by retracing the preceding argument, making use of the Galileo instead of the Lorentz transformation. With

$$\begin{aligned} v_x &= v_{x'} + u \\ v_y &= v_{y'} \end{aligned} \tag{1-21}$$

we obtain

$$\begin{aligned} {}^2v_x^{\text{in}} &= u \\ {}^2v_y^{\text{in}} &= -\alpha_{\text{in}} \\ {}^2v^{\text{in}} &= (u^2 + \alpha_{\text{in}}^2)^{1/2} \end{aligned}$$

and

$$\begin{aligned} {}^2v_x^{\text{fin}} &= u \\ {}^2v_y^{\text{fin}} &= \alpha_{\text{fin}} \\ {}^2v^{\text{fin}} &= (u^2 + \alpha_{\text{fin}}^2)^{1/2} \end{aligned}$$

Conservation of the x component of momentum requires

$$0 + m([u^2 + \alpha_{\text{in}}^2]^{1/2}) \cdot u = 0 + m([u^2 + \alpha_{\text{fin}}^2]^{1/2}) \cdot u$$

or

$$\alpha_{\text{in}} = \alpha_{\text{fin}} = \alpha$$

as before. Conservation of the y component requires

$$m(\alpha) \cdot \alpha - m([u^2 + \alpha^2]^{1/2})\alpha = -m(\alpha) \cdot \alpha + m([u^2 + \alpha^2]^{1/2})\alpha$$

or

$$m(\alpha) = m([u^2 + \alpha^2]^{1/2})$$

The inertial mass is evidently independent of the speed

$$m(0) = m(u)$$

and

$$\mathbf{p} = \mu\mathbf{v}$$

This proves the result quoted in Section 1-4 between Eqs. (1-24) and (1-25).

We see clearly how the change of kinematics from the Galileo to the Lorentz transformation requires the change of dynamics from $\mathbf{p} = \mu\mathbf{v}$ to $\mathbf{p} = \mu(1 - v^2/c^2)^{-1/2}\mathbf{v}$.

The law of conservation of momentum is, for contact or instantaneous action-at-a-distance forces, equivalent to "action equals reaction" ([1], Law III). For the electromagnetic interaction propagating at a speed c in vacuum, momentum is present in the field, and the conservation law holds for the total momentum of particles plus field. (There is a recoil associated with the emission and absorption of electromagnetic radiation. In quantum electrodynamics the field is shown to be equivalent in a certain sense to an assemblage of photons, and the field momentum is simply the momentum of the photons.) It can be shown that in general the law of conservation of momentum for an isolated system is equivalent to the homogeneity of physical space. It is in this sense more basic than electrodynamics, which describes only one of the four kinds of interaction known to us at present. [The other three are the gravitational, the hadronic (nuclear), and the "weak" (leptonic)].

It is remarkable that the conservation law alone suffices to determine the momentum–velocity relation, without reference to the nature of the interactions between the particle and the rest of the world. During the years preceding special relativity, there had been numerous efforts to formulate an electromagnetic theory of matter (J. J. Thomson, Larmor, Lorentz, Poincaré, Abraham, Langevin, etc. For an illuminating discussion, see [37], particularly Sections 26–36, 179–186.) Matter was assumed to consist of charged particles ("electrons") that are really continuous distributions of electricity. The momentum and energy of the electron are the momentum and energy of its electromagnetic field as given by Maxwell's theory. Inertia is a manifestation of back emf inside the electron. When the charge distribution is accelerated as a whole, electric fields are induced that give a net force on the elements of the electron equal to the inertial reaction. It was possible to calculate for specific model charge distributions the relation between momentum and velocity. Lorentz showed that if the charge distribution undergoes the $(1 - v^2/c^2)^{1/2}$ contraction when in motion with respect to the ether, the relation is of the form Eq. (1-33). The subsequent experimental proof of this equation is, however, not a triumph of the electromagnetic theory of matter, for we have just seen that the relation has a more general basis. It shows merely that Lorentz's model is consistent with special relativity.[2] The efforts of classical electron theory to provide an extended spatial model for the elementary charged particles on the basis of classical electrodynamics must be regarded as definitive failures. The theory never could explain the stability of the particles, and the quantization of electric charge and angular momentum and proper mass. Certainly the self-electromagnetic field contributes to the momentum and energy of the elementary particles, but only a quantal and relativistic theory unifying the hadronic, weak, and electromagnetic interactions can be expected to provide a satisfactory framework in which to place the elementary particles.

3-4. COVARIANT EQUATION OF MOTION OF A CHARGED PARTICLE

The vector equation of motion of a charged particle

$$
e\left(\mathscr{E} + \frac{\mathbf{v}}{c} \times \mathscr{B}\right) = \frac{d}{dt}(\mu\gamma\mathbf{v}) \tag{3-5}
$$

can be written in spacetime coordinate notation and put in manifestly covariant form. The right-hand member is [Eqs. (2-27) and (2-60a)],

$$
\frac{1}{\gamma d\tau}d(\mu w_j) = \frac{\mu}{\gamma}\frac{dw_j}{d\tau} \qquad (j = 1, 2, 3)
$$

The left-hand side is

$$
\frac{e}{\gamma c}\mathscr{F}_{j\lambda}w_\lambda \qquad \begin{matrix}(j = 1, 2, 3)\\(\lambda = 1, 2, 3, 4)\end{matrix}
$$

---

[2] To quote Lorentz ([38], p. 125): "The formula for momentum is a general consequence of the principle of relativity, and a verification of that formula is a verification of the principle and tells us nothing about the nature of mass or of the structure of the electron."

with

$$\mathscr{F}_{\mu\nu} = \begin{pmatrix} 0 & \mathscr{B}_z & -\mathscr{B}_y & -i\mathscr{E}_x \\ -\mathscr{B}_z & 0 & \mathscr{B}_x & -i\mathscr{E}_y \\ \mathscr{B}_y & -\mathscr{B}_x & 0 & -i\mathscr{E}_z \\ i\mathscr{E}_x & i\mathscr{E}_y & i\mathscr{E}_x & 0 \end{pmatrix} \qquad (2\text{-}54)$$

and

$$w_4 = \gamma ic \qquad (2\text{-}60\text{b})$$

For example,

$$\mathscr{F}_{2\lambda}w_\lambda = -\mathscr{B}_z \cdot \gamma v_x + 0 \cdot \gamma v_y + \mathscr{B}_x \cdot \gamma v_z - i\mathscr{E}_y \cdot \gamma ic$$

$$= \gamma c \left(\mathscr{E} + \frac{\mathbf{v}}{c} \times \mathscr{B} \right)_y$$

On multiplication by γ, the equation becomes

$$\frac{e}{c} \mathscr{F}_{j\lambda}w_\lambda = \mu \frac{dw_j}{d\tau} \qquad \begin{array}{l}(j = 1, 2, 3) \\ (\lambda = 1, 2, 3, 4)\end{array} \qquad (3\text{-}11\text{a})$$

It equates the spacelike components of two four-vectors

$$\frac{e}{c} \overset{\approx}{\mathscr{F}} \tilde{w} \quad \text{and} \quad \mu \frac{d\tilde{w}}{d\tau}$$

Since the equality is an identity holding in every coordinate system, the timelike components must also be equal in every coordinate system:

$$\frac{e}{c} \mathscr{F}_{4\lambda}w_\lambda = \mu \frac{dw_4}{d\tau} \qquad (\lambda = 1, 2, 3, 4) \qquad (3\text{-}11\text{b})$$

and

$$\boxed{\frac{e}{c} \overset{\approx}{\mathscr{F}} \tilde{w} = \mu \frac{d\tilde{w}}{d\tau}} \qquad (3\text{-}11\text{c})$$

In effect, equality of the spacelike components in every inertial frame implies equality of the four-vectors. Designating them $\tilde{A}$ and $\tilde{B}$, we note that the special Lorentz transformation Eq. (2-49b) gives

$$A_1 = \gamma(A_1' - i\beta A_4') \qquad B_1 = \gamma(B_1' - i\beta B_4')$$

$$A_2 = A_2' \qquad\qquad\quad B_2 = B_2'$$

$$A_3 = A_3' \qquad\qquad\quad B_3 = B_3'$$

$$A_4 = \gamma(i\beta A_1' + A_4') \qquad B_4 = \gamma(i\beta B_1' + B_4')$$

Because $A_1 = B_1$ and $A_1' = B_1'$, it follows from the first line of equations that

$$A_4' = B_4'$$

Equality of all the corresponding components in R' implies equality of the four-vectors.

The derivation of Eq. (3-11c) illustrates a powerful technique: Write an equation in an obviously covariant form [such as Eq. (3-11a)]. The transformation properties of tensors then guarantee that the equality holds for all the spacetime components if it holds for any.

The timelike component [Eq. 3-11b)] of the equation of motion is

$$\frac{e}{c}i\gamma(\mathscr{E} \cdot \mathbf{v}) = \mu i c \frac{d\gamma}{dt}$$

or

$$(e\mathscr{E} \cdot \mathbf{v}) = \frac{d}{dt}(\mu c^2 \gamma) \tag{3-12}$$

The left-hand side is the rate at which the electromagnetic field does work on the particle:

$$e\mathscr{E} \cdot \mathbf{v} = e\left(\mathscr{E} + \frac{\mathbf{v}}{c} \times \mathscr{B}\right) \cdot \mathbf{v} = \mathbf{F} \cdot \mathbf{v}$$

Since the energy of the particle is defined [Eq. (1-6)] by

$$dE = \mathbf{F} \cdot d\mathbf{r}$$

the left-hand side of Eq. (3-12) is dE/dt. Equation (3-12) thus gives the relativistic *energy–velocity relation*

$$dE = d(\mu c^2 \gamma) = d\left(\frac{\mu c^2}{\sqrt{1 - v^2/c^2}}\right) \tag{3-13}$$

Integrating, we see that the energy must equal $\mu c^2 \gamma$ to within an additive constant. We shall find shortly that this constant is zero.[3]

Just as the spatial projection [Eq. (3-11a)] of the four-vector equation of motion [Eq. (3-11c)] implies the vector equation of motion,

$$\mathbf{F} = \frac{d\mathbf{p}}{dt} = \frac{d}{dt}(\mu \gamma \mathbf{v})$$

so the temporal projection [Eq. (3-11b)] implies the theorem of work

$$\mathbf{F} \cdot \mathbf{v} = \frac{dE}{dt} = \frac{d}{dt}(\mu c^2 \gamma)$$

The four projected equations are not independent because the four components of $\tilde{w}$ are not independent ($\tilde{w}^2 = -c^2$). We shall see in the next section that, just as in Newtonian mechanics, the theorem of work can be derived from the vector equation of motion.

Einstein's argument that all forces must transform in the same way leads as before to the conclusion that the energy–velocity relation [Eq. (3-13)] as well as the momentum–velocity relation [Eq. (1-33)] must have general validity, regardless of what kind of interaction is changing the motion. The general covariant equation of motion will be presented in Section 3-7.

3-5. ENERGY, MOMENTUM, AND VELOCITY

We have just seen that the relativistic transformation properties of the electromagnetic field lead, for a charged particle, to the energy–velocity relation

$$dE = d[\mu c^2 \gamma(v)] \tag{3-13}$$

as well as to the momentum–velocity relation

$$\mathbf{p} = \mu \gamma(v)\mathbf{v} \tag{1-33}$$

It is worth verifying that Eq. (3-13) is also a direct consequence of the law of motion:

$$\mathbf{F} = \frac{d\mathbf{p}}{dt} \tag{1-5}$$

---

[3] There was also an additive constant in the momentum. Equation (3-5) shows only that $\Delta \mathbf{p} = \Delta(\mu \gamma \mathbf{v})$. We have, however, imposed the requirement that p vanishes when v vanishes.

$$\mathbf{p} = \mu\gamma\mathbf{v} \tag{1-33}$$

for any kind of interaction.

We have

$$dE = \mathbf{F} \cdot d\mathbf{r} = \frac{d\mathbf{p}}{dt} \cdot d\mathbf{r} = d\mathbf{p} \cdot \mathbf{v} \tag{3-14}$$

Since we know the relation [Eq. (1-33)] between $\mathbf{p}$ and $\mathbf{v}$, we can evaluate dE.

First we prove a useful geometrical lemma, depending only on the parallelism of $\mathbf{p}$ and $\mathbf{v}$:

$$d\mathbf{p} \cdot \mathbf{v} = v\, dp$$

In words, the scalar product of velocity and infinitesimal momentum change equals the speed times the infinitesimal change in magnitude of the momentum. Referring to Fig. 3-4, which shows $\mathbf{p}$ and $d\mathbf{p}$ adding to give $\mathbf{p} + d\mathbf{p}$,

$$d\mathbf{p} \cdot \mathbf{v} = |d\mathbf{p}| \cdot v \cdot \cos\varphi = v\, dp$$

Thus,

$$dE = v\, dp \tag{3-15}$$

If the momentum does not change in magnitude there is no change in the energy. A transverse force, which changes only the direction of $\mathbf{p}$, does no work and leaves E constant.

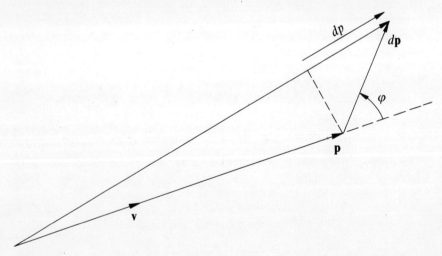

Figure 3-4.

Using now the momentum–velocity relation Eq. (1-33),

$$dp = d\left[\mu v\left(1 - \frac{v^2}{c^2}\right)^{-1/2}\right] = \mu\, dv\left(1 - \frac{v^2}{c^2}\right)^{-3/2}$$

so that by Eq. (3-15),

$$dE = \mu v\, dv\left(1 - \frac{v^2}{c^2}\right)^{-3/2} = -\frac{\mu c^2}{2}\left(1 - \frac{v^2}{c^2}\right)^{-3/2} d\left(1 - \frac{v^2}{c^2}\right)$$

$$= \mu c^2\, d\left[\left(1 - \frac{v^2}{c^2}\right)^{-1/2}\right] = \mu c^2\, d\gamma$$

We have thus arrived at Eq. (3-13):

$$dE = d(\mu c^2 \gamma) \tag{3-13}$$

It is a corollary of the law of motion.

The kinetic energy, defined as the increment of energy from a state of rest, is

$$K = \int_{\gamma=1} dE = \mu c^2 \gamma - \mu c^2 = \mu c^2(\gamma - 1) \tag{3-16}$$

For the nonrelativistic limit, series expansion of γ gives

$$K = \mu c^2\left(1 + \frac{1}{2}\frac{v^2}{c^2} + \frac{3}{8}\frac{v^4}{c^4} + \cdots - 1\right)$$

$$= \frac{1}{2}\mu v^2\left(1 + \frac{3}{4}\frac{v^2}{c^2} + \cdots\right) \approx \frac{1}{2}\mu v^2$$

Every expression in relativistic mechanics must, of course, reduce to the corresponding nonrelativistic one as $c = \infty$.

Since

$$m = \mu\gamma \tag{1-2}$$

Eq. (3-13) can be written

$$dE = d(mc^2) \tag{3-17a}$$

Integrating, we get

$$\Delta E = c^2 \cdot \Delta m \tag{3-17b}$$

With a work ΔE done on the body is associated an increase Δm of its inertial mass, with constant of proportionality c^2. This is our first encounter with the inertia of energy, a fundamental consequence of special relativity. Because m is by definition p/v, the practical significance of the energy–inertia relation Eq. (3-17a) is simply

$$dE = v\, dp \tag{3-15}$$

with p and v related by

$$p = \mu\left(1 - \frac{v^2}{c^2}\right)^{-1/2} v \tag{1-33}$$

Equation (3-13) is Eq. (3-15) with dp expressed in terms of v and dv by means of Eq. (1-33). If instead we express v in terms of p, by Eq. (1-33), we obtain the following useful relation between dE and dp:

$$(c^2 p^2 + \mu^2 c^4)^{1/2}\, dE = c^2 p\, dp \tag{3-18}$$

Equations (3-16) and (1-33) show how the energy and momentum of the particle increase without limit as $v \to c$. An accelerator giving the particle an increment of energy ΔE increases its velocity by the amount corresponding to

$$\Delta\gamma = \frac{\Delta E}{c^2 \mu}$$

Since

$$\frac{v}{c} = \left(1 - \frac{1}{\gamma^2}\right)^{1/2}$$

v remains less than c no matter how large γ becomes. The relation between velocity and energy is just such as to keep a material ($\mu \neq 0$) particle from ever reaching the speed of light. The speed of light appears in the theory as an unattainable limiting velocity.

The Stanford 2-mile linear accelerator, with its 18-GeV electrons, provides the highest speed as yet attained in a terrestrial device. The γ of 3.5×10^4 corresponds to

$$\frac{v}{c} = 1 - \frac{1}{2\gamma^2} - \cdots = 1 - \frac{1}{2.5 \times 10^9}$$

The successful operation of this accelerator has confirmed relativistic dynamics in every detail.

3-6. FOUR-MOMENTUM

The momentum $\mathbf{p} = \mu\gamma\mathbf{v}$ is, as we have remarked in the preceding section, equal to $\mu\mathbf{w}$:

$$p_j = \mu w_j \qquad (j = 1, 2, 3) \tag{3-19a}$$

Since μ is invariant, p_j transforms like w_j. The momentum is the spacelike part of the four-vector $\mu\tilde{w}$. This four-vector is called the *four-momentum*, $\tilde{p}$:

$$\tilde{p} = \mu\tilde{w} \tag{3-19b}$$

The timelike component is

$$p_4 = \mu w_4 = \mu i c\gamma = \frac{i}{c}(\mu c^2 \gamma) \tag{3-19c}$$

or

$$p_0 = \frac{p_4}{i} = \frac{1}{c}(\mu c^2 \gamma) \tag{3-19cc}$$

We have seen [Eq. (3-13)] that the increment of $\mu c^2 \gamma$ is the increment of energy of the particle. We now set the constant of integration equal to zero,[4] and define the *energy* of the particle as $\mu c^2 \gamma$:

$$\boxed{E = \mu c^2 \gamma = \frac{\mu c^2}{\sqrt{1 - v^2/c^2}}} \tag{3-20}$$

---

[4] It is shown in Appendix E that the principle of relativity requires this choice. Otherwise, the energy of the particle in its rest frame would depend on the relative velocity of the laboratory.

Thus,

$$E = cp_0 = \frac{c}{i} p_4 \tag{3-21}$$

The energy is equal, within a universal constant factor, to the timelike component of the four-momentum. The four-momentum

$$\tilde{p} = \left(\mathbf{p}, \frac{i}{c} E \right) = \left\{ \frac{E}{c}, \mathbf{p} \right\} \tag{3-22}$$

combines the momentum and energy into one spacetime four-vector. The components transform like those of every four-vector. In the special Lorentz transformation of Eqs. (2-49), with β and γ referring to the transformation velocity u, not the particle velocity v,

$$
\begin{aligned}
p_1' &= \gamma(p_1 + i\beta p_4) \\
p_2' &= p_2 \\
p_3' &= p_3 \\
p_4' &= \gamma(-i\beta p_1 + p_4)
\end{aligned}
\tag{3-23a}
$$

or

$$
\begin{aligned}
p_0' &= \gamma(p_0 - \beta p_1) \\
p_1' &= \gamma(p_1 - \beta p_0) \\
p_2' &= p_2 \\
p_3' &= p_3
\end{aligned}
\tag{3-23b}
$$

Expressing the components in terms of momentum and energy, we write these equations as

$$
\boxed{
\begin{aligned}
E' &= \gamma(E - \beta c p_x) \\
p_x' &= \gamma\left(p_x - \frac{\beta}{c} E \right) \\
p_y' &= p_y \\
p_z' &= p_z
\end{aligned}
}
\tag{3-23c}
$$

We have here the transformation law of momentum and energy. Energy and momentum are intertwined, like time and position. The momentum components transverse to the direction of relative motion are invariant; the longitudinal component and the energy transform together like longitudinal position and time.

In the nonrelativistic limit,

$$p = \mu v$$

$$E = \mu c^2$$

and Eqs. (3-23c) reduce to

$$p'_x = p_x - \mu u$$

$$p'_y = p_y$$

$$p'_z = p_z$$

as expected.

EXERCISE   Derive the transformation law for force from the four-vector character of four-momentum.

SOLUTION   We seek the transformation of $d\mathbf{p}/dt$, knowing that $\mathbf{p}$ is the spacelike part of a four-vector and t is the timelike part of a four-vector.

Since the Lorentz transformation is linear, $d\tilde{p}$ transforms in the same way as $\tilde{p}$, and $d\tilde{x}$ in the same way as $\tilde{x}$. Thus,

$$dE' = \gamma(dE - \beta c \, dp_x)$$

$$dp'_x = \gamma\left(dp_x - \frac{\beta}{c} dE\right)$$

$$dp'_y = dp_y \tag{3-23d}$$

$$dp'_z = dp_z$$

Let $R^\dagger$ be the instantaneous rest frame and take the $x, x^\dagger$ axes parallel to the particle's velocity. Then β and γ refer to the particle's velocity in R. Equations (3-23d) inverted give

$$dp_x = \gamma\left(dp_x^\dagger + \frac{\beta}{c} dE^\dagger\right)$$

$$dp_y = dp_y^\dagger$$

$$dp_z = dp_z^\dagger$$

whereas the SLT of $d\tilde{x}$ gives, because $v^\dagger = 0$,

$$dt = \gamma \, dt^\dagger + \frac{\beta}{c} \, dx^\dagger \;\; = \gamma \, dt^\dagger$$

Thus,

$$F_x = \frac{dp_x}{dt} = \frac{dp_x^\dagger}{dt^\dagger} + \frac{\beta}{c} \frac{dE^\dagger}{dt^\dagger} = F_x^\dagger + \frac{\beta}{c} \frac{dE^\dagger}{dt^\dagger}$$

$$F_y = \frac{dp_y}{dt} = \frac{dp_y^\dagger}{\gamma \, dt^\dagger} = \gamma^{-1} F_y^\dagger$$

$$F_z = \frac{dp_z}{dt} = \frac{dp_z^\dagger}{\gamma \, dt^\dagger} = \gamma^{-1} F_z^\dagger$$

Now

$$\frac{dE^\dagger}{dt^\dagger} = \mathbf{F}^\dagger \cdot \mathbf{v}^\dagger = 0$$

so

$$\begin{aligned} F_x &= F_x^\dagger \\ F_y &= \gamma^{-1} F_y^\dagger \\ F_z &= \gamma^{-1} F_z^\dagger \qquad \text{Q.E.D.} \end{aligned} \qquad (3\text{-}2c)$$

The square of $\tilde{p}$

$$\tilde{p}^2 = p_\lambda{}^2 = \mathbf{p} \cdot \mathbf{p} - \frac{E^2}{c^2} = p^2 - \frac{E^2}{c^2}$$

is readily evaluated in the rest frame, where $p^\dagger = 0$.

$$\tilde{p}^2 = (p^\dagger)^2 - \frac{(E^\dagger)^2}{c^2} = -\frac{E^{\dagger 2}}{c^2}$$

and, since $E^\dagger = \mu c^2$ [Eq. (3-19)],

$$\boxed{\tilde{p}^2 = -\mu^2 c^2} \qquad (3\text{-}24)$$

It can, of course, be evaluated in any frame. Thus, in R, Eqs. (3-19) give

$$\tilde{p}^2 = \mu^2 \gamma^2 \beta^2 c^2 - \mu^2 c^2 \gamma^2 = -\mu^2 c^2$$

The four-momentum is timelike, and its invariant length is, to within a factor $\pm ic$, the proper mass. Conversely, the proper mass, defined as the inertial mass in the rest frame, is essentially a four-scalar and deserves the name *invariant mass*. The energy, $E = \mu c^2 \gamma$, and the inertial mass, $m = \mu \gamma$, are equal to one another to within the factor c^2, and are essentially the timelike component of the four-momentum. The momentum is the spacelike component of the four-momentum.

Comparing Eqs. (3-20) and (3-16), we see that

$$E = K + \mu c^2 \tag{3-25}$$

The energy equals the increment of energy from rest (K) plus a term proportional to the rest mass called the _rest energy_. The proper mass is essentially positive (momentum points same way as velocity), and the energy can range from the positive value μc^2 on up. Under weak field conditions, as in a spectrometer or an accelerator, the rest energy is locked up in the particle and is a mere additive constant. But in inelastic transitions (Chapter 4) involving destruction and creation of particles, it becomes strikingly manifest as it gets converted into other forms of energy.

The equations relating p, E, and v are consistent with the relativistic transformation relations, and, in fact, the Lorentz transformation equations for the momentum and energy—Eqs. (3-23c)—can be used to derive them. The essential features of the four-momentum are that it is timelike, so that a reference frame (rest frame $R^\dagger$) exists in which the spacelike part $p^\dagger$ vanishes, and that in this frame the timelike component $p_0^\dagger = E^\dagger/c$, is positive. With the R' frame taken to be the rest frame $R^\dagger$ and the $x, x^\dagger$ axes taken parallel to the velocity, the quantities β and γ refer to v as well as u, and

$$E = \gamma(E^\dagger + \beta c p_x^\dagger) = \left(\frac{E^\dagger}{c^2}\right) c^2 \gamma$$

$$p_x = \gamma\left(p_x^\dagger + \frac{\beta}{c} E^\dagger\right) = \left(\frac{E^\dagger}{c^2}\right) \gamma \beta c$$

$$p_y = p_y^\dagger = 0$$

$$p_z = p_z^\dagger = 0$$

We see that

$$\mathbf{p} = \left(\frac{E^\dagger}{c^2}\right) \gamma \mathbf{v}$$

$$E = \left(\frac{E^\dagger}{c^2}\right) c^2 \gamma \tag{3-26a}$$

and

$$\mathbf{p} = \left(\frac{E}{c^2}\right)\mathbf{v} \qquad (1\text{-}3b)$$

This last equation is the physical essence of the slogan "$E = mc^2$" because m is defined as p/v. The proper mass μ, defined as the inertial mass in the rest frame,

$$\mu = \lim_{v \to 0} \frac{p}{v}$$

equals by Eq. (1-3b) the energy in the rest frame divided by c^2

$$\mu = \frac{E^\dagger}{c^2} \qquad (3\text{-}26b)$$

in agreement with Eq. (3-20), and Eqs. (3-26a) become

$$p = \mu\gamma v \qquad (1\text{-}33)$$

$$E = \mu c^2 \gamma \qquad (3\text{-}20)$$

The expression Eq. (3-24) for the invariant length squared of p

$$p^2 - \frac{E^2}{c^2} = -\mu^2 c^2 \qquad (3\text{-}27a)$$

holds in every inertial reference frame (see Eq. (1-4)). It can be represented by a right triangle (Fig. 3-5) with E/c as hypotenuse and p and μc as sides. The three sides can, of course, be expressed in energy units by multiplying through by c, or in mass units by dividing through by c. In the extreme relativistic range ($v \approx c$),

$$p = \mu\gamma\beta c \approx \mu\gamma c = \frac{E}{c}$$

and the triangle is as sketched in Fig. 3-6a, the μc side being relatively short and the p side almost as long as the E/c hypotenuse. In the nonrelativistic region ($v \ll c$),

$$p = \mu\gamma\beta c \approx \mu\beta c$$

$$E = \mu\gamma c^2 \approx \mu c^2$$

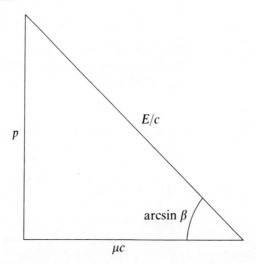

Figure 3-5. Right triangle of four-momentum. The lengths of the sides are in units of momentum.

and the triangle is as shown in Fig. 3-6b. The p side is relatively short and the E/c hypotenuse is barely longer than the μc side. The kinetic energy is small in comparison with μc^2:

$$K = \mu c^2(\gamma - 1) \approx \mu c^2 \frac{\beta^2}{2} \approx \frac{p^2}{2\mu}$$

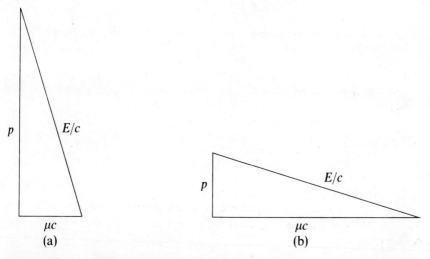

Figure 3-6. Right triangle of four-momentum in extreme relativistic case (a) and nonrelativistic case (b).

the familiar kinetic energy–momentum relation of nonrelativistic mechanics. In general,

$$\frac{p}{E} = \frac{\mu\gamma\beta c}{\mu\gamma c^2} = \frac{\beta}{c} = \frac{v}{c^2} \tag{3-28}$$

[in agreement with Eq. (1-3b)], so that β is the sine of the angle opposite the momentum side of the triangle (Fig. 3-5). The quantity

$$\gamma = (1 - \beta^2)^{-1/2} = \frac{E}{\mu c^2}$$

is evidently the secant of this angle.

A special class of particle—the photon and the various kinds of neutrinos—represents the limiting case $v = c$, $\mu = 0$. The momentum and energy are finite and satisfy the equation

$$p = \frac{E}{c}$$

giving [Eq. (3-28)] $v = c$ and [Eq. (3-27a)] $\mu = 0$. With $\mu = 0$, the triangle is reduced to a line of length p.

The expression for the invariant length squared of the four-momentum [Eq. (3-27a)] leads to

$$E = (c^2 p^2 + \mu^2 c^4)^{1/2} \tag{3-27b}$$

which gives the energy in terms of the momentum and the proper mass. Equation (3-18) can then be written

$$E \, dE = c^2 p \, dp \tag{3-29}$$

an equation also obtainable from Eqs. (3-15) and (3-28), or by differentiating Eq. (3-27a).

3-7. GENERAL COVARIANT EQUATION OF MOTION

The general law of motion

$$\mathbf{F} = \frac{d\mathbf{p}}{dt} \tag{1-5}$$

can be written [Eqs. (2-27)]

$$\gamma F_j = \frac{dp_j}{d\tau} \qquad (j = 1, 2, 3) \tag{3-30a}$$

The right-hand side is the spacelike part of a four-vector. Its timelike part is

$$\frac{dp_4}{d\tau} = \frac{i}{c}\frac{dE}{d\tau} = \frac{i\gamma}{c}\frac{dE}{dt} = \frac{i\gamma}{c}(\mathbf{F}\cdot\mathbf{v}) \qquad (3\text{-}30\text{b})$$

Equations (3-30a and b) can be written as one spacetime equation

$$\mathscr{K}_\lambda = \frac{dp_\lambda}{d\tau} \qquad (3\text{-}30\text{c})$$

or

$$\tilde{\mathscr{K}} = \frac{d\tilde{p}}{d\tau} \qquad (3\text{-}30\text{d})$$

The four-vector $\tilde{\mathscr{K}}$ is called the *four-force* or Minkowski force. Its components are

$$\mathscr{K}_1 = \gamma F_x$$
$$\mathscr{K}_2 = \gamma F_y$$
$$\mathscr{K}_3 = \gamma F_z \qquad (3\text{-}31)$$
$$\mathscr{K}_4 = \frac{i}{c}\gamma(\mathbf{F}\cdot\mathbf{v})$$

or

$$\mathscr{K}_0 = \frac{1}{c}\gamma(\mathbf{F}\cdot\mathbf{v})$$

The spacelike part is γ times the physical force. The timelike part is proportional to γ times the rate at which the physical force does work. $\tilde{\mathscr{K}}$ is a spacelike four-vector unless $F = 0$, in which case all its components vanish.

Just as in a general sense force is defined by Eq. (1-5), so four-force is defined by Eq. (3-30d). The spatial projection of this four-vector equation gives the law of motion, the temporal projection the theorem of work. The four component equations are not independent; the components of $\tilde{p}$ are constrained by the identity

$$\tilde{p}^2 = -\mu^2 c^2 \qquad (3\text{-}24)$$

In the absence of force, all components of $\tilde{\mathscr{K}}$ are zero, and the four-momentum is constant. Both the momentum vector and the energy are constants of the motion.

Since $\tilde{p} = \mu \tilde{w}$, Eq. (3-30) can be written

$$\tilde{\mathscr{K}} = \mu \frac{d\tilde{w}}{d\tau} = \mu \frac{d^2\tilde{x}}{d\tau^2} \tag{3-32}$$

In this form, the covariant equation of motion is like "$\mathbf{F} = m\mathbf{a}$" with proper time replacing time as the independent variable, four-force replacing force, proper mass replacing inertial mass, and four-position ($\tilde{x}$) replacing position.

An example of the law of motion Eq. (3-30d) is, of course, Eq. (3-11) for the motion of a charged particle, where

$$\tilde{\mathscr{K}} = \frac{e}{c} \tilde{\tilde{\mathscr{F}}} \tilde{w} \tag{3-33}$$

3-8.   MOMENTUM SPACE AND MOMENTUM DISTRIBUTIONS. CROSS SECTIONS

Experiments in atomic, nuclear, and particle physics are often concerned with the momentum distribution of the particles involved. It is useful to consider a three-dimensional momentum space based on our Cartesian $Oxyz$ system in R; the axes are p_x, p_y, p_z (Fig. 3-7). In this space we can, if we wish, introduce spherical polar coordinates p, χ, ψ with the x axis

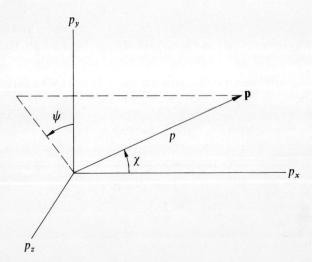

Figure 3-7.   Cartesian and spherical polar coordinates in momentum space in R.

as polar axis

$$p_x = p \cos \chi$$
$$p_y = p \sin \chi \cos \psi \qquad (3\text{-}34\text{a})$$
$$p_z = p \sin \chi \sin \psi$$

Similarly, we can plot the R' momentum components—p'_x, p'_y, p'_z, in an R' momentum space—and introduce spherical polar coordinates there

$$p'_x = p' \cos \chi'$$
$$p'_y = p' \sin \chi' \cos \psi' \qquad (3\text{-}34\text{b})$$
$$p'_z = p' \sin \chi' \sin \psi'$$

The transformation equations of four-momentum in a SLT [Eqs. (3-23c)] can be written

$$E' = \gamma(E - \beta cp \cos \chi)$$
$$p' \cos \chi' = \gamma\left(p \cos \chi - \frac{\beta}{c}E\right)$$
$$p' \sin \chi' \cos \psi' = p \sin \chi \cos \psi \qquad (3\text{-}35)$$
$$p' \sin \chi' \sin \psi' = p \sin \chi \sin \psi$$

These equations contain implicitly our earlier results on the change of velocity in a Lorentz transformation (Section 2-8). For example, the last two give

$$\tan \psi' = \tan \psi$$

or

$$\psi' = \psi \qquad (3\text{-}36\text{a})$$

in agreement with the third of Eqs. (2-43); and

$$p' \sin \chi' = p \sin \chi$$

The combination of this relation with the second Eq. (3-35) gives

$$\tan \chi' = \frac{p' \sin \chi'}{p' \cos \chi'} = \frac{p \sin \chi}{\gamma[p \cos \chi - (\beta/c)E]} = \gamma^{-1}\frac{\sin \chi}{\cos \chi - [(\beta/c)(E/p)]} \qquad (3\text{-}36\text{b})$$

Since

$$\frac{E}{cp} = \frac{c}{v}$$

this last equation is

$$\tan \chi' = \gamma^{-1} \frac{\sin \chi}{\cos \chi - u/v} \qquad (3\text{-}36c)$$

in agreement with the second of Eqs. (2-43) (interchange primed and un-primed quantities and reverse the sign of u). The results on change of direction in a special Lorentz transformation have thus been rederived. It is left to the reader (Problem 3-6) to derive the result [Eq. (2-43)] for the change of speed.

In terms of momentum components parallel and perpendicular to the relative velocity of the two frames, Eqs. (3-35) are

$$E' = \gamma(E - \beta c p_{\parallel})$$

$$p'_{\parallel} = \gamma\left(p_{\parallel} - \frac{\beta}{c}E\right) \qquad (3\text{-}37)$$

$$\mathbf{p}'_{\perp} = \mathbf{p}_{\perp}$$

These equations apply to the general Lorentz transformation; $\parallel$ and $\perp$ refer to the relative velocity of the two frames.

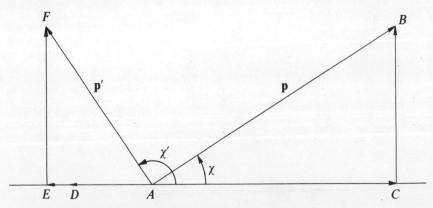

Figure 3-8.   Momentum $\mathbf{p}$ in R and momentum $\mathbf{p}'$ in R' shown in abstract momentum space. In this case, $\beta = 3/5$ and $E/pc = 2$. $\overrightarrow{AC}$ is $\mathbf{p}_{\parallel}$, $|DC|$ is $\beta E/c$, $\overrightarrow{AE}$ is $\gamma \times \overrightarrow{AD}$. $\overrightarrow{CB}$ is $\mathbf{p}_{\perp} = \mathbf{p}'_{\perp} = \overrightarrow{EF}$.

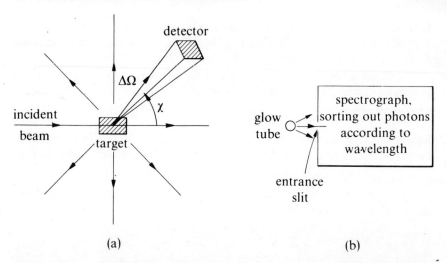

(a)   (b)

Figure 3-9.   Measurement of a distribution function. In (a) one determines the distribution of the scattering angles χ, ψ, using a detector that subtends a certain solid angle $\Delta\Omega$. In (b), one determines the distribution of the wavelength of the photons entering the spectrometer. It is on the basis of this example that the term *spectrum* or spectral function is also used for distribution function.

It is convenient to introduce an abstract momentum space in which momentum vectors from R and R' are plotted together. The relation between $\mathbf{p}$ and $\mathbf{p}'$ in this space is shown in Fig. 3-8. To find $\mathbf{p}'$ knowing $\mathbf{p}$, we start with $\overrightarrow{AB}$. From its longitudinal part $\overrightarrow{AC}$ we subtract $\overrightarrow{DC}$, of length $\beta E/c$. The resulting vector $\overrightarrow{AD}$ is now expanded by the factor γ to get the longitudinal part of $\mathbf{p}'$, $\overrightarrow{AE}$. The transverse momentum is invariant and is equal to $\overrightarrow{CB} = \overrightarrow{EF}$. Thus $\overrightarrow{AF}$ is the momentum vector in R', $\mathbf{p}'$. In the example shown, the particle's direction is forward in R, backward in R'.

Most experiments and the corresponding calculations are concerned with *distribution functions* of the physical variables. One repeats an identical measurement again and again in order to determine this function to the desired precision. The physical variable might, for example, be a component of the momentum of a particle, such as the polar angle χ (angle of scattering of a beam of particles in Fig. 3-9a), or the wavelength of the photon from an atom excited in a glow tube (Fig. 3-9b), or the distance traveled by an unstable system until it makes a transition (Section 2-7). Calling the measured variable x, we define the distribution function $f(x)$ by the statement that the number of cases in which the value of x is in the range x_1 to x_2 is

$$\int_{x_1}^{x_2} dx \cdot f(x)$$

In words, $f(x)\,dx$ is the number of cases in which the variable has a value in the "bin" from x to $x + dx$. One can normalize to one event, in which case $f(x)$ is the *probability density* of x.

Distribution functions of four-momentum are, of course, measured in the laboratory frame. They often find their most direct physical interpretation in other frames, such as, for example, the rest frame of one of the particles involved. Since we know how four-momentum transforms, it is a straightforward matter to calculate how its distribution transforms. The number of events in a certain set is invariant; it is only the boundaries of the set that change in the transformation. If the transformation takes x into y

$$y = y(x)$$

the new distribution function $g(y)$ is determined by

$$g(y)\,dy = f(x)\,dx$$

or

$$g(y) = f(x)\left|\frac{dx}{dy}\right| \qquad (3\text{-}38a)$$

Where x appears in the right-hand side it is to be expressed in terms of y. Numbers of events are essentially positive, so we always have positive distribution functions and positive ranges of the variables—hence, the absolute value sign. The corresponding expression for the case of a distribution of several independent variables is

$$g(y_1, y_2, \ldots) = f(x_1, x_2, \ldots)\frac{dx_1\,dx_2\cdots}{dy_1\,dy_2\cdots} = f(x_1, x_2, \ldots)\cdot\left|\frac{\partial(x_1, x_2, \ldots)}{\partial(y_1, y_2, \ldots)}\right|$$

$$(3\text{-}38b)$$

The last factor on the right-hand side is the absolute value of the functional determinant (Jacobian) of the transformation. For example, in the transformation between Cartesian coordinates (p_x, p_y, p_z) and spherical polar coordinates (p, χ, ψ) in momentum space, we find from Eqs. (3-34a) that

$$g(p, \chi, \psi) = f(p_x, p_y, p_z)\cdot p^2 \sin\chi$$

Here the result is evident geometrically from consideration of the volume element in momentum space, a box of sides dp_x, dp_y, dp_z for Cartesian coordinates (Fig. 3-10a) and a box of sides $p \sin\chi\,d\psi$, $p\,d\chi$, dp for polar coordinates (Fig. 3-10b).

It is convenient to introduce the solid angle of the cone of directions considered,

$$d\Omega = \sin \chi \, d\chi \, d\psi \qquad (3\text{-}39a)$$

The volume element in momentum space is

$$d^3\mathbf{p} = dp_x \, dp_y \, dp_z = p^2 \, dp \sin \chi \, d\chi \, d\psi = p^2 \, dp \, d\Omega$$

Note that the second order infinitesimal $d\Omega$ is

$$d\Omega = |d \cos \chi| \cdot d\psi \qquad (3\text{-}39b)$$

The distribution function can be zero everywhere except in a small region, where it is very large. The limiting case, where there is a unique value of the dynamical variable, x_0, is described by the causal distribution or *delta function* (Fig. 3-11a). The delta function is zero everywhere except at $x = x_0$.

$$\delta(x - x_0) = 0 \qquad x \neq x_0$$

At x_0 it goes infinite in such a way that

$$\int_{\text{all } x} dx \cdot \delta(x - x_0) = 1$$

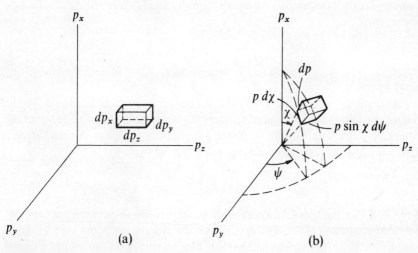

(a)   (b)

Figure 3-10.

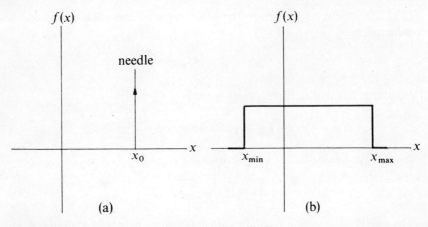

Figure 3-11.

It might, for example, be the limit, as $\varepsilon \to 0$, of a rectangle whose base extends from $x_0 - \varepsilon/2$ to $x_0 + \varepsilon/2$ and whose height is $1/\varepsilon$. When the delta function is a factor in an integrand, integration with respect to x picks out the specified value x_0:

$$\int_{x < x_0}^{x > x_0} dx \cdot h(x)\delta(x - x_0) = h(x_0)$$

At the other extreme is the flat or *uniform distribution* (Fig. 3-11b), where all values of x in its range of values are equally likely. With normalization to one event,

$$f(x) = \frac{1}{x_{max} - x_{min}}$$

Illustrations of both kinds of distribution function are provided by the distribution in the rest frame of the parent, of the momentum of either of the daughter particles resulting from the two-body decay of a particle of zero spin—for example, the pion decay

$$\pi \to \mu + \nu_\mu$$

Since the two particles fly apart with equal and opposite momentum (Fig. 3-12), conservation of energy requires that the momentum must have a unique magnitude p_0. Since the parent system has no spin, no direction in the rest frame is singled out and all directions of emission are equally likely. The probability of p having a value in dp is

$$\delta(p - p_0)\, dp$$

and the probability of the direction being in $d\Omega$ is

$$\frac{1}{4\pi}\, d\Omega$$

The total solid angle range in three dimensions is 4π sr. In terms of spherical polar angles, the last expression is [Eq. (3-39b)]

$$\frac{1}{4\pi}\, |d \cos \chi|\, d\psi$$

The probability density relative to the variables $(p, \cos \chi, \psi)$ is thus

$$f(p, \cos \chi, \psi) = \frac{1}{4\pi}\delta(p - p_0) \tag{3-40a}$$

To find the probability density relative to (p_x, p_y, p_z), we use Eq. (3-38b), noting that, by Eqs. (3-34a),

$$\frac{\partial(p, \cos \chi, \psi)}{\partial(p_x, p_y, p_z)} = \frac{1}{p^2}$$

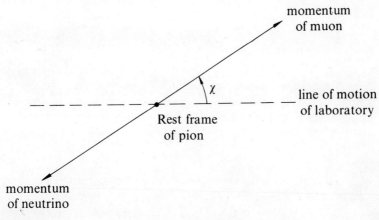

Figure 3-12.

Thus,

$$g(p_x, p_y, p_z) = \frac{1}{4\pi} \frac{\delta([p_x^2 + p_y^2 + p_z^2]^{1/2} - p_0)}{p_x^2 + p_y^2 + p_z^2} \qquad (3\text{-}40b)$$

The transformation of distribution functions between different reference frames is treated in just the same way. The number of events in an experiment is invariant, and the distribution function must transform so as to compensate for the change of the infinitesimal element of integration. For a four-vector with independent components, the Jacobian of the SLT [Eqs. (2-49b)] is unity.

$$\frac{\partial(A_1', A_2', A_3', A_4')}{\partial(A_1, A_2, A_3, A_4)} = \begin{vmatrix} \gamma & 0 & 0 & i\beta\gamma \\ 0 & 1 & 0 & 0 \\ 0 & 0 & 1 & 0 \\ -i\beta\gamma & 0 & 0 & \gamma \end{vmatrix} = 1 \qquad (3\text{-}41)$$

The volume element in spacetime is invariant. For the four-momentum, however, the components are not independent [Eq. (3-24)]. One can take this constraint into account by incorporating

$$\delta(\tilde{p}^2 + \mu^2 c^2)$$

as a factor in the four-dimensional integrand. A more elementary approach leading to the same result is to eliminate the fourth component from the transformation equations by using

$$E = (c^2 p^2 + \mu^2 c^4)^{1/2} \qquad (3\text{-}27b)$$

This gives

$$p_x' = \gamma\left[p_x - \frac{\beta}{c}(c^2 p_x^2 + c^2 p_y^2 + c^2 p_z^2 + \mu^2 c^4)^{1/2} \right]$$

$$p_y' = p_y$$

$$p_z' = p_z$$

and thus

$$\frac{\partial(p_x', p_y', p_z')}{\partial(p_x, p_y, p_z)} = \frac{\partial p_{x'}}{\partial p_x} = \gamma - \beta\gamma c p_x (c^2 p_x^2 + c^2 p_y^2 + c^2 p_z^2 + \mu^2 c^4)^{-1/2}$$

$$= \frac{\gamma E - \beta\gamma c p_x}{E} = \frac{E'}{E} \qquad (3\text{-}42)$$

For a momentum distribution specified by $f(p_x, p_y, p_z)$ in R and by $f'(p'_x, p'_y, p'_z)$ in R', there is, therefore, the relation

$$f(p_x, p_y, p_z)E = f'(p'_x, p'_y, p'_z)E' \qquad \text{(3-43a)}$$

Another way of describing this result is to state that for a particle,

$$\frac{d^3\mathbf{p}}{E}$$

is invariant:

$$\frac{d^3\mathbf{p}}{E} = \frac{dp_x\, dp_y\, dp_z}{E.} = \frac{dp'_x\, dp'_y\, dp'_z}{E'} = \frac{d^3\mathbf{p}'}{E'} \qquad \text{(3-43b)}$$

In a chain of coordinate transformations, the Jacobian of the resultant transformation is the product of the successive Jacobians. A rotation of axes has a Jacobian of $+1$ and an inversion has a Jacobian of -1, so the relations Eqs. (3-41), (3-42), and (3-43) hold for the general Lorentz transformation.

Using spherical polar coordinates in R and R', we can write Eq. (3-43b) as

$$\frac{p^2\, dp\, d\Omega}{E} = \frac{p'^2\, dp'\, d\Omega'}{E'} \qquad \text{(3-44b)}$$

and Eq. (3-29) permits us to convert to

$$p\, dE\, d\Omega = p'\, dE'\, d\Omega' \qquad \text{(3-45b)}$$

The corresponding equations for the distribution functions are

$$g(p, \Omega)\frac{E}{p^2} = g'(p', \Omega')\frac{E'}{p'^2} \qquad \text{(3-44a)}$$

and

$$h(E, \Omega)\frac{1}{p} = h'(E', \Omega')\frac{1}{p'} \qquad \text{(3-45a)}$$

The number of events in an experimental distribution depends, of course, on the duration of the experiment and its specific conditions. Where the experiment is a study of a binary collision, the *cross section* is a convenient measure of the probability of the process under study. It is found by dividing the number of events by the number of particles that were incident per unit area transverse to the direction of relative motion. Numbers of particles and transverse areas are invariant, thus the cross section is invariant. All that changes in a Lorentz transformation are the variables that specify the process.

The letter σ is conventionally used for cross section. It is an area of order of magnitude 10^{-16} cm^2 in common atomic processes, 10^{-24} cm^2 in common nuclear processes, and 10^{-27} cm^2 in common particle processes. The final state in the process is a function of several continuous variables, such as the momentum components of one of the particles. Thus, we can write

$$\sigma = \iint dp\, d\Omega \left(\frac{d^2\sigma}{dp\, d\Omega} \right) \tag{3-46a}$$

The integration is over the entire range of momenta contributing to the specified process. The integrand in this example

$$\frac{d^2\sigma}{dp\, d\Omega}$$

is called the *differential cross section* per unit (magnitude of) momentum per unit solid angle (a second-order differential). According to Eqs. (3-44),

$$\frac{d^2\sigma}{dp\, d\Omega} \frac{E}{p^2} = \frac{d^2\sigma'}{dp'\, d\Omega'} \frac{E'}{p'^2} \tag{3-47}$$

Alternatively, one can use energy and direction as the independent variables:

$$\sigma = \iint dE\, d\Omega \left(\frac{d^2\sigma}{dE\, d\Omega} \right) \tag{3-46b}$$

The integrand

$$\frac{d^2\sigma}{dE\, d\Omega}$$

is called the differential cross section per unit energy per unit solid angle. Equations (3-45) give

$$\frac{d^2\sigma}{dE\, d\Omega} \frac{1}{p} = \frac{d^2\sigma'}{dE'\, d\Omega'} \frac{1}{p'} \tag{3-48}$$

Equations (3-47) and (3-48) are useful because the quantal calculations of differential cross sections are normally evaluated in reference frames other than the laboratory.

EXERCISE   A pencil beam of pions of momentum 10.0 GeV/c decays in vacuum, providing a line source of neutrinos for use in experiments. In the rest frame of the pions, the neutrino momentum distribution is given by Eq. (3-40) with $p_0 = 29.8 \times 10^{-3}$ GeV/c. Calculate the momentum distribution of the neutrinos in the laboratory frame.

   A round bubble chamber of area 1.0 m^2 is located 50 m downstream from the decay region. Calculate the momenta and number of neutrinos striking the chamber.

SOLUTION   Let R' be the rest frame of the pions and R the laboratory frame. The statement that the momentum is "10.0 GeV/c" means that $cp = 10.0$ GeV ($= 10.0 \times 10^9$ eV). The proper mass of the pion is 1.40×10^{-1} GeV/c^2, meaning

$$\mu c^2 = 1.40 \times 10^{-1} \text{ GeV}$$

Therefore,

$$\beta\gamma = \frac{p}{\mu c} = \frac{pc}{\mu c^2} = \frac{10.0}{0.140} = 71.4$$

and

$$\gamma = [1 + (\beta\gamma)^2]^{1/2} = 71.4$$

$$\beta = \left[1 + \frac{1}{(\beta\gamma)^2}\right]^{-1/2} = 1 - \frac{1}{2(\beta\gamma)^2} + \cdots = 1 - 0.981 \times 10^{-4}$$

Equations (3-44a) and (3-40a) give for the momentum distribution of the neutrino ($E = cp$, $E' = cp'$),

$$f(p, \Omega) = \frac{p^2 E'}{p'^2 E} f'(p', \Omega') = \frac{p}{p'} \frac{1}{4\pi} \delta(p' - p_0)$$

The energy transformation [Eq. (3-35)] gives (again using $E = cp$, $E' = cp'$)

$$p' = \gamma p(1 - \beta \cos \chi) \tag{3-49}$$

Thus,

$$f(p, \Omega) = \frac{1}{4\pi} \frac{1}{\gamma(1 - \beta \cos \chi)} \delta(\gamma[1 - \beta \cos \chi]p - p_0)$$

The delta function involves both p and $\cos \chi$. Let us integrate over the solid angle to find the momentum distribution regardless of direction:

$$\iint d\psi \, d(\cos \chi) f(p, \Omega)$$

$$= \frac{1}{4\pi} \int_0^{2\pi} d\psi \int_{-1}^1 d(\cos \chi) \frac{1}{\gamma(1 - \beta \cos \chi)} \delta(\gamma[1 - \beta \cos \chi]p - p_0)$$

In the second integral on the right-hand side, p is held fixed while $\cos \chi$ is varied. With the formal substitution

$$w = p\gamma(1 - \beta \cos \chi) \qquad dw = -p\beta\gamma \, d(\cos \chi)$$

the expression becomes

$$\frac{1}{2} \int_{w = p\gamma(1 - \beta)}^{p\gamma(1 + \beta)} \frac{dw}{p\beta\gamma} \frac{p}{w} \delta(w - p_0) = \frac{1}{2\beta\gamma p_0}$$

which is independent of p. The momentum distribution of the neutrinos in the laboratory is flat. The probability of a momentum in dp is

$$\frac{1}{2\beta\gamma p_0} dp$$

for p ranging between its minimum value (neutrino emitted backward in R')

$$p_{min} = \gamma p'[1 + \beta(\cos \chi')_{min}] = p_0\gamma(1 - \beta)$$

$$= 29.8 \times 10^{-3} \times 71.4 \times 0.981 \times 10^{-4}$$

$$= 29.8 \times 10^{-3} \times 7.01 \times 10^{-3} = 2.09 \times 10^5 \text{ eV/c}$$

and its maximum (neutrino emitted forward in R')

$$p_{max} = \gamma p'[1 + \beta(\cos \chi')_{max}] = p_0\gamma(1 + \beta)$$

$$= 29.8 \times 10^{-3} \times 71.4 \times 2 = 4.26 \text{ GeV/c}$$

Of course,

$$p_{max} - p_{min} = 2\beta\gamma p_0$$

An elementary way of getting this result is to differentiate the Lorentz transformation of energy

$$E = \gamma(E' + \beta p' \cos \chi') = \gamma E_0 + \beta\gamma p_0 \cos \chi'$$

obtaining

$$dE = \beta\gamma p_0 \, d(\cos \chi')$$

A flat distribution in $\cos \chi'$ thus results in a flat distribution in E, ranging from

$$E_{min} = \gamma(E_0 - \beta p_0)$$

to

$$E_{max} = \gamma(E_0 + \beta p_0)$$

This is a general property of *two-body isotropic decays*, independent of proper mass: The energy spectrum of each decay particle in the laboratory is flat. For a neutrino or photon, the momentum spectrum is essentially the same as the energy spectrum.

The momentum and angle in the laboratory are, of course, correlated, for p' in Eq. (3-49) equals p_0:

$$p_0 = \gamma p(1 - \beta \cos \chi)$$

The more forward the neutrino ($\cos \chi$ nearly one) the larger the momentum. The bubble chamber subtends a cone of half-angle $\chi = 11.3$ mrad. The neutrinos striking it range in momentum from 4.26 GeV/c on axis ($\chi = 0$) to

$$p = \frac{p_0}{\gamma(1 - \beta \cos \chi)} = \frac{29.8 \times 10^{-3}}{71.4(1 - 1 + 1.617 \times 10^{-4})} = 2.58 \text{ GeV/c}$$

at the edge ($\chi = 11.3$ mrad). They comprise

$$\frac{4.26 - 2.58}{4.26 - 2.09 \times 10^{-4}} = 39\%$$

of the neutrinos produced.

3-9. MOTION OF A CHARGED PARTICLE IN A CONSTANT ELECTROMAGNETIC FIELD

The law of motion of a charged particle

$$\frac{e}{c}\tilde{\mathscr{F}}\tilde{w} = \frac{d\tilde{p}}{d\tau} \tag{3-50}$$

permits us to determine the motion in any assigned field $\tilde{\mathscr{F}}$. The spatial projection gives the familiar law of evolution in time:

$$e\left(\mathscr{E} + \frac{\mathbf{v}}{c}\times\mathscr{B}\right) = \frac{d\mathbf{p}}{dt} = \frac{d}{dt}\left[\mu\left(1 - \frac{v^2}{c^2}\right)^{-1/2}\mathbf{v}\right] \tag{3-5}$$

The temporal projection is that useful corollary, the theorem of work:

$$e(\mathscr{E}\cdot\mathbf{v}) = \frac{dE}{dt} = \frac{d}{dt}\left[\mu c^2\left(1 - \frac{v^2}{c^2}\right)^{-1/2}\right] \tag{3-12}$$

There is a wealth of important applications: for example, the motion of a cosmic proton in the magnetic field surrounding the earth [39], the motion of an ion in the electromagnetic field produced by the coils and plates of an accelerator [40], the motion of an electron in a vacuum tube device [41]. The entire subject of particle optics is built in a straightforward way on Eq. (3-50). (For sufficiently rapid acceleration, one must add a radiative reaction force; see [42], Chapter 17; [35], Chapter 9, or [43], Chapter 21.)

We have already treated (in Sections 2-11 and 2-12) the special cases of longitudinal motion in a constant electric field and transverse motion in a magnetic or electric field. We shall now take up the general motion in a purely magnetic field and the orbit in a Coulomb field (hydrogenic atom). (For further examples, the reader is referred to [42], Chapter 12, pp. 8–10, or [35], Chapter 3, pp. 6–9.) A recent presentation of "beamology" (particle optics) is that of Banford (see [44]).

Charged Particle in Magnetic Field

With $\mathscr{E} = 0$, Eq. (3-5) reads

$$\frac{e}{c}\mathbf{v}\times\mathscr{B} = \frac{d\mathbf{p}}{dt} = \frac{d}{dt}\left[\mu\left(1 - \frac{v^2}{c^2}\right)^{-1/2}\mathbf{v}\right] \tag{3-51}$$

The force is transverse to the velocity and therefore to the momentum. The momentum remains constant in magnitude, and so does the velocity; only the direction of the motion changes.

The same conclusions follow from Eq. (3-12):

$$0 = \frac{dE}{dt} = \frac{d}{dt}\left[\mu\left(1 - \frac{v^2}{c^2}\right)^{-1/2}\right]$$

Since v is constant, Eq. (3-51) can be written

$$\frac{e}{c}\mathbf{v} \times \mathcal{B} = \mu\left(1 - \frac{v^2}{c^2}\right)^{-1/2}\frac{d\mathbf{v}}{dt} = \mu\gamma\mathbf{a} \qquad (3\text{-}52)$$

which is just

$$\mathbf{F} = m\mathbf{a}$$

with m given by the relativistic inertial mass $\mu\gamma$, not the proper mass μ. The pure magnetic field cannot change v, and m is constant at the value determined by the speed at the instant the field at the particle became purely magnetic. One can take over the results of the nonrelativistic treatment provided one substitutes the correct expression for the constant inertial mass. This holds even at extreme relativistic speeds, provided the field is purely magnetic.

An alternative mathematical technique, which is completely equivalent, is to express $\mathbf{v}$ in terms of $\mathbf{p}$

$$\mathbf{v} = \frac{c\mathbf{p}}{(p^2 + \mu^2 c^2)^{1/2}}$$

and use Eq. (3-51) to determine the momentum as a function of time:

$$e(p^2 + \mu^2 c^2)^{-1/2}\mathbf{p} \times \mathcal{B} = \frac{d\mathbf{p}}{dt}$$

or

$$\frac{ec}{E}\mathbf{p} \times \mathcal{B} = \frac{d\mathbf{p}}{dt} \qquad (3\text{-}53)$$

Of course, E is a constant of the motion, as well as p and v.

In Section 2-12, we saw that for a transverse uniform magnetic field the motion is uniform circular with angular velocity

$$\omega = \frac{e\mathscr{B}}{\mu\gamma c} = \frac{ec\mathscr{B}}{E} \tag{3-54a}$$

and radius

$$\rho = \frac{c\mu w}{e\mathscr{B}} = \frac{cp}{e\mathscr{B}} \tag{3-54b}$$

For motion parallel to the field the force is zero; therefore, the velocity is constant. We shall see shortly that for an arbitrary initial direction the motion is helical (Fig. 3-13), uniform progression parallel to the field combined with uniform rotation about it.

The extension to an inhomogeneous magnetic field is immediate. Consider a particular phase of the motion. Choose local axes with Oz parallel to the local magnetic field and the y–z plane containing $\mathbf{p}$ (or $\mathbf{v}$) (Fig. 3-14). Denote by λ the elevation angle of $\mathbf{p}$ above the plane perpendicular to the magnetic field. It is the angle between $\mathbf{p}$ and the y axis. The components of the law of motion [Eq. (3-53)] give

$$p_z = \text{const} = p \sin \lambda$$

$$p_y = \text{const} = p \cos \lambda$$

$$\frac{dp_x}{dt} = \frac{ec}{E} p\mathscr{B} \cos \lambda$$

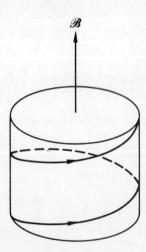

Figure 3-13.

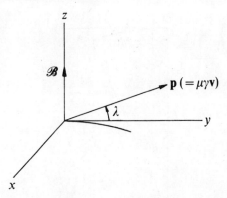

Figure 3-14.

There is transverse acceleration, in the $+x$ direction if the charge is positive, with rate of turning of the momentum vector (Fig. 3-15) given by

$$\frac{d\theta}{dt} = \frac{dp_x/dt}{p_y} = \frac{ecp\mathscr{B}\cos\lambda}{Ep\cos\lambda} = \frac{ec\mathscr{B}}{E} \qquad (3\text{-}55a)$$

and radius of curvature (Fig. 3-16)

$$\rho = \frac{v_y\, dt}{d\theta} = \frac{c^2 p_y/E}{ec\mathscr{B}/E} = \frac{cp\cos\lambda}{e\mathscr{B}} \qquad (3\text{-}55b)$$

The motion consists of a straight component parallel to the field, with speed

$$v_\parallel = v\sin\lambda = \frac{c^2 p}{E}\sin\lambda \qquad (3\text{-}56a)$$

and a curved motion in the plane perpendicular to the field with speed

$$v_\perp = v\cos\lambda = \frac{c^2 p}{E}\cos\lambda \qquad (3\text{-}56b)$$

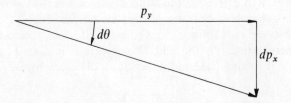

Figure 3-15.

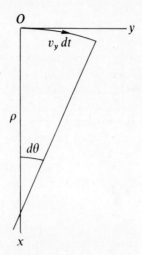

Figure 3-16.

and curvature

$$\frac{1}{\rho} = \frac{e\mathscr{B}}{c}\frac{1}{p\cos\lambda} = \frac{e\mathscr{B}}{c}\frac{1}{p_\perp} \tag{3-56c}$$

If the magnetic field is homogeneous, the motion is uniform translation along the field combined with uniform rotation in a circle at right angles to it. The orbit (Fig. 3-13) is a circular helix with the field as axis. The radius has the constant value given by Eq. (3-50c); the angular velocity is the "cyclotron" frequency [Eqs. (3-55a) or (3-54a)]. If the field does not vary too rapidly with position, as is indeed the case in a bubble chamber or beam-deflecting magnet, the orbit is but little distorted from this simple screw form. The cyclotron frequency about the "guiding center" is proportional to the field strength and inversely proportional to the constant energy or inertial mass. It is independent of the pitch λ.

The observation of orbits in a magnetic field is the basic technique used in particle physics to determine the momentum (cf. Fig. 1-4). With a bubble chamber, spark chambers, or trains of counters we can observe the particle's orbit. The observed curvature ρ^{-1} gives, by Eq. (3-56c), the component of momentum perpendicular to the field. The observed dip angle λ then permits us to calculate the magnitude and direction of the momentum.

$$p = \frac{e}{c}\mathscr{B}\rho\frac{1}{\cos\lambda}$$

Strictly speaking, only the ratio of momentum to charge, or *magnetic rigidity*, is measured by observation of the orbit in a known field. The value of e is found from other observations, usually of the rate of energy loss. In the study of cosmic rays, we use the bending of the incoming charged particles by the earth's magnetic field as a means of determining their magnetic rigidity. We so determine the distribution of (p/e) for the cosmic ray primaries, but we cannot from magnetic deflection alone tell a proton of momentum 10 GeV/c from an alpha particle of momentum 20 GeV/c or a uranium nucleus of momentum 920 GeV/c.

EXERCISE   Determine the motion of a charged particle in a constant homogeneous electric field.

SOLUTION   (See [35], Sections 3–6.) Place the z axis along the field and the y–z plane so as to contain the initial velocity (Fig. 3-17). The momentum (and therefore velocity) vector remains in this plane. The law of motion

$$e\mathscr{E} = \frac{dp_z}{dt}$$

$$0 = \frac{dp_y}{dt}$$

$$0 = \frac{dp_x}{dt}$$

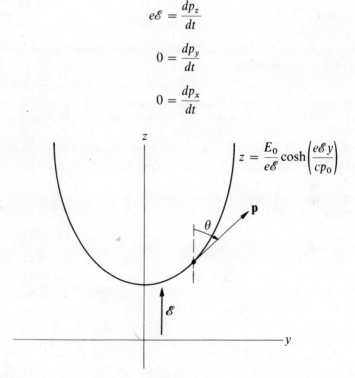

$$z = \frac{E_0}{e\mathscr{E}} \cosh\left(\frac{e\mathscr{E}\, y}{c p_0}\right)$$

Figure 3-17.

gives immediately the momentum as a function of the time

$$p_z = p_0 \cos \theta_0 + e\mathscr{E}t$$

$$p_y = p_0 \sin \theta_0$$

$$p_x = 0$$

Instead of expressing the law of motion in terms of v_z and v_y and going through a rather cumbersome integration, we have determined by inspection the momentum as a function of time. The momentum determines the velocity. We place the origin of time at the instant that the motion is perpendicular to the field. Then $\cos \theta_0 = 0$, $\sin \theta_0 = 1$, and

$$p = (p_0{}^2 + e^2 \mathscr{E}^2 t^2)^{1/2}$$

The inclination to the axis is

$$\tan \theta = \frac{p_y}{p_z} = \frac{p_0}{e\mathscr{E}t}$$

As t increases, the direction of motion approaches that of the electric field. The speed is given by

$$\frac{v}{c} = \frac{p}{(p^2 + \mu^2 c^2)^{1/2}} = \left[\frac{p_0{}^2 + e^2 \mathscr{E}^2 t^2}{p_0{}^2 + e^2 \mathscr{E}^2 t^2 + \mu^2 c^2} \right]^{1/2}$$

which approaches 1 as $t = \infty$. Combining this expression with that for θ, we find

$$\frac{dz}{dt} = v_z = v \cos \theta = \frac{ce\mathscr{E}t}{(p_0{}^2 + e^2 \mathscr{E}^2 t^2 + \mu^2 c^2)^{1/2}} = \frac{c^2 e\mathscr{E}t}{(E_0{}^2 + c^2 e^2 \mathscr{E}^2 t^2)^{1/2}}$$

$$\frac{dy}{dt} = v_y = v \sin \theta = \frac{cp_0}{(p_0{}^2 + e^2 \mathscr{E}^2 t^2 + \mu^2 c^2)^{1/2}} = \frac{c^2 p_0}{(E_0{}^2 + c^2 e^2 \mathscr{E}^2 t^2)^{1/2}}$$

where E_0 is the initial energy. These expressions integrate immediately to

$$z = \frac{1}{e\mathscr{E}} (E_0{}^2 + c^2 e^2 \mathscr{E}^2 t^2)^{1/2} + \text{const}$$

$$y = \frac{cp_0}{e\mathscr{E}} \sinh^{-1} \left(\frac{ce\mathscr{E}t}{E_0} \right) + \text{const}$$

Choosing the origin of coordinates to make both constants of integration zero, we find on eliminating t that

$$z = \frac{E_0}{e\mathscr{E}} \cosh \left(\frac{e\mathscr{E}y}{cp_0}\right)$$

The orbit is a hyperbolic cosine curve. If $p_0 \ll E_0/c$, it is well approximated near its vertex by the nonrelativistic parabolic orbit

$$z = \frac{E_0}{e\mathscr{E}} + \frac{1}{2}\frac{\mu e\mathscr{E}}{p_0^2}y^2$$

Note that the kinetic energy increase is given exactly by

$$K - K_0 = E - E_0 = (E_0^2 + e^2\mathscr{E}^2 t^2)^{1/2} - E_0 = e\mathscr{E}(z - z_0)$$

For the case of longitudinal motion considered in Section 2-11, we set $p_0 = 0$. The velocity becomes

$$\frac{v}{c} = \frac{ce\mathscr{E}t}{(c^2 e^2 \mathscr{E}^2 t^2 + E_0^2)^{1/2}}$$

which is Eq. (2-72), and the position equations become

$$z = \frac{1}{e\mathscr{E}}(E_0^2 + c^2 e^2 \mathscr{E}^2 t^2)^{1/2}$$

$$y = 0$$

which are Eq. (2-73). The world line is a hyperbola in the (z, t) plane.

Charged Particle in a Coulomb Field (Hydrogenic Atom)

A hydrogenic atom consists of a nucleus of charge Ze_0 and an electron of charge $-e_0$. Examples are hydrogen $(Z = 1)$, singly ionized helium $(Z = 2)$, 25-times ionized iron $(Z = 26)$, and muonium $(\mu^+ e^-, Z = 1)$. The hydrogenic atom is the fundamental object of atomic physics. It provides the basis for understanding the entire periodic table. In the approximation that motion of the nucleus can be neglected, an exact classical relativistic solution has been worked out. In this case we have to deal with the motion of the electron in a fixed Coulomb field

$$\mathscr{E} = \frac{Ze_0}{r^3}\mathbf{r} \qquad \mathscr{B} = 0 \qquad\qquad (3\text{-}57)$$

where **r** is the radius vector from the nucleus to the electron. Corrections for motion of the nucleus, treated as a small perturbation, can be made later on.

This classical solution to an atomic problem is of some interest because of the remarkable way in which classical and quantal mechanics correspond. In the "old quantum theory" (Bohr, Ehrenfest, Sommerfeld, 1913–1925) remarkably precise results for the energy levels and transition probabilities of bound states were obtained by treating atomic systems classically and then applying specific "quantum conditions," one for each periodicity of the classical motion. Sommerfeld applied this procedure to the hydrogenic atom using classical relativistic dynamics, and derived exactly the same energy levels that were later shown to follow from Dirac's relativistic wave equation. It is a very remarkable coincidence. The experimental data are very close to the theoretical predictions, differing only by a tiny effect (the Lamb–Rutherford shift) that has been explained by relativistic quantum electrodynamics.

The fine structure of the energy levels can also be calculated as a perturbation in the Schroedinger nonrelativistic theory, by correcting for the relativistic dependence of inertial mass on velocity and the spin–orbit interaction (computed relativistically). The same result is obtained as from the completely relativistic treatments, whether classical or quantal. In the latter (Dirac equation), the electron is completely characterized by its charge and proper mass. The physical consequences of spin fall out of the formalism without explicit introduction of spin at the beginning.

Sommerfeld's calculation differs from Bohr's only in the use of the relativistic expressions for energy and momentum in terms of velocity. Bohr's theory gives the structure of the energy levels, Sommerfeld's the fine structure. We follow Sommerfeld's presentation ([45], Chapter V, Section 1).

The force on the electron is

$$\mathbf{F} = -\frac{Ze_0^2}{r^3}\mathbf{r} \tag{3-58}$$

which is the negative gradient of

$$V = -\frac{Ze_0^2}{r}$$

Hence,

$$dE = \mathbf{F} \cdot d\mathbf{r} = -dV$$

and V can be interpreted as the potential energy of the atom, with zero level corresponding to infinite separation. The energy of the electron plus the potential energy of the electron's interaction with the external field is a constant of the motion.

$$dE + dV = 0$$

or, by Eq. (3-13),

$$\mu c^2 \left(1 - \frac{v^2}{c^2}\right)^{-1/2} - \frac{Ze_0^2}{r} = \text{const} = W \qquad (3\text{-}59)$$

We shall refer to this equation as the energy integral.

Another conservation law that still applies is that of angular momentum. It results from the central nature of the Coulomb force [Eq. (3-58)]. In effect, let us form the moment of the law of motion

$$\mathbf{r} \times \mathbf{F} = \mathbf{r} \times \frac{d\mathbf{p}}{dt}$$

$$= \frac{d}{dt}(\mathbf{r} \times \mathbf{p}) - \frac{d\mathbf{r}}{dt} \times \mathbf{p}$$

But $\mathbf{p}$ is parallel to $\mathbf{v} = d\mathbf{r}/dt$, so the second term of the right-hand member vanishes; the left-hand member vanishes because $\mathbf{F}$ is parallel to $\mathbf{r}$. We have the familiar result

$$\mathbf{L} = \mathbf{r} \times \mathbf{p} = \text{const} \qquad (3\text{-}60)$$

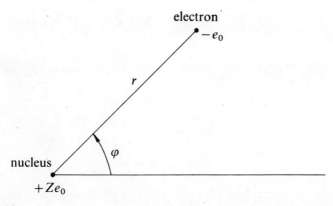

Figure 3-18.

In applying this angular momentum integral, we must use the relativistic expression for the momentum, Eq. (1-33).

Since $\mathbf{r} \times \mathbf{p}$ is a constant vector, the motion must be in a plane. We use polar coordinates (r, φ) in this plane, with the origin at the nucleus (Fig. 3-18). The magnitude of the angular momentum is

$$L = r\mu\gamma r\frac{d\varphi}{dt} = \mu\left(1 - \frac{v^2}{c^2}\right)^{-1/2} r^2\frac{d\varphi}{dt} \tag{3-61}$$

Between the first integrals of energy [Eq. (3-59)] and angular momentum [Eq. (3-61)], we can eliminate the time, obtaining the differential equation of the orbit. Solving Eq. (3-61) for dt

$$dt = \mu\gamma r^2 \frac{d\varphi}{L}$$

and substituting in

$$v^2 = \left(\frac{dr}{dt}\right)^2 + \left(r\frac{d\varphi}{dt}\right)^2$$

we get

$$\frac{\gamma^2 v^2}{c^2} = \frac{L^2}{\mu^2 c^2}\left[\left(\frac{1}{r^2}\frac{dr}{d\varphi}\right)^2 + \frac{1}{r^2}\right]$$

From Eq. (3-59),

$$\gamma = \frac{W + Ze_0{}^2/R}{\mu c^2}$$

Since

$$1 + \gamma^2\beta^2 = \gamma^2$$

the equations combine to give

$$1 + \frac{L^2}{\mu^2 c^2}\left[\left(\frac{1}{r^2}\frac{dr}{d\varphi}\right)^2 + \frac{1}{r^2}\right] = \left(\frac{W + Ze_0^2/r}{\mu c^2}\right)^2 \tag{3-62}$$

which is the differential equation of the path. With the substitution

$$s = \frac{1}{r}$$

it becomes

$$1 + \frac{L^2}{\mu^2 c^2}\left[\left(\frac{ds}{d\varphi}\right)^2 + s^2\right] = \left(\frac{W + Ze_0^2 s}{\mu c^2}\right)^2$$

Differentiating with respect to φ, we get

$$\left(\frac{2}{\mu c^2}\frac{ds}{d\varphi}\right)\frac{L^2}{\mu}\left(\frac{d^2 s}{d\varphi^2} + s\right) = \left(\frac{2}{\mu c^2}\frac{ds}{d\varphi}\right)\left(\frac{W + Ze_0^2 s}{\mu c^2}\right)Ze_0^2$$

Except at extrema of r, where $(ds/d\varphi) = 0$, we can divide through by the common factor, obtaining

$$\frac{d^2 s}{d\varphi^2} + s\left(1 - \frac{Z^2 e_0^4}{c^2 L^2}\right) = \frac{Ze_0^2 W}{c^2 L^2} \tag{3-63}$$

This differential equation has a simple form, the same as that of a simple harmonic oscillator with a constant driving term (of course, the independent variable is the angle φ, not the time t). The coefficient of s

$$\lambda^2 = 1 - \frac{Z^2 e_0^4}{c^2 L^2} \tag{3-64}$$

differs only slightly from unity. It is unity in the nonrelativistic limit ($c = \infty$). The smaller the angular momentum the more λ^2 differs from 1. We know that L is actually quantized in multiples of $\hbar = h/2\pi$. Its smallest possible value is $\hbar$, so that

$$\lambda^2_{\text{min}} = 1 - \left(\frac{Ze_0^2}{\hbar c}\right)^2 = 1 - Z^2 \alpha^2$$

where α is Sommerfeld's famous fine-structure constant, "1/137."

$$\alpha = \frac{e_0^2}{\hbar c} = 7.2972 \times 10^{-3} = (137.04)^{-1}$$

It specifies the strength of the electromagnetic coupling between fundamental particles, relating the quantum of electric charge to the quantum of angular momentum and the speed of light. The difference from $\lambda^2 = 1$ is on the order $(Z/137)^2$ or 53 parts in a million for hydrogen. In terms of the precision of atomic spectroscopy, this amounts to a very large effect. Indeed, until recently the fine structure of hydrogen and singly ionized helium provided more precise confirmation of the relativistic expressions for momentum and energy than did all the experiments on direct deflection of electron beams in evacuated tubes.

The solution of

$$\frac{d^2 s}{d\varphi^2} + \lambda^2 s = \frac{Ze_0{}^2 W}{c^2 L^2} = D \qquad (3\text{-}65)$$

is

$$s = A\cos(\lambda\varphi) + B\sin(\lambda\varphi) + \frac{1}{\lambda^2} D$$

where A and B are constants of integration. Choose the line from which φ is measured so that $\varphi = 0$ at a perihelion (Fig. 3-19). Since

$$\frac{ds}{d\varphi} = -A\sin(\lambda\varphi) + B\lambda\cos(\lambda\varphi)$$

this requires $B = 0$. Thus,

$$s = \frac{1}{r} = A\cos(\lambda\varphi) + \frac{1}{\lambda^2} D \qquad (3\text{-}66)$$

In the nonrelativistic limit ($\lambda = 1$), this is the equation in polar coordinates of an ellipse with a focus at the origin, of eccentricity A/D, and distance from focus to directrix $1/A$ (Fig. 3-20). It is the familiar Kepler ellipse,

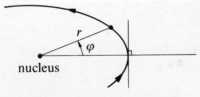

Figure 3-19. $\varphi = 0$ at a perihelion (minimum r).

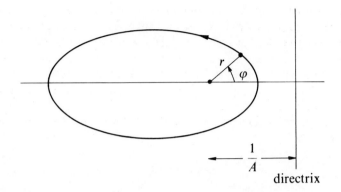

Figure 3-20.

or Bohr orbit. The departure of λ from 1 means that in the relativistic case the ellipse does not quite close. When φ increases by 2π, the argument of the cosine, $\lambda\varphi$, increases by a little less than 2π. We have to go a little further in φ to get to the corresponding value of r. The ellipse turns slowly around, making the orbit a rosette (Fig. 3-21). It is customary to call this motion of the ellipse a "precession." The advance of the aphelion in one revolution is evidently given by

$$\lambda(2\pi + \Delta\varphi) = 2\pi$$

or   (3-67)

$$\Delta\varphi = 2\pi\left(\frac{1 - \lambda}{\lambda}\right)$$

Since $\lambda \lesssim 1$, the ellipse works its way around very slowly relative to the rate at which it is being traced out. The only effect of relativistic dynamics on the orbit is to change the eternal motion in one fixed ellipse into a motion osculated by a slowly rotating ellipse. In the old quantum theory a quantum number is associated with every periodicity. That of orbital revolution gives n, the quantum number determining the gross values of the energy levels. The precession frequency gives another quantum number, associated with angular momentum, determining the fine structure of the energy levels.

One can apply the same orbit calculation to the problem of a planet attracted by the sun, replacing the Coulomb force law by Newton's law of gravitation. All we need to do is replace Ze_0^2 by GMm in Eq. (3-58). For the planet's mass m we may use a constant value, because the change in it has only a higher order effect. Now the elliptical orbit of the innermost

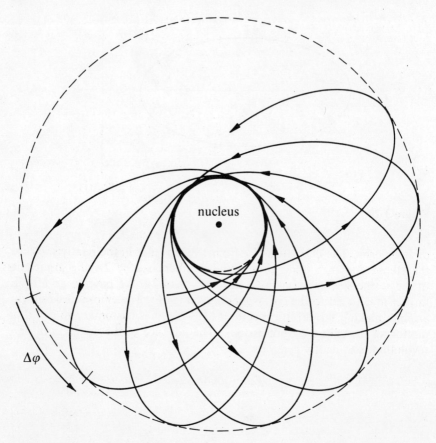

nucleus

$\Delta\varphi$

Figure 3-21. (After Sommerfeld [45]). As the electron goes around its ellipse, the ellipse turns slowly in the same sense, by an amount $\Delta\varphi$ per revolution. The period of this "precession" determines the fine structure of the energy levels.

planet, Mercury, is observed to be precessing in the positive sense at the rate of 43 sec of arc per century. The determination of this precession rate is very difficult, because it requires subtraction of much larger disturbing effects of other bodies on the motions of both Earth and Mercury. But the astronomers agree on the result of the analysis, which they believe to be correct to better than 5%.

Applying Eq. (3-67) for $\Delta\varphi$ to Mercury, we find a predicted precession of 7 sec/century. Thus this special relativity effect, resulting from the momentum–velocity–energy relationship ("variation of inertial mass with velocity"), is not large enough to account for the disturbance of the orbit of Mercury that is observed, although it is in the correct sense. General relativity predicts exactly the entire effect observed.

PROBLEMS

The units eV, eV/c, and eV/c^2 used in some of the problems are described at the beginning of the worked exercise of Section 3-8 and more fully in Section 4-2.

3-1. Generalizing the argument of Section 3-2, use the transformation law of the electromagnetic field [Eq. (2-55)] and the fact that the field of a point charge in its rest frame is purely Coulomb

$$\mathscr{E}^\dagger = \frac{er^\dagger}{(r^\dagger)^3} \qquad \mathscr{B}^\dagger = 0$$

(origin $O^\dagger$ at the charge) to calculate the field of a point charge moving with constant velocity **v** in the laboratory. The answer should be

$$\mathscr{E} = \gamma \frac{e(\mathbf{r} - \mathbf{v}t)}{s^3} \qquad \mathscr{B} = \frac{\mathbf{v}}{c} \times \mathscr{E}$$

with

$$s^2 = (\mathbf{r} - \mathbf{v}t)^2 + (\gamma^2 - 1)(r_{\parallel} - vt)^2$$

Sketch curves showing the longitudinal and transverse electric field at a fixed point as a function of time. Compare the small velocity ($v \ll c$) and large velocity ($v \approx c$) cases.

3-2. Use the result of the preceding problem to calculate the force on an initially stationary electron ($e = -e_0$) when a fast charged ($e = \pm ze_0$) particle flies by at "impact parameter" b. In this calculation, you may neglect the displacement and velocity of the electron during the collision as well as the change of velocity of the fast particle.

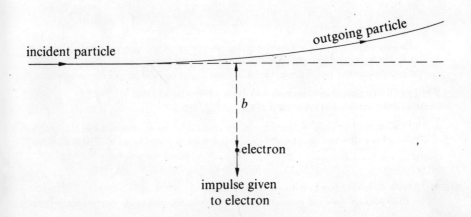

By integrating the force with respect to time, show that the electron acquires transverse momentum of magnitude

$$\frac{2ze_0^2}{bv}$$

Show that a nonrelativistic treatment ($c = \infty$) gives the same result for the impulse given to the electron, although the effective collision duration and the peak transverse force are different.

Note that the energy loss of the fast particle, equal to the kinetic energy acquired by the electron, is proportional to z^2/b^2v^2 and is independent of the particle's proper mass.

3-3. Suppose that in the preceding problem the fast particle is a muon with a magnetic moment in its rest frame of magnitude $m = eh/2\mu c = 4.49 \times 10^{-23}$ erg–G and oriented perpendicular to the plane of the muon orbit and the electron. In the rest frame of the muon, the magnetic dipole field, which can simply be added to the Coulomb field, is

$$\mathscr{B}^\dagger = -\text{grad}\left[\frac{\mathbf{m} \cdot \mathbf{r}^\dagger}{(r^\dagger)^3}\right] \qquad \mathscr{E}^\dagger = 0$$

(origin $O^\dagger$ at the muon). Again using the transformation law Eq. (2-55), determine the electric field in the laboratory, and thence the force and momentum change of the electron. The latter should turn out to be

$$\frac{2me_0}{cb^2}$$

Note that with the given value of m, b has to be of the order of magnitude 10^{-13} cm for the magnetic dipole impulse to be comparable with the Coulomb impulse.

3-4. We showed in Section 3-3 that in reference frame R

$$\mathbf{p} = \mu\mathbf{v}\left(1 - \frac{v^2}{c^2}\right)^{-1/2} \tag{1-33}$$

with μ a constant characterizing the particle. Solving for μ in terms of p and v, verify that μ is invariant:

$$\frac{p}{v(1 - v^2/c^2)^{-1/2}} = \frac{p'}{v'(1 - v'^2/c^2)^{-1/2}}$$

which physicist I is at rest. Rederive the result by requiring conservation of momentum in R', the frame in which physicist II is at rest. The principle of special relativity requires, of course, that the law of conservation of momentum be valid in all inertial frames.

3-6. Derive from the transformation equations of four-momentum [Eq. (3-35)] the transformation equation of speed [first of Eqs. (2-43)].

3-7. In the worked exercise of Section 3-8, we calculated the momentum spectrum in the laboratory of the neutrinos from the decay in flight of a pencil beam of 10.0 GeV/c pions:

$$\pi \to \mu + \nu_\mu$$

Do the corresponding calculation for the muons.

In experiments on the neutrinos, one blocks out the muons by interposing between the decay region and the neutrino target a shield of dense material in which the muons are brought to rest. The stopping process involves a cumulative energy–momentum loss by ionization and excitation of the atoms of the shield. In iron (density $7.9 \, g \, cm^{-3}$), the muon momentum decreases at the average rate $1.4 \times 10^{-2} \, GeV/c \, cm^{-1}$. How many meters of iron shielding are needed to stop all the muons? How far into the iron does the slowest muon penetrate? (The pions that do not decay interact in the first few 0.1 m of iron.)

3-8. In the $\pi \to \mu + \nu_\mu$ disintegration, the velocity of the muon in the pion rest frame is $0.270c$. Using the left subscript position to indicate the reference frame and the left superscript position the particle, we have

$$^{\mu}_{\pi}\beta = 0.270$$

The decay angle $_\pi\theta$ is the angle in the pion rest frame between the muon momentum and the pion momentum (meaning the opposite direction to that of the laboratory motion as seen in the pion frame). The decay angle $_L\theta$ is the angle in the laboratory between the muon momentum and the pion momentum. All values of $_\pi\theta$ are possible.

Show that the value of $_\pi\theta$ giving the largest $_L\theta$ is

$$_\pi\theta(\text{max}) = -\frac{^{\mu}_{\pi}\beta}{^{\pi}_{L}\beta}$$

at which

$$\cot[_L\theta(\text{max})] = {}^{\pi}_{L}\gamma\left[\left(\frac{^{\pi}_{L}\beta}{^{\mu}_{\pi}\beta}\right)^2 - 1\right]^{1/2}$$

Evaluate for $^{\pi}_{L}p = 0.100$, 1.00, 10.0, and 100 GeV/c. What happens at very low pion momentum, say 0.01 GeV/c?

3-9. In an electrostatic velocity separator (Wien filter), a homogeneous vertical electric field

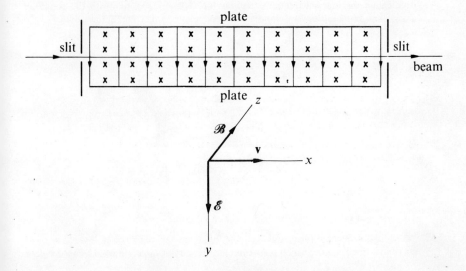

exerts a vertical force on the horizontally moving charged particle, which is balanced by the oppositely directed vertical force exerted by a homogeneous, horizontal, transverse, magnetic field. For particles of the correct velocity, the forces are equal and opposite. These particles are not deflected and thus pass through the slits.

With a 10-m long separator and an electric field $\mathscr{E} = 60 \text{ kV cm}^{-1}$ ($= 200$ Gaussian units of field strength) we wish to select protons of 8.0 GeV/c momentum. What is the velocity of the protons relative to that of light? Give the departure from unity to three significant figures. What magnetic field strength is required? How long does it take a proton to go from one slit to the other?

Consider a "wrong" particle—a pion of the same momentum. What is its velocity relative to that of light? Give the departure from unity to three significant figures. What is its time of flight to the second slit? What is its transverse displacement there? Its transverse momentum? What angle does it make with the proton beam?

3-10. In the preceding problem, transform to the rest frame of a proton, using Eq. (2-55) for the field. What is the electric field? The magnetic field? The force on the proton? The time of passage between the two slits?

Find the velocity of a pion in this frame, the force on it, and its transverse displacement from the first slit to the second.

3-11. In the CERN muon storage ring used for measuring the gyromagnetic ratio (Section 5-4) of the muon [69], a magnet provides a vertical field in a horizontal ring-shaped vacuum tank. Muons of momentum 1.37 GeV/c revolve inside in an orbit of radius 250 cm until they decay (the forward-emitted decay electrons are detected). What is the value of the magnetic field, in gauss? What is the angular velocity of the muons? What is the average number of times that a muon orbits the tank before decaying? (See the Particle Table for muon proper lifetime.)

The same for a muon of less momentum, moving on a circle of radius 247.5 cm.

If muons are injected at an average rate of one every 2×10^{-8} sec, what is the number of muons in the ring after several milliseconds? For how long should injection continue to reach one-half this value?

3-12. Determine the motion of a charged particle in a constant homogeneous electromagnetic field with $\mathscr{E}$ and $\mathscr{B}$ parallel. (See [35], p. 59, for solution.)

Integrate the four differential equations for $x(\tau)$, $y(\tau)$, $z(\tau)$, $t(\tau)$, with the condition $w_\mu^2 = -c^2$. With appropriate choices of axis directions, origin, and initial value of τ,

$$x = \frac{a}{e\mathscr{B}/\mu c} \sin\left(\frac{e\mathscr{B}}{\mu c}\tau\right)$$

$$y = \frac{a}{e\mathscr{B}/\mu c} \cos\left(\frac{e\mathscr{B}}{\mu c}\tau\right)$$

$$z = \frac{(c^2 + a^2)^{1/2}}{e\mathscr{E}/\mu c} \cosh\left(\frac{e\mathscr{E}}{\mu c}\tau\right)$$

$$ct = \frac{(c^2 + a^2)^{1/2}}{e\mathscr{E}/\mu c} \sinh\left(\frac{e\mathscr{E}}{\mu c}\tau\right)$$

The orbit is a spiral around the field direction of radius $a/(e\mathscr{B}/\mu c)$.

3-13. Verify the statement at the end of Section 3-9 that special relativity and Newton's law of gravitation predict a perihelion precession for Mercury of 7 sec/century. The needed data

. are

$$\text{semiaxis major} = 0.387 \times 1.496 \times 10^{13} \text{ cm}$$
$$\text{period} = 88 \text{ days}$$
$$\text{eccentricity} = 0.206$$

You will need to use Kepler's laws to calculate GMm/L.

Chapter 4

Transitions of a System. Conservation of Four-Momentum. The Mass-Energy Relation. Relativistic Kinematics.

4-1. SCATTERING PROCESSES

In the preceding chapter we considered motion of a particle in a weak field. The interaction changed the particle's motion, but the magnitude of its four-momentum, measured by the proper mass, remained constant; the invariant mass was invariable. This is just the situation in Newtonian mechanics: The world is made up of permanent particles, each with a constant "quantity of matter." They approach or recede, but never lose their identity. Their number is an absolute constant. The study of atomic, nuclear, and particle physics has, however, brought to light cases in which interactions result in the disappearance of some particles and the appearance of others. Some examples are:

Radiation by an excited atom. The system initially consists of one excited atom of energy E_i. After the interaction with the radiation field, the system consists of one atom of lower energy E_f, and one photon of energy $E_i - E_f$. The atom in state i was destroyed; the atom in state f and the photon were created.

Nuclear transmutation. For example,

$$\alpha + {}_7N^{14} \rightarrow p + {}_8O^{17}$$

Initially the system consists of an alpha particle and a nitrogen-14 atom, finally of a proton and an oxygen-17 atom. Even though this interaction is really only a rearrangement of neutrons and protons, the proper masses of the two bodies making up the system are different in the initial and final states.

202

Particle production. A typical high energy reaction would be

$$\pi^- + p \to \Lambda^0 + K^0$$

The pion and proton are annihilated; the Λ hyperon and K meson are created.

Relativistic mechanics is capable of describing such phenomena. The treatment of the dynamics must, of course, be quantal, and little can be said classically about what happens during the transition. But before and after, the system consists of independent free particles, which we do know how to describe, and certain conservation laws—such as that of four-momentum—persist through the interaction, relating the final to the initial state.

In this chapter we shall consider such "scattering" processes. Initially, the system consists of one or more free particles; finally, the system consists of one or more free particles. If the final and initial states contain the same number of particles of exactly the same proper mass, charge, spin, and so on, the scattering process is called elastic; otherwise it is termed inelastic.

The creation and destruction of particles renders meaningless the traditional way of distinguishing between simple and complex systems on the basis of number of particles ("dynamics of a particle, dynamics of systems of particles"). A system is characterized by its energy, momentum, angular momentum, electric charge, and so on. It may be, for example, a neutron at one time, a proton, electron, and neutrino at another,

$$n \to p + e^- + \bar{\nu}_\beta$$

The permanence is in its conserved quantities, which move between one partition in the initial state and another one in the final state.

When we describe a system as consisting at a certain time of particles $A, B, C, \ldots$, we must bear in mind that any one of these particles may itself be a complex structure. For example, A may be an atom, consisting of electrons, neutrons, and protons. By treating A as a particle we are assuming that under the circumstances its particle parameters—charge, proper mass, spin, magnetic moment, and so forth—are constant in time. Because of the fact of *quantization*, these parameters never change gradually but only brusquely, in transitions that can be described as destruction of one particle with creation of others. In the weak field of a mass spectrometer a $C^{12}(H^1)_4^+$ ion moves as a whole like a particle of constant charge e_0 and constant proper mass 16 amu. It could not be described so simply if it were in an intense laser beam. Along the same lines, our formalism does not foreclose the possibility that particles that presently appear to be simple (such as the electron) may turn out to have structure.

4-2. RELATIVISTIC UNITS

Before proceeding, let us introduce a system of units that will save writing c's.

In nonrelativistic mechanics there is no intrinsic scale, and nothing about the theory favors one set of units over another. In special relativity, on the other hand, the speed of light in vacuum c is built into the formulas, playing the role of a limit on all velocities. It is natural and convenient to measure velocity in terms of this limiting value—that is, to use units in which $c = 1$. In such a system, the ratio of the units of length and time is fixed. If, for example, the unit of length is the centimeter, the unit of time is the time required for light to travel one centimeter in vacuum. If the unit of time is the year, the unit of length is the light-year.

With such *relativistic units*, c drops out of the formulas. The coordinates of an event are

$$x_\mu = (x, y, z, it) \qquad \mu = 1, 2, 3, 4 \tag{2-48a}$$

$$x_0 = t \tag{2-48b}$$

and for the square of the four-interval between two events [Eq. (2-8)], we have

$$\Delta \tilde{x}^2 = \Delta r^2 - \Delta t^2 = -\Delta s^2$$

The four-velocity has the components

$$w_\mu = (\gamma v_x, \gamma v_y, \gamma v_z, i\gamma) \qquad \mu = 1, 2, 3, 4 \tag{2-60a,b}$$

$$w_0 = \gamma \tag{2-60bb}$$

with

$$\gamma = (1 - v^2)^{-1/2}$$

and the four-momentum has the components

$$p_\mu = (p_x, p_y, p_z, iE) \qquad \mu = 1, 2, 3, 4 \tag{3-22}$$

$$p_0 = E$$

with

$$\mathbf{p} = \mu \gamma \mathbf{v} \tag{1-33}$$

$$E = \mu \gamma \tag{3-20}$$

The invariant length squared of the four-momentum is

$$\tilde{p}^2 = p^2 - E^2 = -\mu^2 \qquad (3\text{-}24)$$

For dimensional checks (invariance with respect to specific choice of units), one merely treats x, y, z, t as having the same dimension, and similarly μ, p, E.

A widely used relativistic system, which we shall call the "practical" one, has the quantum of electric charge incorporated in its definitions. The reader is already familiar with it from the worked exercise of Section 3-8. It uses the electron volt eV (or a decimal multiple thereof—10^0 for atomic, 10^6 for nuclear, 10^9 for particle physics) for energy, momentum, and inertial mass. One electron volt is the energy acquired by a particle of electronic charge e_0 in falling through a potential difference of 1 V $[= (10^8/c)$ Gaussian units of potential difference]. Denoting the numerical value of a physical quantity by its algebraic symbol with the unit attached as subscript, we have

$$(e_0)_{esu} = 4.803 \times 10^{-10}$$

$$(c)_{cm/sec} = 2.9979 \times 10^{10}$$

and, therefore, the unit of energy

$$1\text{ eV} = \frac{(e_0)_{esu} \cdot 1}{(c)_{cm/sec} \times 10^{-8}}\text{ erg} = 1.602 \times 10^{-12}\text{ erg}$$

Thus, an energy E satisfies

$$E_{eV} = \frac{(c)_{cm/sec} \times 10^{-8}}{(e_0)_{esu}} E_{erg} = (1.602 \times 10^{-12})^{-1} E_{erg} \qquad (4\text{-}1a)$$

We measure momentum p by evaluating cp in electron volts. The unit is called "eV/c."

$$p_{eV/c} = (cp)_{eV} = \frac{(c)_{cm/sec} \times 10^{-8}}{(e_0)_{esu}}(cp)_{erg} \qquad (4\text{-}1b)$$

Inertial mass (μ or m) is measured by evaluating $c^2 \times (\mu$ or $m)$ in electron volts. The unit is called "eV/c^2."

$$\mu_{eV/c^2} = (c^2\mu)_{eV} = \frac{(c)_{cm/sec} \times 10^{-8}}{(e_0)_{esu}}(c^2\mu)_{erg} \qquad (4\text{-}1c)$$

The covariant equation

$$[E_{erg}]^2 = [(cp)_{erg}]^2 + [(c^2\mu)_{erg}]^2 \tag{3-27a}$$

becomes

$$[E_{eV}]^2 = [p_{eV/c}]^2 + [\mu_{eV/c^2}]^2 \tag{4-2}$$

confirming the elimination of c from the equations in these units. The rather pedantic "/c" or "/c²" in the names of the units serve the useful function of telling the reader that the quantity under discussion is not an energy but a momentum or a mass.

If in the practical system one chooses to measure length in centimeters, the unit of time (the "shake" or "cm/c") has such a size that

$$t_{cm/c} = (ct)_{cm} = (c)_{cm/sec}t_{sec} = \text{"3"} \times 10^{10}t_{sec} \tag{4-3}$$

where "3" stands for the measured number 2.997925. For the motion of a particle of charge ze_0 in an electromagnetic field $(\mathscr{E}, \mathscr{B})$, the cgs–Gaussian law of motion [Eqs. (D-1) and (1-5)]

$$\frac{d\mathbf{p}_{g\,cm/sec}}{dt_{sec}} = (ze_0)_{esu}\left(\mathscr{E}_{esu} + \frac{\mathbf{v}_{cm/sec}}{c_{cm/sec}} \times \mathscr{B}_G\right)$$

becomes, by Eqs. (4-1b) and (4-3),

$$\frac{d\mathbf{p}_{eV/c}}{dt_{cm/c}} = (c)_{cm/sec} \times 10^{-8} \times z(\mathscr{E}_{esu} + \boldsymbol{\beta} \times \mathscr{B}_G)$$

$$= \text{"300"}z(\mathscr{E}_{esu} + \boldsymbol{\beta} \times \mathscr{B}_G) \tag{4-4a}$$

where "300" is a symbol for the experimental constant $(c)_{cm/sec} \times 10^{-8} = 299.79$ and $\boldsymbol{\beta}$ is the particle velocity normalized to that of light. This equation simplifies finally to the practical equation

$$\frac{d\mathbf{p}_{eV/c}}{dt_{cm/c}} = z(\mathscr{E}_{V/cm} + \boldsymbol{\beta} \times \text{"300"}\mathscr{B}_G) \tag{4-4b}$$

The very useful equation for the curvature in a magnetic field

$$p_{g\,cm/sec}\cos\lambda = \frac{(ze_0)_{esu}}{(c)_{cm/sec}}\mathscr{B}_G\rho_{cm} \tag{3-55b}$$

becomes

$$p_{\mathrm{eV/c}} \cos \lambda = z\text{``300''}\mathscr{B}_{\mathrm{G}} \rho_{\mathrm{cm}}$$

The equations of this book are statements about numbers. It is usually not necessary to attach subscripts indicating the units used, as we have done in this paragraph. The presence or absence of c in a particular unsubscripted equation tells whether traditional length–mass–time (LMT) units or relativistic units are implied. If an equation is written for the latter, it can be converted to the former by inserting such c's as are needed to make the dimensions consistent (c has dimensions L/T). Thus,

$$\mathbf{p} = E\mathbf{v} \qquad \text{(relativistic units)} \tag{1-3b}$$

corresponds to

$$\mathbf{p} = \left(\frac{E}{c^2}\right)\mathbf{v} \qquad \text{(LMT)} \tag{1-3b}$$

or

$$c\mathbf{p} = E\frac{\mathbf{v}}{c} \qquad \text{(LMT)} \tag{1-3b}$$

or

$$\frac{\mathbf{p}}{c} = \left(\frac{E}{c^2}\right)\frac{\mathbf{v}}{c} \qquad \text{(LMT)} \tag{1-3b}$$

One replaces t by ct, v by v/c, p by cp, and μ (or m) by $c^2\mu$ (or $c^2 m$). The electric and magnetic units depend on the mechanical units. If one has held to Gaussian units while changing the mechanical units, the change from practical relativistic to cgs–Gaussian units requires on the right-hand side of Eq. (4-4a) only replacement of "300" by e_0.

4-3. FOUR-MOMENTUM OF A COMPLEX SYSTEM. INERTIA OF ENERGY

We need to develop the formalism for treating a complex system as a single particle. We shall use capital letters for dynamical variables of the aggregation and lowercase letters for those of the parts. The left superscript position will be used for labeling.

The *momentum* of the system is the vector sum of the momenta of the component parts:

$$\mathbf{P} = \sum_i {}^i\mathbf{p} \tag{4-6a}$$

If the system is a collection of free particles, the summation in Eq. (4-6a) is over the momenta of the particles. If the particles are not moving uniformly, it is necessary for comparison with experiment to include in the sum the momenta of the fields through which they interact. The momentum of an atom includes the momentum of the electromagnetic field coupling the nucleus and electrons, as well as the momenta of these particles alone. Furthermore, the nucleus is itself a complex system, and its momentum is the sum of the momenta of the nucleons that make it up and of the fields, hadronic and electromagnetic, that couple them. Moreover, the nucleons themselves are complex, and their momentum is the sum of momenta of a core and mesonic clouds and of the fields through which they interact. In this endless regress it is not possible to make a clear-cut distinction between particle momentum and field momentum; one can only observe the total momentum of an experimentally identifiable system. We make the physical assumption that this total momentum can be written as a sum of terms having the transformation properties of particle momentum. Einstein ([31], pp. 444-445) showed this to hold for electromagnetic field momentum. The experimental confirmations of the formulas to be derived from Eq. (4-6a) indicate that the fields involved in nuclear and particle binding also have this property.

The *four-momentum* $\tilde{P}$ of the system is that four-vector whose spacelike component is the momentum of the system. The special Lorentz transformation (SLT) of four-vectors [Eqs. (2-49bb)] reads

$$\begin{aligned}
P_0' &= \gamma(P_0 - \beta P_1) \\
P_1' &= \gamma(P_1 - \beta P_0) \\
P_2' &= P_2 \\
P_3' &= P_3
\end{aligned} \tag{4-7a}$$

whereas, of course, for each term in the sum,

$$\begin{aligned}
{}^ip_0' &= \gamma({}^ip_0 - \beta\,{}^ip_1) \\
{}^ip_1' &= \gamma({}^ip_1 - \beta\,{}^ip_0) \\
{}^ip_2' &= {}^ip_2 \\
{}^ip_3' &= {}^ip_3
\end{aligned} \tag{4-7b}$$

Adding the Eqs. (4-7b) for the different constituents, and comparing with Eqs. (4-7a), we see that Eq. (4-6a) requires

$$P_0 = \sum_i {}^i p_0 \tag{4-6b}$$

and that therefore,

$$\tilde{P} = \sum_i {}^i \tilde{p} \tag{4-6c}$$

We saw [Eqs. (3-21)] that p_0 is the energy of the particle

$$p_0 = e \tag{3-21}$$

We call P_0 the *energy* of the system

$$E = P_0 \tag{4-8}$$

We shall see that when the system interacts with the outside world, the decrease of P_0 is indeed equal to the increase of energy of the outside world. Equation (4-6b) is now

$$E = \sum_i {}^i e \tag{4-6bb}$$

Again, we note that a simple summation over the particles is only meaningful if they are free. In general, the energies of the coupling fields must be included in the sum.

The center of mass frame, or center of momentum frame, or *rest frame of the system* as a whole, R^*, is defined as that reference frame in which the momentum of the system is zero.

$$P^* = 0 \tag{4-9}$$

That such a frame exists follows from the timelike character of P [Eq. 4-6c]. While the system as a whole is at rest in this frame, the constituent particles may be in violent motion. We call the velocity of this frame the velocity V of the system. To find V, apply a SLT from R to R' with $\mathbf{u}_R(R') = \boldsymbol{\beta}$ parallel to **P**. Equations (4-6) and (4-7) give

$$E' = \sum_i {}^ie' = \sum_i \gamma({}^ie - \beta\, {}^ip_1) = \gamma \sum_i {}^ie - \gamma\beta \sum_i {}^ip_1$$

$$= \gamma(E - \beta P_1) = \gamma(E - \beta P)$$

$$P'_1 = \sum_i {}^ip'_1 = \sum_i \gamma({}^ip_1 - \beta\, {}^ie) = \gamma \sum_i {}^ip_1 - \gamma\beta \sum_i {}^ie$$

$$= \gamma(P_1 - \beta E) = \gamma(P - \beta E)$$

$$P'_2 = \sum_i {}^ip'_2 = \sum_i {}^ip_2 = P_2 = 0$$

$$P'_3 = \sum_i {}^ip'_3 = \sum_i {}^ip_3 = P_3 = 0$$

or, written in a more compact form,

$$E' = \gamma(E - \beta P)$$
$$P'_\parallel = \gamma(P - \beta E) \qquad\qquad (4\text{-}10)$$
$$P'_\perp = 0$$

For P' to be zero [Eq. (4-9)], it is necessary and sufficient that

$$P - \beta E = 0$$

We see that the velocity of the system ($V = \beta$) is its momentum divided by its energy

$$\mathbf{V} = \frac{\mathbf{P}}{E} \qquad\qquad (4\text{-}11)$$

This equation expresses the famous equivalence of energy and inertial mass. Writing Eq. (4-11) as

$$\boxed{\mathbf{P} = E\mathbf{V}} \qquad\qquad (4\text{-}11)$$

we see that the inertial mass, defined as P/V, equals the energy.

$$M = \frac{P}{V} = E \qquad\qquad (4\text{-}12)$$

In LMT units one has here

$$E = Mc^2 \quad \text{(LMT)} \tag{4-12}$$

The import of the equivalence is that *all* kinds of energy $^i e$—rest energy, kinetic energy, field energy—contribute to the inertial mass of the system on an equal basis. Inertia is a general attribute of energy. Energy in all its forms, with due regard to sign, contributes to inertial mass. A stretched or compressed spring has greater inertial mass than when it is undistorted. Warming up a body increases its inertial mass. A metallic cavity has inertial mass from the energy of the electromagnetic field that it embraces. An atom that falls to a level of lower energy has less inertial mass than before.

When we speak of inertia, we mean momentum as related to velocity. The energy–inertial mass equivalence is really the momentum–energy–velocity relationship of Eq. (4-11). Our derivation shows it to be a direct mathematical consequence of the fact that momentum and energy are components of one timelike four-vector.

Inertial mass does not have an arbitrary additive constant. It is determined by the measurable quantities P and V. Correspondingly, there must be (a formal proof on other grounds has been given in Appendix E) a definite zero level of energy without arbitrariness. This is unlike nonrelativistic mechanics, where only changes of energy can be measured. Relativity provides an absolute zero of energy corresponding to the absence of particles and fields. When the field energy can be written as potential energy (as for a hydrogenic atom considered in the frame in which the nucleus is at rest), the zero level of potential energy must be taken to correspond to negligible mutual interaction—that is, infinite separation.

For a real system, momentum and velocity point the same way; the inertial mass is positive. Thus the energy is positive in every inertial frame. This is a constraint on the interactions occurring in nature. It is satisfied by electromagnetism, where the energy density

$$\frac{1}{8\pi}(\mathscr{E}^2 + \mathscr{B}^2)$$

is positive definite. It will have to be satisfied in the future theory of elementary particles and fields.

Energy and inertial mass are different physical quantities, but they go together, with the simple *universal* proportionality relation

$$E = M \tag{4-12}$$

This relation permits us to dispense with one or the other in our formalism. We can specify the timelike component of four-momentum by giving either the energy or the inertial mass. We are familiar since Galileo with a similar situation as between inertial mass and gravitational mass. So long as the universal proportionality of the two is accepted, it is pedantic to distinguish between them in our terminology. The same "10-kg" body can be specified equally well by the statement that in a particular gravitational field the force on it is 10 times that on the standard body ("it weighs 10 kg") and by the statement that its inertial mass is 10 times that of the standard body ("its mass is 10 kg"). Similarly, the same information about a proton is conveyed by the statement that its energy in the laboratory frame is 8.00 GeV as by the statement that its inertial mass in the laboratory frame is 8.00 GeV/c^2 or 14.3×10^{-24} g.

The invariant aspect of the four-momentum is specified[1] by the *proper mass* or *invariant mass*, $^s\mu$. Defined as the inertial mass in the rest frame of the system, it is, by Eq. (4-12), equal to the *energy in the rest frame*

$$^s\mu = E^* \qquad (4-13)$$

This is essentially the invariant length of the four-momentum, since

$$\tilde{P}^2 = P^2 - E^2 = P^{*2} - E^{*2} = -E^{*2}$$

Thus,

$$P^2 - E^2 = -(^s\mu)^2 \qquad (4-14)$$

corresponding exactly to Eq. (3-27a) for a particle. The same right-triangle mnemonic (Fig. 3-5) can be used, with E the hypotenuse, P and $^s\mu$ the sides. Again, the square of a particle's four-momentum can be specified equally well by stating either that its rest energy is 938.256 MeV or that its proper mass is 938.256 MeV/c^2 or 1.67252×10^{-24} g.

For the transformation from R^* to R, Eqs. (4-10) give

$$E = \gamma(E^* + \beta P^*) = \gamma E^*$$
$$P = \gamma(P^* + \beta E^*) = \gamma \beta E^* \qquad (4-15)$$

with

$$\beta = \frac{P}{E} \quad \text{and} \quad \gamma = (1 - \beta^2)^{-1/2}$$

---

[1] Inasmuch as capital μ is indistinguishable from capital m, we use $^s\mu$ for the proper mass of the system.

Hence,

$$\boxed{\begin{aligned} \mathbf{P} &= {}^s\mu\gamma\mathbf{V} = {}^s\mu(1 - V^2)^{-1/2}\mathbf{V} \\ E &= {}^s\mu\gamma = {}^s\mu(1 - V^2)^{-1/2} \end{aligned}} \qquad (4\text{-}16)$$

These equations correspond exactly to Eqs. (1-33) and (3-20) for a particle. In terms of the four-velocity $\tilde{W}$ of the system,

$$\tilde{W} = (\gamma\mathbf{V}, i\gamma) = \{\gamma, \gamma\mathbf{V}\}$$

the Eqs. (4-16) are

$$\tilde{P} = {}^s\mu\tilde{W} \qquad (4\text{-}17)$$

Evidently a requirement that an external field must satisfy to be considered "weak" is that it leave E^* constant as it accelerates the system.

When the system is observed as a set of separated fragments, the invariant mass, or energy of the system in its rest frame, is determined by measuring in any frame the momenta and energies of the parts, and using

$$\mathbf{P} = \sum_i {}^i\mathbf{p} \qquad (4\text{-}6a)$$

$$E = \sum_i {}^ie \qquad (4\text{-}6bb)$$

in

$$({}^s\mu)^2 = E^2 - P^2 \qquad (4\text{-}14)$$

For example, in the scattering process shown in Fig. 1-4, the invariant mass of the system consisting of positive pion (2), proton (3), and negative pion (5) is given by

$$\begin{aligned} ({}^{235}\mu)^2 &= ({}^2e + {}^3e + {}^5e)^2 - ({}^2\mathbf{p} + {}^3\mathbf{p} + {}^5\mathbf{p})^2 \\ &= (7.82)^2 - (7.46)^2 \\ &= (2.31)^2 \end{aligned}$$

or

$$^{235}\mu = 2.31 \text{ GeV/c}^2$$

For a bound system, such as an atom or a nucleus, one finds the invariant mass in the same way as for a particle, by determining two of the three quantities P, E, V and using Eqs. (4-14) or (4-16). The experimental techniques available for use on a system of charge q are electric acceleration (Section 3-9, worked exercise), giving $K/q = (E - \mu)/q$; magnetic deflection [Eqs. (2-78)], giving P/q; electric deflection [Eqs. 2-78)] giving PV/q; and time of flight or Cerenkov radiation angle, giving V. Note that every combination involves some acceleration of the system, showing the kinship of the modern mass spectrometers with the macroscopic mass measuring systems ($m = F/a$) demonstrated in elementary physics lectures.

4-4. CONSERVATION OF FOUR-MOMENTUM

In relativistic mechanics, energy and momentum are tied together as components of the four-momentum, and one cannot be conserved without the other.

Consider a transition of a system. Let us again use the right superscript position to identify initial (in) and final (fin) values. The change in four-momentum

$$\Delta \tilde{P} = \tilde{P}^{\text{fin}} - \tilde{P}^{\text{in}} \tag{4-18a}$$

has the energy component in R

$$\Delta E = E^{\text{fin}} - E^{\text{in}} \tag{4-18b}$$

and the momentum component in R

$$\Delta \mathbf{P} = \mathbf{P}^{\text{fin}} - \mathbf{P}^{\text{in}} \tag{4-18c}$$

Conservation of momentum in R means that

$$\Delta \mathbf{P} = 0 \tag{4-19}$$

Conservation of energy in R means that

$$\Delta E = 0 \tag{4-20}$$

We shall prove that Eq. (4-19) implies Eq. (4-20), and conversely.

Consider the components of $\Delta \tilde{P}$ in R and in another frame R' moving with some constant velocity $\boldsymbol{\beta}$ relative to R. Since $\boldsymbol{\beta}$ is a constant, the same transformation law holds for $\Delta \tilde{P}$ as for $\tilde{P}$. We write Eqs. (4-7a) in more compact form

$$E' = \gamma(E - \beta P_{\parallel})$$
$$P'_{\parallel} = \gamma(P_{\parallel} - \beta E) \qquad (4\text{-}21)$$
$$P'_{\perp} = P_{\perp}$$

so that

$$\Delta E' = \gamma\,(\Delta E - \beta\,\Delta P_{\parallel})$$
$$\Delta P'_{\parallel} = \gamma\,(\Delta P_{\parallel} - \beta\,\Delta E) \qquad (4\text{-}22)$$
$$\Delta P'_{\perp} = \Delta P_{\perp}$$

If momentum is conserved in one inertial frame, it must, according to the principle of special relativity, be conserved in every other: $\Delta P = 0$ implies that $\Delta P' = 0$. Equations (4-22) become, in this case,

$$\Delta E' = \gamma\,\Delta E$$
$$0 = -\gamma\beta\,\Delta E$$

That is,

$$\Delta E = \Delta E' = 0$$

Energy is conserved in all inertial frames if momentum is. Conversely, if energy is conserved ($\Delta E = \Delta E' = 0$), Eqs. (4-22) become

$$0 = -\gamma\beta\,\Delta P_{\parallel}$$
$$\Delta P'_{\parallel} = \gamma\,\Delta P_{\parallel}$$
$$\Delta P'_{\perp} = \Delta P_{\perp}$$

giving

$$\Delta P_{\parallel} = 0 = \Delta P'_{\parallel}$$

Since the transformation velocity $\boldsymbol{\beta}$ can be given any direction, the last equations require that

$$\Delta P = 0 = \Delta P'$$

Momentum is conserved in all inertial frames if energy is. Thus, conservation of energy and momentum are unbreakably bound together. One cannot be conserved and not the other. The laws of conservation of momentum and conservation of energy are one grand law of conservation of four-momentum,

$$\boxed{\Delta \tilde{P} = 0} \tag{4-23}$$

valid for any system free of external influence.

It is different in nonrelativistic mechanics, where momentum is always conserved but energy is only conserved in those special collisions called "elastic." An example of elastic scattering is provided by two steel balls, bouncing apart with the same total kinetic energy as before the collision. If, however, one of the balls is plastic, there is a loss of macroscopic kinetic energy, the collision is "inelastic." Total energy is, of course, conserved, but some of it disappears from the mechanical budget by going into internal degrees of freedom as heat and potential energy of deformation. It is assumed nonrelativistically that this internal energy has no effect on the momentum–velocity relation of the ball as a whole,

$$\mathbf{P} = \mu \mathbf{V}$$

with μ the same before and after the deformation. Relativistically, on the other hand, it is the total energy of the ball that enters in its momentum–velocity relation

$$\mathbf{P} = E \mathbf{V} \tag{4-21}$$

so that we cannot leave the "dissipated" energy out of the mechanical reckoning. Present experimental technique does not permit comparison of the two theories at the level of colliding balls. Nuclear and particle physics experiments have, however, established the correctness of the relativistic formulas. We discuss these matters in greater detail in the next section.

The law of conservation of four-momentum is a pillar of contemporary physics. It appears to hold exactly without exception. For example, an apparent violation in radioactive beta decay led to the speculative hypothesis (Pauli, 1930) of an undetected neutral particle of small proper mass, later named "neutrino." Twenty-three years later the reality of the neutrino was established beyond any doubt by the detection of a reaction produced by it [46].

4-5. REST ENERGY. NONCONSERVATION OF PROPER MASS

Nonrelativistic particle mechanics is not cognizant of the energy associated with proper mass. In the small-velocity, weak-field range of phenomena on which it is based, rest energy is never converted into other forms and thus it escapes detection. In astro-, nuclear, and particle physics, on the other hand, there are numerous examples of conversion of rest energy into other forms. In nature there is free trade between all the additive terms in the energy: particle rest energies $^i\mu$; particle kinetic energies $(^ie - {}^i\mu)$; and field (including potential) energies ^je

$$E = \sum_{\substack{i \\ \text{particles}}} {}^i\mu + \sum_{\substack{i \\ \text{particles}}} (^ie - {}^i\mu) + \sum_{\substack{j \\ \text{fields}}} {}^je \qquad (4\text{-}24)$$

Inelastic transitions can occur in which the partition of E among these terms changes, and, in particular, the sum of the proper masses in the final state is different from its value in the initial state. The journalese description of such processes as "conversion of mass into energy" is misleading; an accurate statement would be "conversion of rest energy into other forms of energy." Energy and inertial mass go hand-in-hand, changing together. One does not get converted into the other. Always

$$e = m \qquad E = M \qquad\qquad \text{(1-3a)} \quad (4\text{-}12)$$

and

$$e^* = \mu \qquad E^* = {}^s\mu \qquad\qquad \text{(3-26b)} \quad (4\text{-}13)$$

These processes rather exemplify the additive character of the terms making up the energy in Eq. (4-24). If E is constant, a decrease of the total proper mass is balanced by a compensating increase in the total kinetic and field energies.

$$\Delta E = \Delta\left(\sum_i {}^i\mu\right) + \Delta\left[\sum_i (^ie - {}^i\mu)\right] + \Delta\left(\sum_i {}^je\right) \qquad (4\text{-}25)$$

The test of Einstein's prediction of the presence of rest energy terms in the sum of Eq. (4-25) ([47]; see also [31]) had to wait upon the observation of transitions in which the different terms of Eq. (4-25) are measurable with comparable precision. The reactions of atomic chemistry are not suitable. For example, the dissociation energy of the H_2 molecule, 4.48 eV, corresponds to only 4.8×10^{-9} g in 2.016. When finally reactions of light nuclei, with their precisely measurable proper masses, became available for study in the laboratory [48], the prediction was tested ([48]; [49]) on

$$p + {_3}\text{Li}^7 \to \alpha + \alpha$$

The loss of proper mass, determined from mass spectrograph measurements on hydrogen, helium, and lithium ions, was found to be 0.0181 ± 0.0006 amu, corresponding to 16.9 ± 0.6 MeV of energy. The gain of kinetic energy, determined from range measurements on the alpha particles, was 17.0 MeV. Subsequently, comparisons of reaction kinetic energies with proper mass measurements over chains of nuclear reactions have confirmed the Einstein relation with considerably higher precision. The development of "atomic" bombs and reactors has given us examples of conversion of rest energy into kinetic energy in gross amounts, and made "eeh equals em-see-squared" a household phrase. It is likely that nuclear reactions in the core of the sun are responsible for the radiant energy without which our physics would not exist.

An extreme example of nonconservation of proper mass is provided by positron annihilation. It is the inelastic scattering of a negative and positive electron, resulting in a pair of photons

$$e^- + e^+ \to \gamma + \gamma$$

Energy and momentum are conserved in the transition, but the total proper mass changes from 1.022 MeV/c^2 to zero. The destruction of rest energy is compensated by the creation of the energy of the photons. Another example of this type is the transition of the metastable neutral pion into two photons

$$\pi^0 \to \gamma + \gamma$$

where the amount of energy converted from "material" to radiant form is 135 MeV in the rest frame of the pion.

4-6. RADIATIVE TRANSITIONS

Radiative transitions involve the emission or absorption of photons. A photon is a packet of electromagnetic radiation. It has field energy ε and field momentum π, related by

$$\pi = \frac{\varepsilon}{c} \qquad \text{(LMT units)}$$

$$\pi = \varepsilon \qquad \text{(relativistic units)}$$

$$(4\text{-}26)$$

Thus the photon's four-momentum $\tilde{\pi}$ has invariant length zero, corresponding to zero proper mass.

$$\tilde{\pi}^2 = 0 \qquad (4\text{-}27)$$

Suppose an atom of four-momentum $\tilde{P}^{\text{in}}$ emits a photon. The process is represented diagramatically at the top of Fig. 4-1. The final state consists of the daughter atom of four-momentum $\tilde{P}^{\text{fin}}$ and the photon of four-momentum $\tilde{\pi}$. Conservation of four-momentum requires

$$\tilde{P}^{\text{in}} = \tilde{P}^{\text{fin}} + \tilde{\pi} \qquad (4\text{-}28a)$$

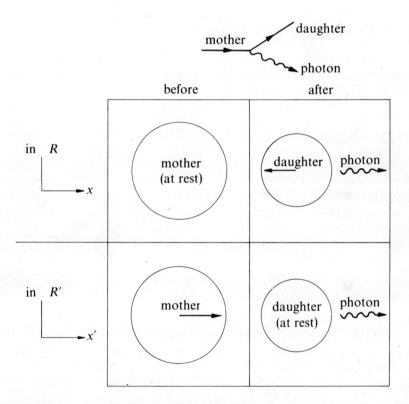

Figure 4-1.   A radiative transition. As indicated diagramatically at the top of the figure, the initial system ("mother") gets replaced by the two-body final system ("daughter" and "photon"). The sketches in the four boxes show the situation before and after, in the rest frame of the mother (R) and the rest frame of the daughter (R').

Thus, the daughter atom has less energy than the mother atom

$$E^{\text{fin}} = E^{\text{in}} - \varepsilon \qquad (4\text{-}28\text{b})$$

and it has a recoil momentum

$$\mathbf{P}^{\text{fin}} - \mathbf{P}^{\text{in}} = -\pi \qquad (4\text{-}28\text{c})$$

Let R (Fig. 4-1) be the rest frame of the mother atom (the entire system). Lay the x axis along the direction of motion of the photon. In this coordinate system, the four-momentum components are, by Eqs. (4-28),

$$\{E^{\text{in}} = \mu^{\text{in}}, 0, 0, 0\} \qquad \text{(mother atom)}$$

$$\{E^{\text{fin}} = \mu^{\text{in}} - \varepsilon, -\varepsilon, 0, 0\} \qquad \text{(daughter atom)} \qquad (4\text{-}29)$$

$$\{\varepsilon, \varepsilon, 0, 0\} \qquad \text{(photon)}$$

Let R' (Fig. 4-1) be the reference frame in which the daughter atom is at rest ($P^{\text{fin}'} = 0$). Its velocity with respect to R is, by Eq. (4-21),

$$\beta = \frac{P_1^{\text{fin}}}{E^{\text{fin}}} = \frac{-\varepsilon}{\mu^{\text{in}} - \varepsilon} \qquad (4\text{-}30)$$

in the x direction. From the transformation Eq. (4-21), the four-momentum components in the coordinate system in R' with axes parallel to the x, y, and z axes in R, are

$$\left\{ \frac{(\mu^{\text{in}})^2 - \mu^{\text{in}}\varepsilon}{[(\mu^{\text{in}})^2 - 2(\mu^{\text{in}})\varepsilon]^{1/2}}, \frac{\varepsilon\mu^{\text{in}}}{[(\mu^{\text{in}})^2 - 2(\mu^{\text{in}})\varepsilon]^{1/2}}, 0, 0 \right\} \qquad \text{(mother atom)}$$

$$\{[(\mu^{\text{in}})^2 - 2(\mu^{\text{in}})\varepsilon]^{1/2}, 0, 0, 0\} \qquad \text{(daughter atom)} \qquad (4\text{-}31)$$

$$\left\{ \frac{\mu^{\text{in}}\varepsilon}{[(\mu^{\text{in}})^2 - 2(\mu^{\text{in}})\varepsilon]^{1/2}}, \frac{\varepsilon\mu^{\text{in}}}{[(\mu^{\text{in}})^2 - 2(\mu^{\text{in}})\varepsilon]^{1/2}}, 0, 0 \right\} \qquad \text{(photon)}$$

Four-momentum is evidently conserved in R' as well as in R. The proper mass of the daughter atom μ^{fin} is its energy in its rest frame

$$E^{\text{fin}'} = [(\mu^{\text{in}})^2 - 2(\mu^{\text{in}})\varepsilon]^{1/2}$$

that is,

$$\mu^{\text{fin}} = \mu^{\text{in}} \left(1 - \frac{2\varepsilon}{\mu^{\text{in}}} \right)^{1/2} \qquad (4\text{-}32)$$

The proper mass of the daughter atom is less than that of the mother. The radiative transition from a higher to a lower state ($\varepsilon > 0$) is associated with a decrease of proper mass. If $\varepsilon \ll \mu$ as in the atomic case (eV's versus GeV's), expansion of Eq. (4-32) gives

$$\mu^{\text{fin}} \approx \mu^{\text{in}} - \varepsilon$$

The loss of proper mass is approximately equal to the radiated energy in the original rest frame. Equation (4-32) can also be written as

$$\frac{\mu^{\text{fin}}}{\mu^{\text{in}}} = \frac{\varepsilon}{\varepsilon'} \quad \qquad (4\text{-}33)$$

ε and ε' being the energy of the photon in the rest frame of the initial and final atom, respectively. In effect [Eq. (4-31)],

$$\varepsilon' = \varepsilon \frac{\mu^{\text{in}}}{[(\mu^{\text{in}})^2 - 2(\mu^{\text{in}})\varepsilon]^{1/2}}$$

Evidently $\varepsilon' > \varepsilon$; the daughter atom recoils opposite to the photon.

In R, the inertial mass of the mother atom is

$$M^{\text{in}} = E^{\text{in}} = \mu^{\text{in}}$$

whereas that of the daughter is [using Eq. (4-30)]

$$M^{\text{fin}} = \frac{P^{\text{fin}}}{\beta} = \frac{-\varepsilon}{-\varepsilon/(\mu^{\text{in}} - \varepsilon)} = \mu^{\text{in}} - \varepsilon$$

Thus,

$$M^{\text{fin}} = M^{\text{in}} - \varepsilon$$

or $\qquad\qquad\qquad\qquad\qquad\qquad\qquad\qquad\qquad\qquad\qquad$ (4-34a)

$$M^{\text{fin}} = M^{\text{in}} - \frac{\varepsilon}{c^2} \quad \text{(LMT)}$$

The inertial mass of the atom is decreased by the mass equivalent of the energy radiated. This is merely the statement that in R the timelike component of the four-momentum of the system is conserved in the transition (mother) $\rightarrow$ (daughter + photon). To bring out its identity with Eq. (4-25), we can write Eq. (4-34a) as

$$0 = M^{fin} - M^{in} + \varepsilon$$
$$= (\mu^{fin} - \mu^{in}) + K^{fin} + \varepsilon$$

The proper mass of the atom is reduced by the kinetic energy of recoil and the photon energy, in the rest frame of the initial state.

As a further check, we note that the timelike component of the four-momentum is also conserved in R'. The inertial mass of the mother atom is [using Eq. (4-30)]

$$M^{in'} = \frac{P^{in'}}{-\beta} = \frac{\mu^{in}(\mu^{in} - \varepsilon)}{[(\mu^{in})^2 - 2(\mu^{in})\varepsilon]^{1/2}}$$

whereas that of the daughter atom is [Eq. (4-31)]

$$M^{fin'} = E^{fin'} = [(\mu^{in})^2 - 2(\mu^{in})\varepsilon]^{1/2}$$

Thus,

$$M^{in'} - M^{fin'} = \frac{\mu^{in}\varepsilon}{[(\mu^{in})^2 - 2(\mu^{in})\varepsilon]^{1/2}} = \varepsilon'$$

or   (4-34b)

$$M^{in'} - M^{fin'} = \frac{\varepsilon'}{c^2} \quad \text{(LMT)}$$

In R', too, the inertial mass of the atom is diminished by the mass equivalent of the energy radiated. In the form of Eq. (4-25),

$$0 = (\mu^{fin} - \mu^{in}) - K^{in'} + \varepsilon'$$

The proper mass of the photon is zero. It is evident from Eq. (4-33) that the sum of the proper masses is not conserved in the transition.

It is also possible to arrive at Eqs. (4-32) and (4-33) from Eq. (4-28a) by simply evaluating the square of the conserved four-momentum. Squaring both sides of Eq. (4-28a),   4-28a ; $\tilde{P}_{in} = \tilde{P}_{in} + \tilde{\pi}$

$$\tilde{P}^{in2} = \tilde{P}^{fin2} + \tilde{\pi}^2 + 2\tilde{P}^{fin}\tilde{\pi}$$

4-14 : $P^2 \cdot E^2 = -(\mu)^2$

or, by Eqs. (4-14) and (4-27),   4-27 ; $\tilde{\pi}^2 = 0$

$$-\mu^{in2} = -\mu^{fin2} + 0 + 2(\mathbf{P}^{fin} \cdot \pi - E^{fin}\varepsilon)$$ (4-35)

The four-scalar product on the right-hand side can be evaluated in any inertial frame. In R,

$$\mathbf{P}^{\text{fin}} = -\boldsymbol{\pi} \qquad \boldsymbol{\pi} = \boldsymbol{\varepsilon}$$

$$E^{\text{fin}} = \mu^{\text{in}} - \varepsilon$$

$-\mu^{\prime H\,2} = -\mu^{fn^2} + 2\left(-\varepsilon^2 - \mu^{\prime\prime}_\varepsilon + \varepsilon^2\right)$

so that Eq. (4-35) becomes

$$-\mu^{\text{in}^2} = -\mu^{\text{fin}^2} - 2\mu^{\text{in}}\varepsilon \qquad\qquad (4\text{-}36\text{a})$$

which is Eq. (4-32). In R',

$$P^{\text{fin}'} = 0$$

$$E^{\text{fin}'} = \mu^{\text{fin}}$$

so that Eq. (4-35) becomes

$$-\mu^{\text{in}^2} = -\mu^{\text{fin}^2} - 2\mu^{\text{fin}}\varepsilon' \qquad\qquad (4\text{-}36\text{b})$$

an equally valid expression. Comparing Eqs. (4-36a and b), we see that

$$\mu^{\text{fin}}\varepsilon' = \mu^{\text{in}}\varepsilon$$

which is Eq. (4-33).

In cases where the initial and final proper masses are known, Eq. (4-32) can be used to calculate the energy of the photon in the parent rest frame

$$\varepsilon = \frac{\mu^{\text{in}^2} - \mu^{\text{fin}^2}}{2\mu^{\text{in}}} \qquad\qquad (4\text{-}37)$$

Thus, in the transition

$$\Sigma^0 \to \Lambda + \gamma$$

discussed in the exercise of Section 2-8, the measured rest energies of Σ^0 and Λ give

$$\mu^{\text{in}} = 1.1926 \text{ GeV}/c^2$$

$$\mu^{\text{fin}} = 1.1156 \text{ GeV}/c^2$$

from which we compute

$$\varepsilon = 74.5 \text{ MeV}$$

The 77.0-MeV loss of rest energy is divided between the photon and the recoil kinetic energy of the Λ so as to maintain zero momentum.

Einstein's argument [47] for a loss of proper mass in radiating was based on consideration of an extremely simple, extremely unlikely, but not impossible, process.[2] A body simultaneously emits in its rest frame (R^*) two equal and opposite wave trains (nowadays called photons). In this way it loses energy without suffering a recoil. The two photons have total energy ε^* and total momentum $\pi^* = 0$.

Consider the process in another reference frame R, in which R^* has velocity $V = \beta$ in the x direction. In R, the mother and daughter system have the same velocity V. The energy of the photons in R is given by the transformation equation (4-21)

$$\varepsilon = \gamma(\varepsilon^* + \beta\pi_\parallel^*) = \gamma\varepsilon^* = \varepsilon^*(1 - V^2)^{-1/2}$$

Conservation of energy requires that the energy of the daughter be less than that of the mother by this amount. Inasmuch as the velocities are the same, μ^{fin} must be less than μ^{in}. In effect,

$$\mu^{\text{in}}(1 - V^2)^{-1/2} = \mu^{\text{fin}}(1 - V^2)^{-1/2} + \varepsilon^*(1 - V^2)^{-1/2}$$

Dividing through by $(1 - V^2)^{-1/2}$, we find

$$\mu^{\text{in}} = \mu^{\text{fin}} + \varepsilon^*$$

In this case of recoilless radiation, the proper mass decreases by the amount corresponding to the energy radiated in the body's rest frame.

Einstein drew the sweeping conclusion: "The mass of a body is a measure of its energy content; if the energy changes by L, the mass changes in the same sense by $L/9 \times 10^{20}$, the energy being measured in ergs, and the mass in grams. It is not impossible that with bodies whose energy content is variable to a high degree (e.g., with radium salts) the theory may be successfully put to the test. If the theory corresponds to the facts, radiation conveys inertia between the emitting and absorbing bodies."

The results of this section are based only on the transformation properties of photon four-momentum. Their applicability extends to all transitions to a two-body final state in which one body has zero proper mass. Besides the photon, the neutrinos have this property. Thus the reaction considered in

---

[2] Einstein's first publications (1901–1902) were on thermodynamics!

the worked exercise of Section 3-8

$$\pi^+ \rightarrow \mu^+ + \nu_\mu$$

is covered. It is also not necessary that the mother and daughter "atoms" be bound systems. Any collection of particles with nonzero invariant mass will do. The formulas of this section apply, for example, to

$$n \rightarrow p + e^- + \bar{\nu}_\beta$$

with μ^{fin} standing for the invariant mass of the proton–electron system and ε for the energy of the antineutrino.

4-7. TWO-BODY FINAL STATE, BOTH OF ZERO PROPER MASS

We consider now cases like

$$e^- + e^+ \rightarrow \gamma + \gamma$$

and

$$\pi^0 \rightarrow \gamma + \gamma$$

(Reactions leading to $\nu + \gamma$ or $\nu + \nu$ have not been observed, and seem to be forbidden by a quantal conservation law.) Since a particle of zero proper mass has no rest frame, the transformation used in the preceding section does not apply.

A diagram of the process is shown in Fig. 4-2a. We abbreviate the proper mass of the initial system to μ. Conservation of four-momentum

$$\tilde{P}^{\text{in}} = {}^1\tilde{\pi} + {}^2\tilde{\pi} \tag{4-38}$$

gives in the rest frame R of the parent (Fig. 4-2b)

$$0 = {}^1\pi + {}^2\pi$$

$$\mu^{\text{in}} = {}^1\varepsilon + {}^2\varepsilon$$

and, since $\pi = \varepsilon$,

$${}^1\varepsilon = {}^2\varepsilon = \frac{\mu}{2} \tag{4-39}$$

The photons share the energy equally and balance off each other's momentum. These features of the reaction have been confirmed to high precision by observations of the annihilation gamma rays from stopped positrons.

The transformation to another reference frame is straightforward. If a photon's momentum π makes an angle χ with the velocity of R' in R, β, Eq. (4-21) gives

$p = EV$

$$\varepsilon' = \gamma(\varepsilon - \beta\pi\cos\chi)$$

$$\pi'\cos\chi' = \gamma(\pi\cos\chi - \beta\varepsilon)$$

$$\pi'\sin\chi' = \pi\sin\chi$$

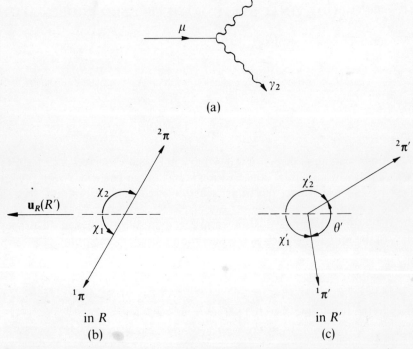

Figure 4-2.   Breakup of a system into two independent parts of zero proper mass (e.g., $\pi^0 \to \gamma + \gamma$). In (a) a diagramatic representation is given. In (b) the momentum vectors $^1\pi$ and $^2\pi$ in the rest frame (R) of the system are drawn from a common origin. This does not mean that the photons are necessarily collinear. (c) The corresponding momentum diagram in another frame R', whose velocity with respect to R is $\mathbf{u}_R(R')$. In R', the momentum vectors of the photons are pulled in the direction of motion of R with respect to R'.

which with

$$\pi = \varepsilon \qquad \pi' = \varepsilon'$$

leads to

$$\varepsilon' = \gamma\varepsilon(1 - \beta \cos \chi)$$
$$\varepsilon' \cos \chi' = \gamma\varepsilon(\cos \chi - \beta) \qquad (4\text{-}40)$$
$$\varepsilon' \sin \chi' = \varepsilon \sin \chi$$

In R', the photon is pulled by the transformation toward the direction in which the parent system (R) was moving (Fig. 4-2c).

The invariant square of both sides of Eq. (4-38) gives

$$\tilde{P}^2 = -\mu^2 = 0 + 0 + 2\,{}^1\tilde{\pi}^2\tilde{\pi} = 2({}^1\boldsymbol{\pi} \cdot {}^2\boldsymbol{\pi} - {}^1\varepsilon^2\varepsilon)$$

With θ denoting the angle between the photon momenta ${}^1\boldsymbol{\pi}$ and ${}^2\boldsymbol{\pi}$, this equation reads

$$\mu^2 = 2\,{}^1\varepsilon^2\varepsilon(1 - \cos \theta)$$

In R, $\cos \theta = -1$, and, by Eqs. (4-39),

$$\mu^2 = 2\frac{\mu}{2}\frac{\mu}{2}2 = \mu^2$$

In R' (any other inertial frame—for example, the laboratory frame),

$$\mu^2 = 2\,{}^1\varepsilon'\,{}^2\varepsilon'(1 - \cos \theta') \qquad (4\text{-}41)$$

This relation between the energies and included angle of the two photons characterizes photons resulting from a two-photon decay. It is used in experimental particle research to recognize photons from π^0 ($\mu = 135 \text{ MeV/c}^2$) or η^0 ($\mu = 549 \text{ MeV/c}^2$) decay against a background of uncorrelated photons.

4-8. TWO-BODY FINAL STATE: GENERAL CASE

A system "0" divides into independent pieces "1" and "2" of nonzero rest mass (Fig. 4-3a). Conservation of four-momentum states that

$$^0\tilde{P} = {}^1\tilde{P} + {}^2\tilde{P} \qquad (4\text{-}42)$$

Let us label the reference frame used in giving a value to a spacetime quantity by a left subscript. Thus, $_i^jE$ is the energy of system j in the rest frame of system i.

In the rest frame of the initial system R_0, Eqs. (4-42) are

$$\mu_0 = {}_0^0E = {}_0^1E + {}_0^2E \tag{4-43a}$$

$$0 = {}_0^1\mathbf{P} + {}_0^2\mathbf{P} \tag{4-43b}$$

The momentum vectors (Fig. 4-3b) are thus equal and opposite, of magnitude, say, q_0. This magnitude determines $_0^1E$ and $_0^2E$ by Eq. (4-14),

$$_0^1E = (q_0{}^2 + \mu_1{}^2)^{1/2} \qquad _0^2E = (q_0{}^2 + \mu_2{}^2)^{1/2} \tag{4-44}$$

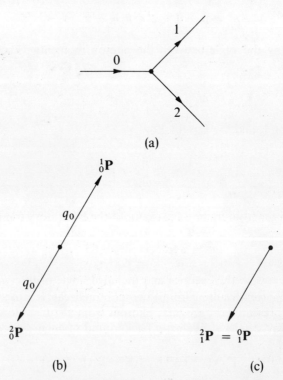

(a)

(b)   (c)

Figure 4-3.   Breakup of a system (0) into two independent parts (1) and (2), at least one of which has nonzero proper mass. (a) Graph representing the transition. (b) Momentum vectors in the rest frame of (0), drawn back to back. (c) Momentum vectors in the rest frame of (1) ($\mu_1 \neq 0$).

They, in turn, are constrained by Eq. (4-43a) to add up to μ_0. This relation determines q_0.

Writing Eq. (4-42) as

$$^2\tilde{P} = {}^0\tilde{P} - {}^1\tilde{P}$$

and forming the square of both sides, we get

$$-\mu_2{}^2 = -\mu_0{}^2 - \mu_1{}^2 - 2({}^0\mathbf{P}\cdot{}^1\mathbf{P} - {}^0E^1E)$$

Evaluating the four-scalar product in R_0, where

$$^0_0P = 0 \qquad {}^0_0E = \mu_0$$

gives

$$^1_0E = \frac{1}{2\mu_0}(\mu_0{}^2 + \mu_1{}^2 - \mu_2{}^2) \tag{4-45a}$$

Similarly, squaring

$$^1\tilde{P} = {}^0\tilde{P} - {}^2\tilde{P}$$

and evaluating in R_0 gives

$$^2_0E = \frac{1}{2\mu_0}(\mu_0{}^2 + \mu_2{}^2 - \mu_1{}^2) \tag{4-45b}$$

Each of Eqs. (4-45) comes from the other by interchange of the labels 1 and 2. They have the form

$$^j_iE = \frac{1}{2\mu_i}(\mu_i^2 + \mu_j^2 - \mu_k^2) \qquad i \to j + k \tag{4-45c}$$

Equation (4-37) is the special case $\mu_j = 0$.

We can evaluate q_0 by using either of Eqs. (4-44). The result is

$$q_0 = \frac{1}{2\mu_0}\{[\mu_0{}^2 - (\mu_1 + \mu_2)^2][\mu_0{}^2 - (\mu_1 - \mu_2)^2]\}^{1/2} \tag{4-46a}$$

The square bracket is a symmetric homogeneous function of $\mu_0{}^2$, $\mu_1{}^2$, $\mu_2{}^2$:

$$q_0 = \frac{1}{2\mu_0}[(\mu_0{}^2)^2 + (\mu_1{}^2)^2 + (\mu_2{}^2)^2 - 2\mu_0{}^2\mu_1{}^2 - 2\mu_1{}^2\mu_2{}^2 - 2\mu_2{}^2\mu_0{}^2]^{1/2}$$

The speed of each fragment in R_0 is

$$\begin{matrix} {}_0^1V = \dfrac{q_0}{{}_0^1E} & \text{or} & \left[1 - \left(\dfrac{\mu_1}{{}_0^1E}\right)^2\right]^{1/2} \end{matrix} \qquad (4\text{-}47a)$$

$$\begin{matrix} {}_0^2V = \dfrac{q_0}{{}_0^2E} & \text{or} & \left[1 - \left(\dfrac{\mu_2}{{}_0^2E}\right)^2\right]^{1/2} \end{matrix} \qquad (4\text{-}47b)$$

The final state velocities are completely determined by conservation of four-momentum, except for the orientation of the direction of motion.

Let us evaluate the dynamical variables also in the rest frame of one of the decay products, say, R_1 (Fig. 4-3c). Using the same trick as for getting Eqs. (4-45), we square Eq. (4-42) and evaluate it in R_1:

$$-\mu_0{}^2 = -\mu_1{}^2 - \mu_2{}^2 + 2({}^1\mathbf{P}\cdot{}^2\mathbf{P} - {}^1E{}^2E)$$

$$= -\mu_1{}^2 - \mu_2{}^2 - 2\mu_1\,{}_1^2E$$

or

$$\begin{matrix} {}_1^2E = \dfrac{1}{2\mu_1}(\mu_0{}^2 - \mu_2{}^2 - \mu_1{}^2) \end{matrix} \qquad (4\text{-}48a)$$

Squaring,

$$^2\tilde{P} = {}^0\tilde{P} - {}^1\tilde{P}$$

and evaluating in R_1, we obtain

$$\begin{matrix} {}_1^0E = \dfrac{1}{2\mu_1}(\mu_0{}^2 + \mu_1{}^2 - \mu_2{}^2) \end{matrix} \qquad (4\text{-}48b)$$

And, of course,

$$^1_1E = \mu_1$$

Typical examples of the type of transition considered in this section are the natural decay of uranium

$$_{92}U^{238} \rightarrow {}_{90}Th^{234} + \alpha$$

or the accelerator reaction

$$p + {}_4Be^9 \rightarrow {}_5B^9 + n$$

where the entire left-hand side is system (0). Because of the relatively small change of proper mass, one can in these cases use the nonrelativistic relation

$$K \approx \frac{P^2}{2\mu}$$

to compute recoil kinetic energy. But the conversion between kinetic energy and rest energy, with the associated change of total proper mass, is a purely relativistic effect, in flagrant contradiction of Newton's assumption that proper mass is conserved. The one case in which there is no conversion of rest energy is that of elastic scattering, such as

$$e^- + Hg \rightarrow e^- + Hg$$

when the mercury atom does not get lifted to an excited state.

4-9. TWO-BODY INITIAL STATE (BINARY COLLISIONS)

Let us now consider cases in which the initial state consists of two independent systems, "1" and "2". The final state "0" may be of one piece or many. We are concerned at present only with its total four-momentum $^0\tilde{P}$. The process is shown graphically in Fig. 4-4. The situation is the same as the one considered in the preceding section (Fig. 4-3a) except that all the arrows are reversed. Conservation of four-momentum again gives

$$^1\tilde{P} + {}^2\tilde{P} = {}^0\tilde{P} \tag{4-42}$$

and all the equations derived from this equation in the preceding section hold. The frame R_0 is the center of mass or center of momentum or rest frame of the colliding particles [Eq. (4-9)], previously labeled by an asterisk; μ_0 is the total energy of the colliding particles in this frame. In the old notation, Eqs. (4-45) are

$$^jE^* = \frac{1}{2E^*}(E^{*2} + \mu_j{}^2 - \mu_k{}^2) \quad \begin{array}{l} j=1 \quad k=2 \\ \text{or} \\ j=2 \quad k=1 \end{array} \tag{4-45d}$$

and Eq. (4-46a) is

$$q_0 = \frac{1}{2E^*}\{[E^{*2} - (\mu_1 + \mu_2)^2][E^{*2} - (\mu_1 - \mu_2)^2]\}^{1/2} \qquad (4\text{-}46\text{b})$$

$$= \frac{1}{2E^*}(E^{*4} + \mu_1{}^4 + \mu_2{}^4 - 2E^{*2}\mu_1{}^2 - 2\mu_1{}^2\mu_2{}^2 - 2\mu_2{}^2E^{*2})^{1/2}$$

It is useful to compute the four-momentum components in the rest frame of one of the colliding particles. Many experiments with high energy particle beams employ stationary targets, so that R_2, say, coincides with the laboratory reference frame.

To find the energies in R_2, we can use the standard procedure: Calculate the square of an appropriate form of the four-momentum conservation law and evaluate it in R_2. Squaring the form Eq. (4-42), we obtain

$$-\mu_1{}^2 - \mu_2{}^2 - 2\mu_2\,{}_2^1E = -\mu_0{}^2$$

or

$$\tfrac{1}{2}E = \frac{1}{2\mu_2}\,(\mu_0{}^2 - \mu_1{}^2 - \mu_2{}^2) \qquad (4\text{-}49\text{a})$$

Squaring the form

$${}^1\tilde{P} = {}^0\tilde{P} - {}^2\tilde{P}$$

gives

$$-\mu_1{}^2 = -\mu_0{}^2 - \mu_2{}^2 + 2\mu_2\,{}_2^0E$$

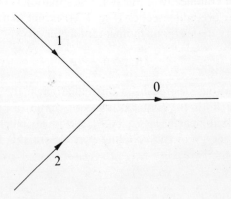

Figure 4-4.

or

$$\substack{0\\2}E = \frac{1}{2\mu_2}\,({\mu_0}^2 + {\mu_2}^2 - {\mu_1}^2) \tag{4-49b}$$

Equations (4-49) are, of course, just Eqs. (4-48) with 2 and 1 interchanged. The energies $\substack{1\\2}E$ or $\substack{0\\2}E$ determine the magnitude of the momentum $\substack{1\\2}P = \substack{0\\2}P$ by the right-triangle relation Eq. (4-14).

In terms of R_2 quantities, μ_0 is given by

$$\begin{aligned}{\mu_0}^2 &= (\substack{1\\2}E + \substack{2\\2}E)^2 - (\substack{1\\2}\mathbf{P} + \substack{2\\2}\mathbf{P})^2 = (\substack{1\\2}E + \mu_2)^2 - (\substack{1\\2}P)^2\\ &= {\mu_1}^2 + {\mu_2}^2 + 2\mu_2\,\substack{1\\2}E \end{aligned} \tag{4-50a}$$

This is, of course, just another way of writing Eq. (4-49a). In terms of the beam kinetic energy

$$\substack{1\\2}K = \substack{1\\2}E - \mu_2$$

Eq. (4-50a) is

$$\mu_0^2 = (\mu_1 + \mu_2)^2 + 2\mu_2\,\substack{1\\2}K \tag{4-50b}$$

To transform all four-momentum components between R_0 and R_2, we need to know the relative velocity

$$\beta = \substack{0\\2}V = \frac{\substack{0\\2}P}{\substack{0\\2}E} = \frac{\substack{1\\2}P + 0}{\substack{1\\2}E + \mu_2} = \frac{\substack{1\\2}P}{\substack{1\\2}E + \mu_2} \tag{4-51}$$

which we can express in terms of either the beam momentum or the beam energy by using

$$(\substack{1\\2}E)^2 = (\substack{1\\2}P)^2 + (\mu_1)^2 \tag{4-14}$$

One-Body Final State

This completely inelastic collision is exemplified macroscopically by the meeting and subsequent sticking together of two putty balls. Microscopically, the process can only occur if an excited bound state of the colliding systems exists having just the right rest energy to permit conservation of both energy and momentum. Consider, for example, the hypothetical sticking together of two hydrogen atoms in an atomic hydrogen glow discharge tube:

$$H + H \rightarrow H_2^*$$

(the asterisk means excited state). In the center of mass frame, the atoms have equal and opposite momentum and the excited molecule is at rest. The rest energy of the molecule must equal the sum of the rest energies and kinetic energies of the atoms. Clearly this is not possible for the ground state of H_2 whose proper mass is 4.48 eV less than the sum of the atomic proper masses. If there is an excited state meeting the energy requirement, it can only be a virtual bound state or "resonance," breaking into pieces sooner or later. In the case of the putty balls, an enormous number of compound excited states are available (thermal excitation), and the mean life for decay from any one of them into a state with two separated putty balls is very, very, very long (like the time needed for a mixture of two gases to separate itself spontaneously).

By way of verifying that heat energy contributes to rest mass, consider the head-on collision of two putty balls (Fig. 4-5). In the center of mass frame, conservation of the spacelike part of four-momentum requires

$$0 = \mu_1(1 - V_1{}^2)^{-1/2}V_1 + \mu_2(1 - V_2{}^2)^{-1/2}V_2 = \mu_0(1 - V_0{}^2)^{-1/2}V_0$$

so that

$$V_0 = 0$$

The combined ball is at rest as a whole. Conservation of the timelike part of four-momentum requires

$$\mu_1(1 - V_1{}^2)^{-1/2} + \mu_2(1 - V_2{}^2)^{-1/2} = \mu_0(1 - V_0{}^2)^{-1/2} = \mu_0$$

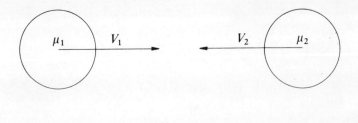

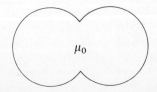

Figure 4-5.   Head-on collision of two putty balls, in the center of mass frame. The balls stick together, forming a system at rest of proper mass μ_0.

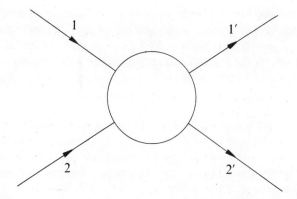

Figure 4-6.

The proper mass of the combined ball is greater than the sum of the proper masses of the component balls, by just the total kinetic energy of the colliding balls:

$$\mu_1 + K_1 + \mu_2 + K_2 = \mu_0$$

This macroscopic kinetic energy gets converted into chaotic internal energy of the combined ball, contributing to μ_0 and thus to the inertia.

Elastic Scattering

At this opposite extreme, the final state consists of two emerging systems with exactly the same particle parameters as the two incident systems. Besides the cases already mentioned, some additional examples are:
(1) Rutherford scattering experiment:

$$\alpha + \text{nucleus} \to \alpha + \text{nucleus in same state}$$

(2) Proton–proton scattering without production of other particles:

$$p + p \to p + p$$

(It is not possible to tell which proton is which. All that matters is that the initial and final state consist of two protons.)
(3) Elastic scattering of a pion and proton:

$$\pi^- + p \to \pi^- + p$$

The reaction can be labeled (Fig. 4-6) $1 + 2 \to 1' + 2'$. So far as four-momenta are concerned, it is $\textcircled{1} + \textcircled{2} \to \textcircled{0}$ followed by $\textcircled{0} \to \textcircled{1'} + \textcircled{2'}$.

The center of mass frame of the system is R_0. In R_0, the initial particles have energies given by Eqs. (4-45) and equal and opposite momenta of magnitude given by Eq. (4-46a). The final particles have energies and momenta given by the corresponding expressions with primes. Since

$$\mu'_1 = \mu_1 \qquad \mu'_2 = \mu_2$$

the corresponding initial and final quantities are equal. In a vector diagram (Fig. 4-7) in which the momenta are drawn from the same point, the momentum vectors are on diameters of a sphere of radius q_0. Only the direction of the (equal and opposite) momenta can change. The angular deflection of either particle is called the scattering angle $_0\chi$. This angle and the corresponding azimuthal angle completely characterize the change of motion in the collision. Quantum mechanics provides the formalism for calculating their distribution functions as determined by the nature of the interaction. Conservation of four-momentum says nothing about this. It only reduces the number of independent variables in the problem by four.

The scattering angle in the center of mass frame is simply related to the four-momentum transfer

$$\tilde{\Delta} = {}^{1'}\tilde{P} - {}^1\tilde{P} = {}^2\tilde{P} - {}^{2'}\tilde{P}$$

This is dynamically significant if the interaction can be thought of as occurring in the two steps indicated in Fig. 4-8a (emission and absorption

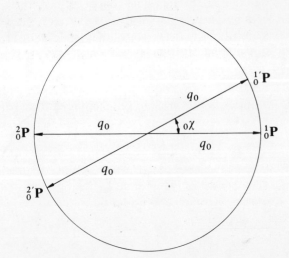

Figure 4-7.   Momentum vectors in center of mass frame in elastic collision $1 + 2 \rightarrow 1' + 2'$.

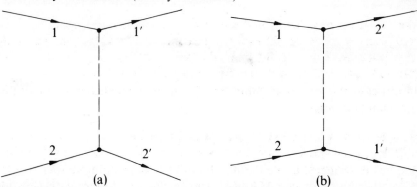

Figure 4-8.

of a virtual particle). The square of this four-vector, the "four-momentum transfer squared," is

$$\tilde{\Delta}^2 = (^{1'}\tilde{P})^2 + (^{1}\tilde{P})^2 - 2(^{1'}\tilde{P})(^{1}\tilde{P}) = -\mu_1{}^2 - \mu_1{}^2 - 2(^{1'}\mathbf{P} \cdot {}^{1}\mathbf{P} - {}^{1'}E^1E)$$

Evaluating the four-scalar product in R_0, we get

$$\tilde{\Delta}^2 = -2\mu_1{}^2 - 2q_0{}^2\cos{}_0\chi + 2(_0^1E)^2 \qquad (4\text{-}52a)$$

$$= 2q_0{}^2(1 - \cos{}_0\chi) \qquad (4\text{-}52b)$$

In a grazing collision ($_0\chi = 0$), the four-momentum transfer squared is zero. In a head-on collision ($_0\chi = \pi$), $\tilde{\Delta}^2$ has its highest possible value, $4q_0{}^2$. Evidently the four-momentum transfer is a spacelike four-vector. The linear relationship between $\tilde{\Delta}^2$ and $\cos{}_0\chi$ implies that equal intervals of $\tilde{\Delta}^2$ correspond to equal intervals of solid angle in the center of mass frame. The invariance of $\tilde{\Delta}^2$ permits us to calculate it in any frame. In R_2,

$$\tilde{\Delta}^2 = (^{2}\tilde{P})^2 + (^{2'}\tilde{P})^2 - 2(^{2}\tilde{P})(^{2'}\tilde{P}) = -\mu_2{}^2 - \mu_2{}^2 - 2(^{2}\mathbf{P} \cdot {}^{2'}\mathbf{P} - {}^{2}E \cdot {}^{2'}E)$$

becomes

$$\tilde{\Delta}^2 = -2\mu_2{}^2 + 2\mu_2\,{}_2^{2'}E = 2\mu_2\,{}_2^{2'}K \qquad (4\text{-}53)$$

The invariant four-momentum transfer squared is seen to be proportional to the laboratory kinetic energy acquired by the struck particle. If the recoil velocity is small,

$$_2^{2'}K \approx \frac{(_2^{2'}P)^2}{2\mu_2}$$

and

$$\tilde{\Delta}^2 \approx (\tfrac{2'}{2}P)^2$$

In this case, the four-momentum transfer squared equals the square of the momentum transfer

$$\tfrac{2'}{2}\mathbf{P} - \tfrac{2}{2}\mathbf{P} = \tfrac{2'}{2}\mathbf{P}$$

In general, Δ^2 increases monotonically with the momentum transfer.

Another four-momentum transfer can be defined, corresponding to the other pairing of final and initial particles in a two-step scattering process (Fig. 4-8b). Calling the four-momentum transfer already discussed $\tilde{\Delta}(1, 1')$,

$$\tilde{\Delta}(1, 1') = {}^{1'}\tilde{P} - {}^{1}\tilde{P} = {}^{2}\tilde{P} - {}^{2'}\tilde{P} = -\tilde{\Delta}(2, 2') \tag{4-54a}$$

the other four-momentum transfer $\tilde{\Delta}(1, 2')$ is

$$\tilde{\Delta}(1, 2') = {}^{2'}\tilde{P} - {}^{1}\tilde{P} = {}^{2}\tilde{P} - {}^{1'}\tilde{P} = -\tilde{\Delta}(2, 1') \tag{4-54b}$$

Evaluating its square,

$$\tilde{\Delta}^2(1, 2') = ({}^{2'}\tilde{P})^2 + ({}^{1}\tilde{P})^2 - 2({}^{2'}\tilde{P})({}^{1}\tilde{P}) = -\mu_2{}^2 - \mu_1{}^2 - 2({}^{2'}\mathbf{P} \cdot {}^{1}\mathbf{P} - {}^{2'}E^1E)$$

in the center of mass frame,

$$\tilde{\Delta}^2(1, 2') = -\mu_2{}^2 - \mu_1{}^2 - 2q_0{}^2 \cos(\pi - {}_0\chi) + 2{}_0^2 E_0^1 E$$

$$= -\mu_2{}^2 - \mu_1{}^2 + 2q_0{}^2 \cos {}_0\chi + 2{}_0^2 E{}_0^1 E \tag{4-55}$$

This equation is also a linear function of the cosine of the scattering angle, but with the opposite sign. It has its greatest value when $\tilde{\Delta}^2(1, 1')$ is smallest. The sum of the two $\tilde{\Delta}^2$'s is independent of the scattering angle, depending only on μ_0 and the proper masses. In fact, adding Eqs. (4-52a) and (4-55) gives, with use of Eq. (4-45a),

$$\tilde{\Delta}^2(1, 1') + \tilde{\Delta}^2(1, 2') = -2\mu_1{}^2 - 2\mu_2{}^2 + \mu_0{}^2 \tag{4-56}$$

To relate $\tilde{\Delta}^2(1, 2')$ to laboratory quantities, we evaluate the square of $\tilde{\Delta}(2, 1')$ [Eq. (4-54b)] in R_2:

$$\tilde{\Delta}^2(1, 2') = \tilde{\Delta}^2(2, 1') = ({}^{1'}\tilde{P} - {}^{2}\tilde{P})^2 = -\mu_1{}^2 - \mu_2{}^2 + 2\mu_2{}_2^{1'}E$$

$$= -(\mu_1 - \mu_2)^2 + 2\mu_2{}_2^{1'}K \tag{4-57}$$

This equation corresponds to Eq. (4-53) for the other $\tilde{\Delta}^2$, relating $\tilde{\Delta}^2(1, 2')$ to the laboratory kinetic energy of the "incident" particle after the collision.

The results of relativistic quantal calculations of the probability of a particular scattering angle are usually expressed in terms of the invariants $\mu_0{}^2$ and $\tilde{\Delta}^2$. Equations (4-50), (4-53), and (4-57) relate these invariants to the respective laboratory quantities: beam energy; recoil kinetic energy; and residual kinetic energy. They hold whatever is the mechanism of interaction, being based only on conservation of four-momentum.

Equation (4-51) permits transformation of the four-momentum components between R_0 and R_2. The restriction of $_0\chi$ to the range $0-\pi$ limits the possible energies and angles in the laboratory frame. The details are left for the exercises (an excellent summary is given in [35], Section 2–5).

EXERCISE A photon has an elastic collision with a free electron (Compton scattering). Determine the relation between the scattering angle and the energy of the scattered photon in the laboratory frame.

SOLUTION Let "1" be the incident photon, "2" the struck electron, "1'" the scattered photon, and "2'" the recoiling electron. Representing again a photon by a wavy line, the reaction is as shown in Fig. 4-9. The formulas of this section apply, with $\mu_1 = \mu_1' = 0$. In the usual experimental situation, the struck electron belongs to a stationary target placed in the x-ray beam, so that R_2 is the laboratory frame. The scattering angle of the photon is $(^{1'}\mathbf{P}, {}^1\mathbf{P})$, suggesting that we evaluate

$$\tilde{\Delta}^2(1, 1') = (^{1'}\tilde{P} - {}^1\tilde{P})^2 = (\tilde{\pi}' - \tilde{\pi})^2$$

in R_2. The result is

$$\tilde{\Delta}^2 = -2(_2\pi')(_2\pi)\cos{}_2\chi + 2(_2\varepsilon')(_2\varepsilon) = 2(_2\varepsilon')(_2\varepsilon)(1 - \cos{}_2\chi)$$

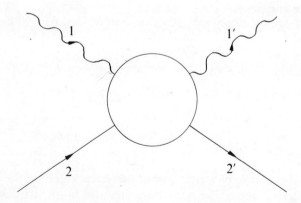

Figure 4-9.

The other evaluation [Eq. (4-53)] of $\tilde{\Delta}^2$ in R_2 gives

$$\tilde{\Delta}^2 = 2\mu_2 {}_2^2 K \qquad (4\text{-}53)$$

Since (conservation of energy)

$$_2\varepsilon + \mu_2 = {}_2\varepsilon' + \mu_2 + {}_2^2 K$$

Eq. (4-53) becomes

$$\tilde{\Delta}^2 = 2\mu_2({}_2\varepsilon - {}_2\varepsilon')$$

Equating the two expressions of $\tilde{\Delta}^2$, we obtain the relation sought:

$$1 - \cos {}_2\chi = \mu_2\left(\frac{1}{{}_2\varepsilon'} - \frac{1}{{}_2\varepsilon}\right)$$

As explained in Appendix F, a plane wave is characterized by an amplitude and phase and by a propagation four-vector $\tilde{k}$ that specifies the direction of propagation, the wavelength λ, and the frequency v. The spacelike part of $\tilde{k}$ is a vector in the direction of propagation of magnitude $2\pi/\lambda$; the timelike part is the circular frequency $\omega = 2\pi v$. The quantal wave–particle relation is the relativistically covariant equation

$$\tilde{P} = \hbar\tilde{k}$$

with $\hbar$ a universal constant ($\hbar = 1.0545 \times 10^{-27}$ erg-sec $= 6.5819 \times 10^{-16}$ eV-sec; it is the original Planck constant h divided by 2π). The timelike component is for a photon

$$\varepsilon = \hbar\omega = \frac{\hbar 2\pi c}{\lambda}$$

and our result takes on the familiar Compton form

$$_2\lambda' - {}_2\lambda = \frac{2\pi\hbar}{\mu_2 c}(1 - \cos {}_2\chi) \qquad \text{(LMT)}$$

giving the change of wavelength of the scattered photon as a function of scattering angle. The scattering angles of both particles and their energies are evidently functions of $_2\chi$ and, thus, of $_0\chi$ or $\tilde{\Delta}^2$.

Sometimes R_2 is not the laboratory frame. In interstellar space, elastic collisions occur between very high energy cosmic-ray particles and thermal photons from stars [50]. In the fixed star reference frame, the collision can entail an enormous loss of energy by the charged particle and, correspondingly, an enormous increase of photon energy. A visible photon is replaced by a gamma ray! The process has recently been observed on earth, using a laser beam aimed head-on at a beam of electrons in an accelerator ([51], [52]).

Inelastic Scattering, Two-Body Final State

Again we have $1 + 2 \to 1' + 2'$, but this time we do not require $\mu'_1 = \mu_1$ and $\mu'_2 = \mu_2$. Numerous examples of such collisions have already been mentioned. For the sake of concreteness we may think now, for example, of

$$\pi^+ + p \to \Sigma^+ + K^+$$

Figure 4-6 applies here as well as to the elastic case, and our two-body formulas Eqs. (4-45) and (4-46) apply, as before, to the initial and the final state, with R_0 being the center-of-mass frame of the system. The only difference is that now

$$\,^{1'}_{0}E = \frac{1}{2\mu_0}(\mu_0{}^2 + \mu_1{}'^2 - \mu_2{}'^2) \neq \,^{1}_{0}E = \frac{1}{2\mu_0}(\mu_0{}^2 + \mu_1{}^2 - \mu_2{}^2) \quad (4\text{-}58a)$$

$$\,^{2'}_{0}E = \frac{1}{2\mu_0}(\mu_0{}^2 + \mu_2{}'^2 - \mu_1{}'^2) \neq \,^{2}_{0}E = \frac{1}{2\mu_0}(\mu_0{}^2 + \mu_2{}^2 - \mu_1{}^2) \quad (4\text{-}58b)$$

$$q'_0 = \frac{1}{2\mu_0}\{[\mu_0{}^2 - (\mu_{1'} + \mu_{2'})^2][\mu_0{}^2 - (\mu_{1'} - \mu_{2'})^2]\}^{1/2}$$

$$\neq q_0 = \frac{1}{2\mu_0}\{[\mu_0{}^2 - (\mu_1 + \mu_2)^2][\mu_0{}^2 - (\mu_1 - \mu_2)^2]\}^{1/2} \quad (4\text{-}59)$$

In the vector diagram of momenta in R_0 (Fig. 4-10), the final state momenta are on a diameter of a sphere of radius q'_0, different from q_0. They are still equal and opposite, and the collision is still completely specified (except for possible spin orientations) by the scattering angle $_0\chi$ and azimuth.

The expression Eq. (4-50) giving the center of mass energy μ_0 in terms of the beam energy still applies, of course. As μ_0 is increased, the radii q_0 and q'_0 increase; as it is decreased, the radii decrease. When q'_0 reaches the value zero (1' and 2' created at rest in the center of mass), one is at the threshold for the reaction. A lower value of μ_0 would not correspond to a real momentum q'_0. A larger value would give the final state particles kinetic in addition to rest energy. From Eq. (4-59) we see that

$$\mu_0(\text{thresh}) = \mu_{1'} + \mu_{2'}$$

By Eq. (4-50b), this value corresponds to a beam kinetic energy of

$$\tfrac{1}{2}K(\text{thresh}) = \frac{1}{2\mu_2}[(\mu_{1'} + \mu_{2'})^2 - (\mu_1 + \mu_2)^2] \qquad (4\text{-}60)$$

The square bracket consists of the sum of the final proper masses, squared, minus the sum of the initial proper masses, squared.

There are again two different four-momentum transfers, depending on which final particle one associates with which initial one. These are (Fig. 4-8a)

$$\tilde{\Delta}(1, 1') = {}^{1'}\tilde{P} - {}^{1}\tilde{P} = {}^{2}\tilde{P} - {}^{2'}\tilde{P} = -\tilde{\Delta}(2, 2')$$

as before, and (Fig. 4-8b)

$$\tilde{\Delta}(1, 2') = {}^{2'}\tilde{P} - {}^{1}\tilde{P} = {}^{2}\tilde{P} - {}^{1'}\tilde{P} = -\tilde{\Delta}(2, 1')$$

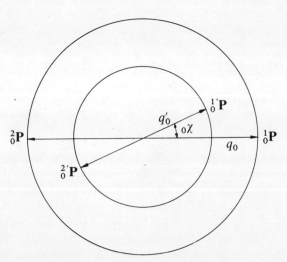

Figure 4-10.   Momentum vectors in center of mass frame in two-body reaction $1 + 2 \rightarrow 1' + 2'$. The radius of the outer circle is q_0; that of the inner circle is q_0'.

The broken lines in the graphs represent some physical coupling mechanism, such as, possibly, exchange of a virtual particle. The squares of these four-momentum transfers are still simply related to the scattering angle in the center of mass frame. Proceeding as in the derivation of Eq. (4-52), we find that

$$\tilde{\Delta}^2(1, 1') = -\mu_{1'}{}^2 - \mu_1{}^2 - 2({}^{1'}\mathbf{P} \cdot {}^1\mathbf{P} - {}^{1'}E^1E)$$

$$= -\mu_{1'}{}^2 - \mu_1{}^2 - 2q_0q_0' \cos({}_0\chi) + 2_0^{1'}E_0^1E \qquad (4\text{-}61)$$

As before, $\tilde{\Delta}^2(1, 1')$ is a linear function of the cosine of the scattering angle. Similarly,

$$\tilde{\Delta}^2(1, 2') = -\mu_1{}^2 - \mu_{2'}{}^2 - 2({}^1\mathbf{P} \cdot {}^{2'}\mathbf{P} - {}^1E^{2'}E)$$

$$= -\mu_1{}^2 - \mu_{2'}{}^2 - 2q_0q_0' \cos(\pi - {}_0\chi) + 2_0^1E_0^{2'}E$$

$$= -\mu_1{}^2 - \mu_{2'}{}^2 + 2q_0q_0' \cos {}_0\chi + 2_0^1E_0^{2'}E \qquad (4\text{-}62)$$

again a linear function of $\cos({}_0\chi)$ but with the opposite sign. A small ${}_0\chi$—that is, a small angle between ${}_0^1\mathbf{P}$ and ${}_0^{1'}\mathbf{P}$—corresponds to small $\tilde{\Delta}^2(1, 1')$ and large $\tilde{\Delta}^2(1, 2')$. Adding Eqs. (4-61) and (4-62) and using Eq. (4-45a), we obtain

$$\tilde{\Delta}^2(1, 1') + \tilde{\Delta}^2(1, 2') = -\mu_1{}^2 - \mu_2{}^2 - \mu_{1'}{}^2 - \mu_{2'}{}^2 + \mu_0{}^2 \qquad (4\text{-}63)$$

Evidently, the sum of the two $\tilde{\Delta}^2$'s is independent of the scattering angle, depending only on μ_0 and the proper masses.

Equations (4-61), (4-62), and (4-63) correspond exactly to Eqs. (4-52), (4-55), and (4-56) of the elastic scattering case. For the expressions giving μ_0^2 and the $\tilde{\Delta}^2$'s in terms of laboratory (R_2) quantities, we have

$$\mu_0{}^2 = (\mu_1 + \mu_2)^2 + 2\mu_2 {}_2^1 K \qquad (4\text{-}50b)$$

as before. Corresponding to Eq. (4-53), we find

$$\tilde{\Delta}^2(1, 1') = ({}^2\tilde{P} - {}^{2'}\tilde{P})^2 = -(\mu_{2'} - \mu_2)^2 + 2\mu_2 {}_2^{2'} K \qquad (4\text{-}64)$$

and corresponding to Eq. (4-57),

$$\tilde{\Delta}^2(1, 2') = ({}^2\tilde{P} - {}^{1'}\tilde{P})^2 = -(\mu_{1'} - \mu_2)^2 + 2\mu_2 {}_2^{1'} K \qquad (4\text{-}65)$$

The dependence of the invariant four-momentum transfers on kinetic energies in R_2 after the interaction is the same as in the elastic case, except for shifts of range resulting from differences of proper mass. Note the linear relationship between laboratory energy in the final state and invariant four-momentum transfer squared, which is characteristic of binary collisions with two-body final states.

EXERCISE   Consider the two-body inelastic process

$$\pi^-_1 + p_2 \to K^+_{1'} + \Sigma_{2'}$$

In a hydrogen bubble chamber experiment [53], the distributions of final-state laboratory momenta were determined for various incident π^- momenta. For π^- incident at 1.9 GeV/c, the distribution is fitted by the simple form

$$\frac{d\sigma}{d\Omega_0} = 7.5(1 - \cos_0\chi) \times 10^{-30} \text{ cm}^2/\text{sr}$$

where Ω_0 is the solid angle in the center of mass (R_0) frame. (See Section 3-8 for the definition of cross section σ. There is a unique magnitude of the momentum, q'_0, corresponding to a delta-function factor in $d^2\sigma/dp_0\,d\Omega_0$).

Assuming the validity of this fit, determine the differential cross section with respect to $\tilde{\Delta}^2(\pi^-, \Sigma^-)$.

SOLUTION   With

$$\mu_1 = 0.1396 \text{ GeV/c}^2$$

$$\mu_2 = 0.9383 \text{ GeV/c}^2$$

$$\mu'_1 = 0.4938 \text{ GeV/c}^2$$

$$\mu'_2 = 1.1974 \text{ GeV/c}^2$$

we find

(Eq. (4-50))	$\mu_0 = 2.115$ GeV/c^2
(Eq. (4-46))	$q_0 = 0.841$ GeV/c
(Eq. (4-45a))	$_0^1 E = 0.854$ GeV
(Eq. (4-45b))	$_0^2 E = 1.261$ GeV
(Eq. (4-59))	$q'_0 = 0.599$ GeV/c
(Eq. (4-58a))	$_0^{1'} E = 0.776$ GeV
(Eq. (4-58b))	$_0^{2'} E = 1.339$ GeV

(Eq. (4-61)) $\tilde{\Delta}^2(1, 1') = 1.061 - 1.008 \cos {}_0\chi \text{ GeV/c}^2$

(Eq. (4-62)) $\tilde{\Delta}^2(1, 2') = 0.830 + 1.008 \cos {}_0\chi \text{ GeV/c}^2$

From this last equation,

$$\cos {}_0\chi = \frac{\tilde{\Delta}^2(1, 2') - 0.830}{1.008}$$

and

$$d(\cos {}_0\chi) = \frac{1}{1.008} d[\tilde{\Delta}^2(1, 2')]$$

Assuming azimuthal symmetry, we obtain

$$\frac{d\sigma}{d\tilde{\Delta}^2} = 2\pi \frac{d\sigma}{d\Omega_0} \frac{d(\cos {}_0\chi)}{d\tilde{\Delta}^2}$$

$$= 85.3 - 46.4 \, \tilde{\Delta}^2 \quad 10^{-30} \text{ cm}^2/(\text{GeV/c})^2$$

The differential cross section is a linear function of $\tilde{\Delta}^2(\pi^-, \Sigma^-)$, with its greatest value $93.6 \times 10^{-30} \text{ cm}^2/(\text{GeV/c})^2$ for "backward" scattering [Σ^- emerging forward; $\tilde{\Delta}^2(\pi^-, \Sigma^-) = -0.178$] and its smallest value 0 for "forward" scattering [K^+ emerging forward, $\tilde{\Delta}^2(\pi^-, \Sigma^-) = 1.838$].

At larger μ_0, the backward peak becomes more pronounced. There is some theoretical basis for fitting the exponential form

$$\frac{d\sigma}{d\tilde{\Delta}^2} \sim \exp[-a(\tilde{\Delta}^2 - \tilde{\Delta}^2_{\min})]$$

By plotting our linear function on semilog paper (Fig. 4-11), we see that this form, with $a = 0.53 \, (\text{GeV/c})^{-2}$, is a fair approximation in the backward hemisphere.

To each value of the scattering angle ${}_0\chi$ or $\tilde{\Delta}^2$ corresponds a definite vector momentum of Σ^- and K^+ in the laboratory. To calculate the differential cross section in the laboratory, we can apply the general procedure of the worked exercise of Section 3-6. As was explained there, a short way of determining the laboratory energy distribution involves making use of the linear relation between laboratory energy on the one hand and $\cos {}_0\chi$ or $\tilde{\Delta}^2$ on the other [Eqs. (4-64) and (4-65)].

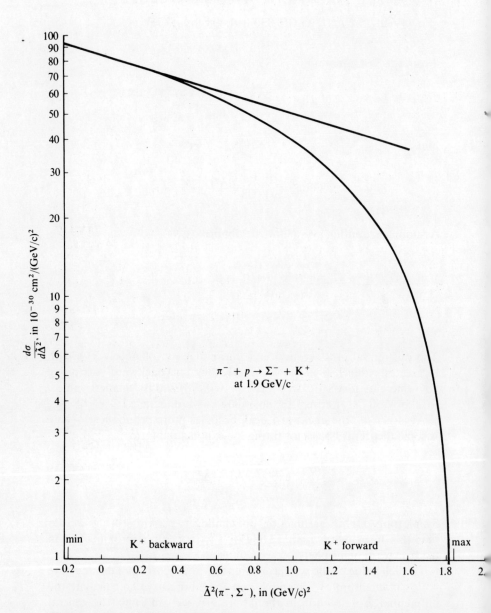

Figure 4-11. Differential cross section for invariant four-momentum transfer squared from π^- to Σ^- in $\pi^- + p \to \Sigma^- + K^+$ at 1.9 GeV/c (calculated from fit given by Dahl *et al.* [53]). The straight line is a fit to the form $(d\sigma/d\tilde{\Delta}^2) \propto \exp[-a(\tilde{\Delta}^2 - \tilde{\Delta}^2_{\min})]$ with $a = 0.53$ (GeV/c)$^{-2}$.

Inelastic Scattering, the General Case. Thresholds

When there are more than two independent systems in the final state

$$1 + 2 \rightarrow 1' + 2' + 3' + \cdots \qquad (4\text{-}66)$$

we can apply the results of the preceding sections by the artifice of grouping the final particles into a two-body system

$$1 + 2 \rightarrow (1' + 2') + (3' + \cdots)$$

and then considering the breakup of each bracket into its components,

$$(1' + 2') \rightarrow 1' + 2'$$

and so on. This procedure is merely an exercise in the algebra of four-momenta with respect to addition, implying nothing at all about the possible existence of short-lived intermediate states.

The threshold for arriving at the particular final state, Eq. (4-66), is

$$\mu_0(\text{thresh}) = \mu_{1'} + \mu_{2'} + \mu_{3'} + \cdots \qquad (4\text{-}67)$$

In effect, the cheapest way to make this set of particles is to make them at rest in the overall center of mass. Any motion there will make μ_0 larger, since

$$\mu_0 = [\mu_{1'}{}^2 + ({}_0^1 P)^2]^{1/2} + [\mu_{2'}{}^2 + ({}_0^2 P)^2]^{1/2} + [\mu_{3'}{}^2 + ({}_0^3 P)^2]^{1/2} + \cdots$$

Now μ_0 only increases slowly with the beam energy in R_2 [Eqs. (4-50)],

$$\mu_0 = (\mu_1{}^2 + \mu_2{}^2 + 2\mu_2 {}_2^1 E)^{1/2} \qquad (4\text{-}50c)$$

and when the target is stationary, R_2 is the laboratory frame. If a proton from the Brookhaven accelerator, of momentum 30 GeV/c, collides with a proton in a hydrogen target, the center of mass energy is, by Eq. (4-50c),

$$\mu_0 = [2\mu(\mu + E)]^{1/2}$$
$$= [2 \times 0.938 \times 30.94]^{1/2} = 7.61 \text{ GeV}$$

The same experiment carried out at Serpukhov, where beam momenta of 76 GeV/c have been attained, would provide a center of mass energy of

$$[2 \times 0.938 \times 76.94]^{1/2} = 12.0 \text{ GeV}$$

An increase of laboratory beam energy by a factor of 2.5 has increased the center of mass energy by a factor of only 1.6. The square-root dependence of μ_0 on E makes it more and more costly to reach high energy states by the technique of collisions with stationary targets. The 200-GeV design value of the Weston accelerator will only provide 19.4 GeV in the proton–proton center of mass.

Motion of the target particle toward the beam particle can increase μ_0 appreciably. In general,

$$\mu_0{}^2 = (^2E + {}^1E)^2 - (^2\mathbf{P} + {}^1\mathbf{P})^2$$

$$= \mu_2{}^2 + \mu_1{}^2 + 2(^2E\,^1E - {}^2\mathbf{P}\cdot{}^1\mathbf{P})$$

$$= (\mu_2{}^2 + \mu_1{}^2 + 2\mu_2\,{}^1E) + 2\,^2K\,^1E - 2\,^2\mathbf{P}\cdot{}^1\mathbf{P} \qquad (4\text{-}68a)$$

Evaluating Eq. (4-68a) in the laboratory (L), we find

$$\mu_0{}^2 = (\mu_2{}^2 + \mu_1{}^2 + 2\mu_2\,{}_L^1E) + 2_L^2K\,{}_L^1E - 2_L^2P\,{}_L^1P\cos({}_L^2\mathbf{P}, {}_L^1\mathbf{P}) \qquad (4\text{-}68b)$$

For oppositely directed colliding particles, the last term makes a positive contribution. It is appreciable if ${}_L^2P$ is not small compared to μ_2. Antiprotons were made at the Berkeley accelerator in 1955 [54] with protons of ${}_L^1P = 5.8$ GeV/c incident on a copper target. The copper nucleus consists of nucleons with an isotropic distribution of momentum ranging up to several hundred MeV/c in magnitude ("Fermi motion"). In a collision of a beam proton with a nuclear proton that happens to be at rest, μ_0 is 3.57 GeV, which falls short of the $4 \times 0.938 = 3.75$ GeV required for

$$p + N \to N + N + \bar{p} + N$$

(N stands for nucleon). If, however, a beam proton interacts with a nuclear proton that is moving toward it with ${}_L^2P = 200$ MeV/c, we find from Eq. (4-68b) that $\mu_0 = 3.92$ GeV—more than enough to make a nucleon–antinucleon pair. With a complex target the distribution function of μ_0 can evidently have appreciable width.

"Clashing beams" provide a means for exciting states of large proper energy. The accelerator produces oppositely directed beams of sufficiently high intensity that interactions can be observed in the region where they cross. The laboratory frame now coincides with R_0, and

$$\mu_0 = [({}_L^2E + {}_L^1E)^2 - ({}_L^2\mathbf{P} + {}_L^1\mathbf{P})^2]^{1/2} = {}_L^2E + {}_L^1E \qquad (4\text{-}69)$$

The 28-GeV proton–proton colliding beam system under construction at CERN should provide final states of 56 GeV proper energy. A system of this type, bringing together electrons and positrons of up to 700 MeV, has recently been successfully used at Novosibirsk [55] to study the reaction

$$e^- + e^+ \to \pi^+ + \pi^-$$

as a function of center of mass energy. A pronounced peak in the cross section when μ_0 is in the 700–800 MeV range gives evidence for an intermediate state with the proper mass, and proper mass spectrum, of the ρ meson:

$$e^- + e^+ \to \rho \to \pi^+ + \pi^-$$

EXERCISE The process ("triplet production")

$$\gamma + e^- \to e^+ + e^- + e^-$$

has been observed with gamma rays striking stationary electrons. What is the threshold photon energy in the laboratory? If the process is to be observed in a head-on collision of a photon from a ruby laser ($_L\varepsilon = 1.79$ eV) and an electron, what is the minimum energy of the electron?

SOLUTION Since

$$\mu_{e+} = \mu_{e-} = \mu_e (= 0.511 \text{ MeV}/c^2)$$

the threshold energy in the center of mass frame is, by Eq. (4-67),

$$\mu_0(\text{thresh}) = 3\mu_e$$

In the case of the stationary electrons in the laboratory,

$$\mu_0{}^2 = (_L\varepsilon + \mu_e)^2 - (_L\pi + 0)^2 = 2 {_L\varepsilon}\mu_e + \mu_e{}^2$$

Equating this equation to $9\mu_e{}^2$, we find

$$_L\varepsilon = 4\mu_e = 2.044 \text{ MeV}$$

For the second situation, we again evaluate $\mu_0{}^2$ in the laboratory

$$\mu_0{}^2 = (_L\varepsilon + {_L}E)^2 - (_L\pi + {_L}\mathbf{P})^2 = 0 + \mu_e{}^2 + 2(_L\varepsilon)(_L E) - 2(_L\pi) \cdot (_L\mathbf{P})$$

$$= \mu_e{}^2 + 2 {_L\varepsilon}(_L E + {_L}P)$$

which, equated to $9\mu_e{}^2$, gives

$$4\mu_e{}^2 = {}_L\varepsilon({}_LE + {}_LP)$$

Since

$$\frac{4\mu_e{}^2}{{}_L\varepsilon} = 5.85 \times 10^{11} \text{ eV} \gg 0.511 \times 10^6 \text{ eV}$$

we have

$${}_LE(\text{thresh}) = 293 \text{ GeV}$$

This is beyond the reach of present-day electron accelerators (the Stanford machine reaches 20 GeV). The process undoubtedly occurs in interstellar space between thermal photons and cosmic electrons [50].

4-10. RELATIVISTIC KINEMATICS

This concludes our discussion of conservation of four-momentum in transitions. The material is often referred to as "relativistic kinematics" [56]. It is basic to the analysis of scattering and decay processes in particle and nuclear physics. When the relations that it predicts between the various dynamical variables are ignored, there result "kinematic biases" in the interpretation of the data. Relativistic kinematics is the framework within which the quantal dynamics must be worked out.

The use of the word "kinematics" may be confusing, because its traditional meaning is the study of conceivable motions without consideration of the forces that determine them—just geometry plus time. Now conservation of four-momentum is a feature of real interactions. On the other hand, we have not made any specific assumptions about the interactions other than that four-momentum is conserved. And this conservation law goes so deep—it can be shown to be equivalent to the homogeneity of space and time—that conservation of four-momentum is in a sense part of the structure of space-time.

PROBLEMS

4-1. A beam of protons of energy 200 GeV passes through a target. The electrons in the target can be regarded as essentially at rest in the laboratory frame. Calculate ${}_L\gamma$ for the frame in which the protons are at rest. What is the energy of a target electron in this frame? What is the total energy of a proton and an electron in their center of mass frame?

4-2. The inverse of the radiative transition considered in Section 4-6 is one in which an atom absorbs a photon, going from a state of lower to one of higher energy. Show that the increase in proper mass of the atomic system is given by Eq. (4-32) with ε the negative of the energy of the photon in the rest frame of the excited atom.

4-3. Show that in the transition

$$\pi^+ \to \mu^+ + \nu_\mu$$

the velocity of the muon in the pion rest frame is given by

$$_\pi^\mu\beta = \frac{\mu_\pi^2 - \mu_\mu^2}{\mu_\pi^2 + \mu_\mu^2}$$

With the values given in the Particle Table, it equals 0.270, giving muon momentum 29.7 MeV/c and kinetic energy 4.12 MeV.

4-4. The π^0 meson usually disintegrates into two photons

$$\pi^0 \to \gamma + \gamma$$

The angular distribution is isotropic in the rest frame of the π^0. Calculate the distribution function of the angle between the two photons in the laboratory. Evaluate and plot it for π^0 energies equal to μc^2, $10\,\mu c^2$, $100\,\mu c^2$, and $1000\,\mu c^2$.

4-5. Is it possible for a photon to change into a pair of positive and negative electrons as it moves through the vacuum?

4-6. The most precise way of determining the proper mass (believed to be zero) of the normally undetected neutrino emitted in the transition $\pi \to \mu + \nu$ is to apply relativistic kinematics to the measured proper masses and energies of the detectable pion and muon. Show that

$$\mu_\nu^2 = (\mu_\pi - \mu_\mu)^2 - 2\mu_\pi {_\pi^\mu K}$$

where $_\pi^\mu K$ is the kinetic energy in the rest frame of the pion.

A recent advance in the precision of x-ray spectroscopy has sharpened the value of μ_μ to 139.577 ± 0.013 MeV/c^2 [R. E. Shafer, *Phys. Rev.* **163**, 1451 (1967)]. With $\mu_\mu = 105.659 \pm 0.002$ MeV/c^2 and $_\pi^\mu K = 4.122 \pm 0.016$ MeV, calculate the neutrino proper mass and its estimated error.

4-7. A polonium nucleus at rest in the laboratory disintegrates into a lead nucleus and an alpha particle:

$$_{84}\text{Po}^{210} \to {_{82}}\text{Pb}^{206} + {_2}\text{He}^4$$

The kinetic energy of the alpha particle is measured to be 5.30 MeV. Calculate (to three significant figures) the "Q" of the reaction, defined as the loss of rest energy

$$Q = \mu(\text{Po}) - \mu(\text{Pb}) - \mu(\alpha) \qquad (\text{Answer: } Q = 5.41 \text{ MeV})$$

The final state is extremely nonrelativistic (both bodies have small velocity), but the observed decrease of proper mass in the transition is a purely relativistic effect.

4-8. A negative pion slowed down in matter usually disappears by reacting with two nucleons of a nucleus, as in

$$\pi^- + p + n \to n + n$$

Calculate the kinetic energy and momentum of each final neutron in the center-of-mass frame. Assume that the initial energy is simply the sum of the three rest energies.

4-9. A K^0 meson breaks up into two pions:

$$K^0 \to \pi^+ + \pi^-$$

Calculate the momentum and energy of the π^+ in the rest frame of the K^0, of the π^+, and of the π^-.

The angular distribution in the K^0 rest frame is isotropic (K^0 has spin zero). Show that in any other frame the energy distribution of the π^+ is flat.

If the K^0 has momentum 3.0 GeV/c in the laboratory, what are the greatest and least values of ${}^{\pi^+}_L E$?

4-10. Consider the head-on collision of identical putty balls (discussed in Section 4-9) in the frame in which one of the balls is initially at rest. What is the speed of the final particle and its proper mass?

4-11. (From University of Illinois Ph.D. Qualifying Examination.) Particle 1 of rest mass μ_1 moving with a speed β strikes particle 2 of rest mass μ_2, which is at rest. Particles 3 and 4 with rest masses μ_3 and μ_4, respectively, are the outgoing particles in the reaction induced by this collision. The reaction is shown schematically.

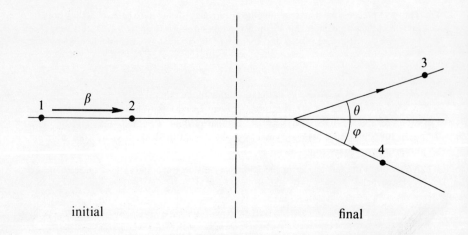

(a) Calculate the minimum total energy that particle 1 must have if $(\mu_3 + \mu_4) > (\mu_1 + \mu_2)$ and the reaction is to take place.

(b) Assume now that the speed (β_{cm}) of the center-of-momentum frame, the speed (β_3), and the angle θ of the emerging particle 3 are known and that these are measured with respect to the laboratory frame. Find the angle θ^* with which particle 3 emerges when viewed in the center of momentum frame.

(c) At what angle, relative to particle 3, does particle 4 emerge as measured in the center-of-momentum frame?

4-12. Show that in elastic scattering, the greatest possible scattering angle of 1 in the initial rest frame of 2 is given by

$$\sin(_2\chi) = \frac{\mu_2}{\mu_1}$$

Evaluate for scattering by an atomic electron of an incident electron, muon, pion, K meson, proton.

4-13. An experiment on proton–proton scattering is planned in which one observes the scattering at 90° in the center of mass frame. With protons of momentum 6 GeV/c incident on a liquid hydrogen target, calculate the energy, momentum, and invariant mass of the initial proton–proton system. Calculate the velocity of the center of mass frame. Calculate the initial momentum and energy of each proton in this frame.

   Calculate the final momentum and energy of each proton in the laboratory frame. What angle does each proton make with the incident beam direction?

4-14. An elastic collision between a particle of unknown proper mass and an electron permits calculation of the unknown proper mass if one can measure the incident momentum and the recoil momentum with sufficient accuracy. Derive the formula expressing the proper mass in terms of the magnitudes of the momenta and the angle that the knocked-on electron makes with the primary.

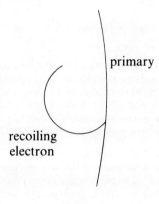

   Applying this technique to cloud chamber photographs of cosmic-ray tracks, Leprince–Ringuet and his coworkers [L. Leprince-Ringuet and M. Lhéritier, *J. Phys. Radium* **7**, 65 (1946)] were able to measure the mass of the muon and to discover the K^+ meson. In the muon picture [reproduced in G. D. Rochester and J. G. Wilson, *Cloud Chamber Photographs of the Cosmic Radiation* (Pergamon, London, 1952), p. 57], the radius of curvature of the primary is 115–120 cm, that of the recoiling electron is 1.05–1.10 cm, and the cosine of the angle is 0.97. The magnetic field is 2650 G. Calculate the proper mass of the incident particle.

4-15. In passing through a bubble chamber, a fast particle has elastic collisions with electrons of the liquid. For energy transfers large compared to atomic binding energies, the electron may be treated as free and at rest in the laboratory.

Consider head-on collisions (maximum possible four-momentum transfer squared) for the case $\mu \gg \mu_e$. Show that if

$$_LE \ll \frac{\mu^2}{\mu_e}$$

the recoil kinetic energy

$$_e^{e'}K(\text{max}) \approx 2\mu_e \left(\frac{_LP}{\mu}\right)^2$$

This equation depends only on the velocity of the fast particle in the laboratory. If, on the other hand,

$$_LE \gg \frac{\mu^2}{\mu_e}$$

show that the greatest possible energy transfer is

$$_e^{e'}K(\text{max}) \approx {_LE}$$

and the minimum residual energy of the incident particle is

$$_L^{1'}E(\text{min}) \approx \frac{\mu^2}{2\mu_e}$$

Thus practically the entire energy of the incident particle is transferred, despite the great difference in proper mass. Show that the two energy regions considered correspond to the incident particle being nonrelativistic or extreme relativistic in the center of mass frame.

Evaluate the maximum possible energy transfer and the minimum residual energy for incident pions, K mesons, and protons of 2, 20, and 200 GeV/c.

4-16. Elastic scattering of optical photons from a laser against high energy electrons from an accelerator is being used at the Stanford Linear Accelerator Center to produce a beam of polarized gamma rays of known energy. Plane polarized photons of $\lambda = 6943\,\text{Å}$ ($_L\varepsilon = 1.78\,\text{eV}$) collide head on with unpolarized electrons of 20 GeV ($_LE = 20\,\text{GeV}$). Calculate the range of energies and angles of the scattered photon in (a) the rest frame of the incident electron and (b) the laboratory frame.

4-17. We continue the analysis of the Compton scattering situation of the preceding problem. In the scattering plane sketched, the polarization vector of the incident photon e is perpendicular to π and that of the scattered photon e' is perpendicular to π'. Denote the angle between e and e' by Θ. In the rest frame of the incident electron (R_2), the differential scattering cross section (averaged over electron spin orientations) is given by the Klein–Nishina formula [W. Heitler, *Quantum Theory of Radiation*, 3rd ed. (Clarendon Press, Oxford, 1954), p. 217].

$$\frac{d\sigma}{d\Omega} = \frac{1}{4} \left(\frac{e_0^2}{\mu_e c^2}\right)^2 \left(\frac{_2\varepsilon'}{_2\varepsilon}\right)^2 \left[\frac{_2\varepsilon}{_2\varepsilon'} + \frac{_2\varepsilon'}{_2\varepsilon} - 2 + 4\cos^2{_2\Theta}\right]$$

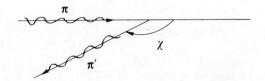

If **e** is in the scattering plane, $_2\Theta = {}_2\chi$ corresponds to no change of polarization in scattering ($\parallel$), whereas $_2\Theta = \pi/2$ corresponds to orthogonal polarization of the scattered photon ($\perp$). If **e** is perpendicular to the scattering plane, $_2\Theta = 0$ corresponds to no change of polarization ($\parallel$), whereas $_2\Theta = \pi/2$ corresponds to orthogonal polarization ($\perp$). The quantities $_2\varepsilon$, $_2\varepsilon'$, and $_2\chi$ are related by the equation derived in the worked example of Section 4-9

$$1 - \cos {}_2\chi = \mu_e\left(\frac{1}{{}_2\varepsilon'} - \frac{1}{{}_2\varepsilon}\right)$$

Calculate the depolarization of the back scattered ($_2\chi = \pi$) laser photons:

$$\frac{(d\sigma/d\Omega)_\perp}{(d\sigma/d\Omega)_\parallel + (d\sigma/d\Omega)_\perp}$$

4-18. A proton of momentum 70 GeV/c with respect to the laboratory collides with a proton in a target nucleus. The target proton happens to be approaching the beam proton head-on with a momentum of 0.20 GeV/c with respect to the laboratory. Determine the invariant mass of the two-proton system. Compare with the result for the case that the target proton is at rest.

Returning to the case of the moving target, calculate the velocity of the center-of-mass frame in the laboratory. Determine the momentum and energy of each proton in this frame.

Among the particles resulting from the collision is a Λ hyperon. In the center-of-mass frame, it is emitted at right angles to the direction of the protons with a momentum of 2.0 GeV/c. It lives for 2.0×10^{-10} sec in its rest frame before breaking up.

How long does it live in the center-of-mass frame? What is its momentum in the laboratory (magnitude and direction)? How far does it move in the laboratory before disintegrating?

4-19. The reaction

$$\pi^- + p \to p + B^-$$
$$\quad\quad\quad \hookrightarrow \pi^- + \omega$$
$$\quad\quad\quad\quad\quad \hookrightarrow \pi^+ + \pi^- + \pi^\circ$$

has been studied at incident pion energies of 5.0 and 7.5 GeV. Derive an expression for the momentum of the outgoing proton in the rest frame of B in terms of the proper masses and the beam energy in the laboratory where the target proton is at rest. The answer is

$$p = \frac{[(\mu_B{}^2 - 2\mu_P E_{\text{beam}} - \mu_\pi{}^2)^2 - 4\mu_B{}^2\mu_P{}^2]^{1/2}}{2\mu_B}$$

Evaluate for the two energies mentioned.

4-20. Particles a and b come together; the result of the interaction is four free particles: c, 1, 2, and 3.

$$a + b \rightarrow c + 1 + 2 + 3$$

Show that in the center of mass frame, c has its greatest momentum when 1, 2, and 3 move with equal velocity in the direction opposite to c, like a single particle of invariant mass $\mu_1 + \mu_2 + \mu_3$.

4-21. Consider the possible reaction

$$\pi^- + p \rightarrow \Omega^- + K^+ + K^0 + K^0$$

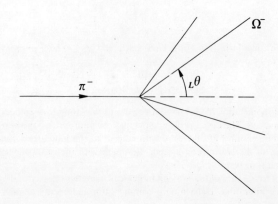

where in the laboratory frame the pion has momentum 7.5 GeV/c and the proton is at rest. What is the largest angle $_L\theta$ in the laboratory frame at which the Ω^- can be produced? [Hint: Use the result of the preceding problem.]

4-22. Suppose a 20 GeV photon from the Stanford linear accelerator strikes a proton in a liquid hydrogen target. It might produce a pair of particles W and $\overline{W}$ of equal proper mass, in the reaction

$$\gamma + p \rightarrow p + W + \overline{W}$$

What is the highest proper mass the "W particle" might have without violating conservation of four-momentum? Assuming the reaction just barely occurs (i.e., 20 GeV is threshold), what are the laboratory momenta of the three particles in the final state?

4-23. Consider the inelastic transition

$$\pi^- + p \rightarrow K^0 + \Lambda^0$$

The target proton is at rest in the laboratory. What is the kinetic energy of the pion at threshold?

In an experiment using pions of momentum 2.50 GeV/c, an example of this transition is observed in which the Λ^0 has momentum 0.600 GeV/c at an angle of 45° with respect to the incident pion. What is the momentum (magnitude and direction) of the K in the laboratory? What is the velocity of the center of mass frame? In this frame, what is the momentum (magnitude and direction) of the Λ? Of the K? You do not need to do the different parts of this problem in the order stated.

Chapter 5

Successive Lorentz Transformations. Motion of Spin

We consider now the compounding of successive reference frame transformations. We need to know what happens to four-vector components when a Lorentz transformation from frame R to R' is followed by one from R' to R''. There are two main applications. One is to the motion of the spin of an accelerated system, such as an atomic electron. Here we need to evaluate quantities in the laboratory frame, in the rest frame of the electron at one instant, and in the rest frame of the electron at another instant. The other is to the analysis of transitions that proceed in stages, such as

$$K^- + p \rightarrow \pi^- + Y_1^{*\,+}$$
$$\quad\quad\quad\quad \llcorner\!\!\rightarrow \Lambda + \pi^+$$
$$\quad\quad\quad\quad\quad\quad \llcorner\!\!\rightarrow p + \pi^-$$

Here one routinely needs to calculate four-momentum components in the laboratory frame, the center of mass frame, the rest frame of the Y_1^* hyperon, and the rest frame of the Λ hyperon.

5-1. TRANSFORMATION MATRICES IN SPACETIME

Let us evaluate some transformation matrices in spacetime. The utility of matrices rests on the fact that compounding transformations corresponds to multiplying matrices. The transformation

$$A'_\mu = a_{\mu\nu} A_\nu \tag{2-52}$$

followed by

$$A''_\lambda = b_{\lambda\mu} A'_\mu$$

258

is evidently

$$A_\lambda'' = c_{\lambda\nu} A_\nu$$

with

$$c_{\lambda\nu} = b_{\lambda\mu} a_{\mu\nu}$$

We have here simply the matrix multiplication law

$$\mathbf{c} = \mathbf{b} \cdot \mathbf{a}$$

that is,

$$
\begin{pmatrix}
c_{11} & c_{12} & c_{13} & c_{14} \\
c_{21} & c_{22} & c_{23} & c_{24} \\
c_{31} & c_{32} & c_{33} & c_{34} \\
c_{41} & c_{42} & c_{43} & c_{44}
\end{pmatrix}
=
\begin{pmatrix}
b_{1\mu} a_{\mu 1} & b_{1\mu} a_{\mu 2} & b_{1\mu} a_{\mu 3} & b_{1\mu} a_{\mu 4} \\
b_{2\mu} a_{\mu 1} & b_{2\mu} a_{\mu 2} & b_{2\mu} a_{\mu 3} & b_{2\mu} a_{\mu 4} \\
b_{3\mu} a_{\mu 1} & b_{3\mu} a_{\mu 2} & b_{3\mu} a_{\mu 3} & b_{3\mu} a_{\mu 4} \\
b_{4\mu} a_{\mu 1} & b_{4\mu} a_{\mu 2} & b_{4\mu} a_{\mu 3} & b_{4\mu} a_{\mu 4}
\end{pmatrix}
$$

$$
=
\begin{pmatrix}
b_{11} & b_{12} & b_{13} & b_{14} \\
b_{21} & b_{22} & b_{23} & b_{24} \\
b_{31} & b_{32} & b_{33} & b_{34} \\
b_{41} & b_{42} & b_{43} & b_{44}
\end{pmatrix}
\cdot
\begin{pmatrix}
a_{11} & a_{12} & a_{13} & a_{14} \\
a_{21} & a_{22} & a_{23} & a_{24} \\
a_{31} & a_{32} & a_{33} & a_{34} \\
a_{41} & a_{42} & a_{43} & a_{44}
\end{pmatrix}
$$

The λ, ν element of $\mathbf{c}$ is the product of corresponding elements of the λ row of $\mathbf{b}$ with the ν column of $\mathbf{a}$. These 4×4 matrices tell us how the components of every four-vector change when the coordinate system undergoes the corresponding transformation.

It is easy to write down the 4×4 matrices for rotation or inversion: One simply borders the relevant 3×3 matrices with zeros. Thus, a rotation of axes through the angle θ about the x_3 axis gives rise to the matrix

$$\text{Rot}_3(\theta) =
\begin{pmatrix}
\cos\theta & \sin\theta & 0 & 0 \\
-\sin\theta & \cos\theta & 0 & 0 \\
0 & 0 & 1 & 0 \\
0 & 0 & 0 & 1
\end{pmatrix}
\tag{5-1}$$

meaning that an arbitrary four-vector $\tilde{A}$ has its components in the two coordinate systems related by

$$A'_1 = \cos \theta \cdot A_1 + \sin \theta \cdot A_2 + 0 \cdot A_3 + 0 \cdot A_4$$
$$A'_2 = -\sin \theta \cdot A_1 + \cos \theta \cdot A_2 + 0 \cdot A_3 + 0 \cdot A_4$$
$$A'_3 = 0 \cdot A_1 + 0 \cdot A_2 + 1 \cdot A_3 + 0 \cdot A_4$$
$$A'_4 = 0 \cdot A_1 + 0 \cdot A_2 + 0 \cdot A_3 + 1 \cdot A_4$$

We have already written down the matrix for the special Lorentz transformation (SLT), Eq. (2-49c). We denote it $L_1(u)$, L standing for Lorentz and the subscript 1 indicating that the x_1, x'_1 axes have been chosen to be parallel to the direction of relative motion of the reference frames,

$$L_1(u) = \begin{pmatrix} \gamma & 0 & 0 & i\beta\gamma \\ 0 & 1 & 0 & 0 \\ 0 & 0 & 1 & 0 \\ -i\beta\gamma & 0 & 0 & \gamma \end{pmatrix} \tag{5-2}$$

A SLT in which the number 3 axes are so honored is described by the matrix

$$L_3(u) = \begin{pmatrix} 1 & 0 & 0 & 0 \\ 0 & 1 & 0 & 0 \\ 0 & 0 & \gamma & i\beta\gamma \\ 0 & 0 & -i\beta\gamma & \gamma \end{pmatrix}$$

The matrix representing a general homogeneous Lorentz transformation is simply the matrix product of matrices like Eqs. (5-1) and (5-2) corresponding to the various transformations carried out one after the other.

Let us evaluate the matrix elements for a pure Lorentz transformation (PLT) or "boost," in which the rotations are taken about the x_3 and x'_3 axes. Such would be the case shown in Fig. 2-10, where the first transformation is $Rot_3(40°)$, the second is $L_1(u)$ with $u = u_R(R')$, and the third is $Rot_3(-40°)$. Denoting the matrix of such a PLT by[1] $P(\theta, u)$,

$$P(\theta, u) = Rot_3(-\theta) \cdot L_1(u) \cdot Rot_3(\theta) \tag{5-3a}$$

---

[1] It might be appropriate to think of P as standing for Poincaré, who emphasized [9] the essential invariance features of the Lorentz transformation. In fact, mathematical physicists call the group of general Lorentz transformations (including shift of origin) the Poincaré group.

The matrix $\mathsf{Rot}_3(\theta)$ converts (A_1, A_2, A_3, A_4) to $(A_\|, A_\perp, A_3, A_4)$. Then $\mathsf{L}_1(u)$ converts these components to $(A'_\|, A'_\perp, A'_3, A'_4)$; and, finally, $\mathsf{Rot}_3(-\theta)$ converts these to (A'_1, A'_2, A'_3, A'_4). Thus, $\mathsf{P}(\theta, u)$ converts the components of $\tilde{A}$ from their values in $Ox_1x_2x_3$ in R to their values in $O'x'_1x'_2x'_3$ in R'.

None of the transformations affects the number 3 component. One could omit the third row and column of each matrix. We shall keep the full 4×4 form for the time being.

Substituting from Eqs. (5-1) and (5-2), we get

$$\mathsf{L}_1(u) \cdot \mathsf{Rot}_3(\theta) = \begin{pmatrix} \gamma \cos\theta & \gamma \sin\theta & 0 & i\beta\gamma \\ -\sin\theta & \cos\theta & 0 & 0 \\ 0 & 0 & 1 & 0 \\ -i\beta\gamma\cos\theta & -i\beta\gamma\sin\theta & 0 & \gamma \end{pmatrix}$$

so that Eq. (5-3a) becomes

$$\mathsf{P}(\theta, u) = \begin{pmatrix} (\gamma - 1)\cos^2\theta + 1 & (\gamma - 1)\sin\theta\cos\theta & 0 & i\beta\gamma\cos\theta \\ (\gamma - 1)\sin\theta\cos\theta & (\gamma - 1)\sin^2\theta + 1 & 0 & i\beta\gamma\sin\theta \\ 0 & 0 & 1 & 0 \\ -i\beta\gamma\cos\theta & -i\beta\gamma\sin\theta & 0 & \gamma \end{pmatrix} \quad (5\text{-}3b)$$

Let us check our manipulations. If $\theta = 0$,

$$\mathsf{P}(0, u) = \mathsf{L}_1(u)$$

If $\theta = \pi/2$,

$$\mathsf{P}\left(\frac{\pi}{2}, u\right) = \mathsf{L}_2(u)$$

If $u = 0$, then $\gamma = 1$ and $\beta = 0$, so that

$$\mathsf{P}(\theta, 0) = \mathsf{I}$$

the identity matrix. The resultant of equal and opposite rotations is unity.

We can without loss of generality limit u to positive values. With $u \geq 0$, $\mathsf{P}(\theta + \pi, u)$ corresponds to the transformation with reversed relative velocity from $\mathsf{P}(\theta, u)$.

For the general PLT, there are two angle parameters, specifying the orientation of $\mathbf{u}_R(R')$ with respect to the $Ox_1x_2x_3$ axis frame—for example, spherical polar angles (θ, φ).

EXERCISE   The weathervane example of Section 2-5 was worked out on the basis of Eq. (2-30c), derived by straightforward use of a SLT. As a check of Eq. (5-3b), derive the same result using a PLT between the vane's rest frame and the ground frame, with number 1 axis in the former along the vane.

SOLUTION   Figure 5-1 shows the situation in each reference frame, the rest frame of the vane $(R^\dagger)$ and the ground frame (R). The vane is a horizontal stick pointing northeast in its rest frame. We determine its motion in R by following two of its points, the center C and the arrowhead A. In $R^\dagger$, we put the origin $O^\dagger$ at C and the $x^\dagger$ axis along the stick. The coordinates of C and A are $(0, 0, 0)$ and $(\lambda, 0, 0)$. The rest frame of the vane is moving east with constant velocity $u = 0.19c$, and θ, the angle between $-\mathbf{u}_R{}^\dagger(R)$ and the vane, is 45°. We put the $y^\dagger$ axis in the plane of $\mathbf{u}_R{}^\dagger(R)$ and the vane.

Going from R to $R^\dagger$ the transformation matrix is $P(-45°, u)$. Since we know the coordinates of C and A in $R^\dagger$, it is more convenient to carry out the reverse transformation from $R^\dagger$ to R, which is $P(135°, u)$, or

$$\begin{pmatrix} (\gamma - 1)/2 + 1 & -(\gamma - 1)/2 & 0 & -i\beta\gamma/\sqrt{2} \\ -(\gamma - 1)/2 & (\gamma - 1)/2 + 1 & 0 & i\beta\gamma/\sqrt{2} \\ 0 & 0 & 1 & 0 \\ i\beta\gamma/\sqrt{2} & -i\beta\gamma/\sqrt{2} & 0 & \gamma \end{pmatrix}$$

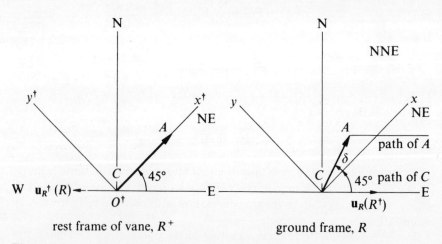

rest frame of vane, R^+ ground frame, R

Figure 5-1.   Weather vane CA pointing northeast in its rest frame points north–northeast in ground frame $(\gamma = 2.4)$.

This means that

$$A_1 = \left(\frac{\gamma - 1}{2} + 1\right)A_1^\dagger - \left(\frac{\gamma - 1}{2}\right)A_2^\dagger - \left(\frac{i\beta\gamma}{\sqrt{2}}\right)A_4^\dagger$$

$$A_2 = -\left(\frac{\gamma - 1}{2}\right)A_1^\dagger + \left(\frac{\gamma - 1}{2} + 1\right)A_2^\dagger + \left(\frac{i\beta\gamma}{\sqrt{2}}\right)A_4^\dagger$$

$$A_3 = A_3^\dagger$$

$$A_4 = \left(\frac{i\beta\gamma}{\sqrt{2}}\right)A_1^\dagger - \left(\frac{i\beta\gamma}{\sqrt{2}}\right)A_2^\dagger + \gamma A_4^\dagger$$

Thus, an event at C has the coordinates in R

$$x_1(C) = -\left(\frac{i\beta\gamma}{\sqrt{2}}\right)x_4^\dagger(C)$$

$$x_2(C) = +\left(\frac{i\beta\gamma}{\sqrt{2}}\right)x_4^\dagger(C)$$

$$x_3(C) = 0$$

$$x_4(C) = \gamma x_4^\dagger(C)$$

Elimination of $x_4^\dagger(C)$ gives, for the motion of C in R,

$$x(C) = x_1(C) = -\left(\frac{i\beta}{\sqrt{2}}\right)x_4(C) = \left(\frac{\beta c}{\sqrt{2}}\right)t(C)$$

$$y(C) = x_2(C) = \left(\frac{i\beta}{\sqrt{2}}\right)x_4(C) = -\left(\frac{\beta c}{\sqrt{2}}\right)t(C)$$

as drawn on the figure. An event at A has in R the coordinates

$$x_1(A) = \left(\frac{\gamma - 1}{2} + 1\right)\lambda - \left(\frac{i\beta\gamma}{\sqrt{2}}\right)x_4^\dagger(A)$$

$$x_2(A) = -\left(\frac{\gamma - 1}{2}\right)\lambda + \left(\frac{i\beta\gamma}{\sqrt{2}}\right)x_4^\dagger(A)$$

$$x_3(A) = 0$$

$$x_4(A) = \left(\frac{i\beta\gamma}{\sqrt{2}}\right)\lambda + \gamma x_4^\dagger(A)$$

Elimination of $x_4^\dagger(A)$ gives

$$x(A) = x_1(A) = \left(\frac{\beta c}{\sqrt{2}}\right)t(A) + \frac{\lambda}{2}\frac{\gamma + 1}{\gamma}$$

$$y(A) = x_2(A) = -\left(\frac{\beta c}{\sqrt{2}}\right)t(A) + \frac{\lambda}{2}\frac{\gamma - 1}{\gamma}$$

as shown on the figure. Points C and A, like every point at rest in $R^\dagger$, move in R with the velocity $\mathbf{u}_R(R')$. We find the length and orientation of the vane in R from the space coordinates of C and A at the same time in R

$$t(A) = t(C)$$

With this condition, the deflection of the pointer in R is

$$\tan \delta = \frac{y(A) - y(C)}{x(A) - x(C)} = \frac{\gamma - 1}{\gamma + 1}$$

in agreement with Eq. (2-30d). With $\gamma = 2.4$,

$$\tan \delta = \frac{1.4}{3.4} = 0.411 = \tan 22.4°$$

The vane points north–northeast in the ground frame.

5-2. COMPOSITION OF REFERENCE-FRAME TRANSFORMATIONS

The discussion of transformation of particle velocity in Section 2-8 has provided the basic formulas, because the velocity of a particle P is the same as the velocity of the reference frame in which P is at rest (we limit our considerations to inertial reference frames, in pure translation with respect to one another). Since numerical indices are easier to handle than primes, let us denote the three frames by R_0, R_1, R_2. The velocity with respect to R_i of a point at rest in R_j is written $\mathbf{u}_i(j)$. The angle in R_i between the velocity of a point at rest in R_j and the velocity of a point at rest in R_k, $\angle[\mathbf{u}_i(j), \mathbf{u}_i(k)]$, is denoted by χ_i. The nonrelativistic vector velocity triangle of Fig. 2-32 gets the labels shown in Fig. 5-2. The relativistic linking of time and longitudinal position leads to the velocity transformation Eqs. (2-42) and (2-43). The relative velocity vectors still lie in a plane in velocity space. One can still draw the triangle of Fig. 5-2 as a way of summarizing the relations between angles and velocities, but one cannot apply Euclidean plane geometry to it. The first of Eqs. (2-43),

$$[u_0(2)]^2 = \frac{[u_0(1)]^2 + [u_1(2)]^2 - 2u_0(1) \cdot u_1(2) \cos \chi_1 - [u_0(1) \cdot u_1(2) \sin \chi_1/c]^2}{[1 - u_0(1) \cdot u_1(2) \cos \chi_1/c^2]^2}$$

(5-4a)

is inconsistent (unless $c = \infty$) with the law of cosines

$$[u_0(2)]^2 = [u_0(1)]^2 + (u_1(2))^2 - 2u_0(1)u_1(2) \cos \chi_1$$

and the second

$$\tan \chi = \left[1 - \left(\frac{u_0(1)}{c}\right)^2\right]^{1/2} \frac{u_1(2) \sin \chi_1}{u_0(1) - u_1(2) \cos \chi_1}$$

(5-4b)

is similarly inconsistent with the trigonometric identity

$$\tan \chi_0 = \frac{u_1(2) \sin \chi_1}{u_0(1) - u_1(2) \cos \chi_1}$$

One can still assume that the angles at any one vertex add to a straight angle; for example,

$$\chi_1 + \angle[\mathbf{u}_1(2), -\mathbf{u}_1(0)] = \pi$$

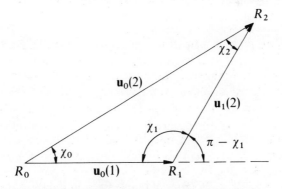

Figure 5-2. Velocity triangle between frames R_0, R_1, and R_2. Ordinary trigonometry is only valid in the nonrelativistic limit.

but one cannot assume that the angles of the triangle add to a straight angle.

$$\pi - \chi_1 \neq \chi_0 + \chi_2$$

It will turn out to be possible to map the triangle so as to permit use of a non-Euclidean geometry.

Composition of Parallel Lorentz Transformations

Before treating the general case, let us consider the simpler one in which the triangle is squeezed down to one dimension. The translations of the reference frames are assumed to be pair-by-pair parallel (Fig. 5-3)

$$\mathbf{u}_0(1)\|\mathbf{u}_0(2) \qquad \mathbf{u}_1(0)\|\mathbf{u}_1(2) \qquad \mathbf{u}_2(0)\|\mathbf{u}_2(1)$$

We locate Cartesian axes in the three frames as follows. In R_0, the x_0 axis is chosen in the direction and sense of $\mathbf{u}_0(1)$. Some arbitrary origin O_0 and orientation of the y_0 axis are selected. In R_1, the x_1 axis is taken collinear with the path of O_0 and in the opposite sense. An origin O_1 is picked on it, and the y_1 axis is oriented to be in the plane swept out in R_1 by the y_0 axis. The origins of t_0 and t_1 are put at the instant of coincidence of O_1 and O_0. Similarly, in R_2 the x_2 axis is collinear with the path of O_1 and oppositely directed. The origin O_2 is the point of this axis coinciding with O_1 when it coincides with O_0. The origin of t_2 is placed at this event. The y_2 axis is in the plane swept out in R_2 by the y_1 axis. Figure 5-4 shows the three axis systems in R_0 at an instant $t_0 > 0$. With this choice of coordinate systems,

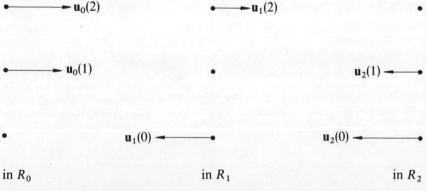

in R_0 in R_1 in R_2

Figure 5-3.   Relative velocities of points fixed in the three frames, shown, left to right, in R_0, R_1, and R_2, and, top to bottom, for points at rest in R_2, R_1, and R_0.

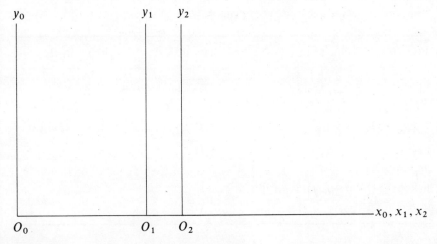

Figure 5-4.   The coordinate axes in R_0.

the transformations under consideration are special Lorentz transformations.

In Eq. (2-42), the velocity components are given in the present notation by

$$v_x = u_0(2) \qquad v'_x = u_1(2) \qquad u = u_0(1)$$

$$v_y = 0 \qquad v'_y = 0$$

$$v_z = 0 \qquad v'_z = 0$$

so that the first equation of Eq. (2-42) reads

$$u_0(2) = \frac{u_0(1) + u_1(2)}{1 + u_0(1)u_1(2)/c^2} \tag{5-5a}$$

in obvious disagreement with the one-dimensional form of Galilean velocity addition [Eq. (1-21)]

$$u_0(2) = u_0(1) + u_1(2)$$

Of course, the result of the SLT from R_0 to R_1 followed by the SLT from R_1 to R_2 is the SLT from R_0 to R_2. It is just that in successive SLT's the relative velocities u do not combine *additively*.

All the relative velocities are displayed in Table 5-1, which was used in drawing Fig. 5-3.

Table 5-1

Point	Velocity in R_0	Velocity in R_1	Velocity in R_2
Fixed in R_0	0	$u_1(0) = -u_0(1)$	$u_2(0) = -u_0(2) =$ $\dfrac{-[u_0(1) + u_1(2)]}{1 + u_0(1)u_1(2)/c^2}$
Fixed in R_1	$u_0(1)$	0	$u_2(1) = -u_1(2)$
Fixed in R_2	$u_0(2) = \dfrac{u_0(1) + u_1(2)}{1 + u_0(1)u_1(2)/c^2}$	$u_1(2)$	0

If we divide Eq. (5-5a) by c, it becomes

$$\beta_0(2) = \frac{\beta_0(1) + \beta_1(2)}{1 + \beta_0(1)\beta_1(2)} \tag{5-5b}$$

which is of the same form as the equation for the hyperbolic tangent of the sum of two arguments

$$\tanh(a_1 + a_2) = \frac{\tanh a_1 + \tanh a_2}{1 + \tanh a_1 \tanh a_2}$$

Like β, the hyperbolic tangent can only take on values in the range -1 to 1. It is thus convenient to introduce a new parameter α to specify the velocity, defined by

$$\alpha = \tanh^{-1}(\beta) \qquad \beta = \tanh \alpha \tag{5-6a}$$

As β ranges from -1 to $+1$, α ranges from $-\infty$ to $+\infty$. It is sometimes called the "rapidity." The nonrelativistic region ($|\beta| \approx 0$) has $|\alpha| \approx 0$. The extreme relativistic region ($|\beta| \approx 1$) has $|\alpha| \gg 1$. Since $\gamma = (1 - \beta^2)^{-1/2}$, we have, identically,

$$\gamma = \cosh \alpha \qquad \beta\gamma = \sinh \alpha \tag{5-6b}$$

The identities involving the hyperbolic functions are helpful in manipulating complicated expressions built up of β's and γ's. The composition law [Eq. (5-5b)] reads

$$\tanh \alpha_0(2) = \frac{\tanh \alpha_0(1) + \tanh \alpha_1(2)}{1 + \tanh \alpha_0(1) \cdot \tanh \alpha_1(2)}$$

so that

$$\alpha_0(2) = \alpha_0(1) + \alpha_1(2) \tag{5-5c}$$

Just as a rotation through an angle θ_1 followed by a rotation about the same axis through an angle θ_2 is equivalent to a rotation through the angle

$(\theta_1 + \theta_2)$, so a SLT of parameter α_1 followed by a SLT, in the same direction, of parameter α_2 is equivalent to a SLT of parameter $(\alpha_1 + \alpha_2)$.

One can verify the last sentence by multiplication of the transformation matrices. By Eq. (5-1),

$$
\text{Rot}_3(\theta_1) = \begin{pmatrix} \cos\theta_1 & \sin\theta_1 & 0 & 0 \\ -\sin\theta_1 & \cos\theta_1 & 0 & 0 \\ 0 & 0 & 1 & 0 \\ 0 & 0 & 0 & 1 \end{pmatrix}
$$

so matrix multiplication gives

$$
\text{Rot}_3(\theta_2) \cdot \text{Rot}_3(\theta_1) =
$$

$$
\begin{pmatrix} \cos\theta_2\cos\theta_1 - \sin\theta_2\sin\theta_1 & \cos\theta_2\sin\theta_1 + \sin\theta_2\cos\theta_1 & 0 & 0 \\ -\sin\theta_2\cos\theta_1 - \cos\theta_2\sin\theta_1 & -\sin\theta_2\sin\theta_1 + \cos\theta_2\cos\theta_1 & 0 & 0 \\ 0 & 0 & 1 & 0 \\ 0 & 0 & 0 & 1 \end{pmatrix}
$$

The identities

$$
\cos(\theta_2 + \theta_1) = \cos\theta_2\cos\theta_1 - \sin\theta_2\sin\theta_1
$$
$$
\sin(\theta_2 + \theta_1) = \sin\theta_2\cos\theta_1 + \cos\theta_2\sin\theta_1
$$

show that this matrix is $\text{Rot}_3(\theta_2 + \theta_1)$.

For the special Lorentz transformations, Eq. (5-2) reads

$$
L_1(\alpha_1) = \begin{pmatrix} \cosh\alpha_1 & 0 & 0 & i\sinh\alpha_1 \\ 0 & 1 & 0 & 0 \\ 0 & 0 & 1 & 0 \\ -i\sinh\alpha_1 & 0 & 0 & \cosh\alpha_1 \end{pmatrix}
$$

Matrix multiplication gives

$$
L_1(\alpha_2) \cdot L_1(\alpha_1) = \begin{pmatrix} \cosh\alpha_2\cosh\alpha_1 + \sinh\alpha_2\sinh\alpha_1 & 0 & 0 & i\cosh\alpha_2\sinh\alpha_1 + i\sinh\alpha_2\cosh\alpha_1 \\ 0 & 1 & 0 & 0 \\ 0 & 0 & 1 & 0 \\ -i\sinh\alpha_2\cosh\alpha_1 - i\cosh\alpha_2\,\text{sihn}\,\alpha_1 & 0 & 0 & \sinh\alpha_2\sinh\alpha_1 + \cosh\alpha_2\cosh\alpha_1 \end{pmatrix}
$$

The identities

$$\cosh(\alpha_2 + \alpha_1) = \cosh \alpha_2 \cosh \alpha_1 + \sinh \alpha_2 \sinh \alpha_1$$

$$\sinh(\alpha_2 + \alpha_1) = \sinh \alpha_1 \cosh \alpha_2 + \cosh \alpha_2 \sinh \alpha_1$$

show that this matrix is $L_1(\alpha_2 + \alpha_1)$. Thus,

$$L_1(\alpha_2 + \alpha_1) = L_1(\alpha_2) \cdot L_1(\alpha_1) \qquad \text{Q.E.D.}$$

This example also gives explicit verification of the previous assertion, that SLT's in the same direction have the group property: The product of two SLT's is a SLT.

The Velocity Parameter ("Rapidity") α and its Hyperbolic Functions

Use of the parameter α instead of u or β or γ makes the equation of a special Lorentz transformation look more like that of a rotation. For a rotation about the z axis,

$$\begin{aligned} x' &= \cos \theta \cdot x + \sin \theta \cdot y \\ y' &= -\sin \theta \cdot x + \cos \theta \cdot y \end{aligned} \qquad \text{(A-2a)}$$

Then the invariance property

$$x'^2 + y'^2 = x^2 + y^2$$

follows from the identity

$$\cos^2 \theta + \sin^2 \theta = 1$$

Analogously, the SLT Eq. (1-31b) now has the form

$$\begin{aligned} ct' &= \cosh \alpha \cdot ct - \sinh \alpha \cdot x \\ x' &= -\sinh \alpha \cdot ct + \cosh \alpha \cdot x \end{aligned} \qquad \text{(1-31f)}$$

and its invariance property

$$c^2 t'^2 - x'^2 = c^2 t^2 - x^2$$

is a consequence of the fundamental identity

$$\cosh^2 \alpha - \sinh^2 \alpha = 1$$

The matrix of the SLT from R to R' is, as we have noted in the example,

$$L_1(\alpha) = \begin{pmatrix} \cosh \alpha & 0 & 0 & i \sinh \alpha \\ 0 & 1 & 0 & 0 \\ 0 & 0 & 1 & 0 \\ -i \sinh \alpha & 0 & 0 & \cosh \alpha \end{pmatrix} \qquad \text{(5-7a)}$$

The four-vector transformation [Eqs. (2-49bb)] is, of course,

$$
\begin{aligned}
A'_0 &= \cosh\alpha \cdot A_0 - \sinh\alpha \cdot A_1 \\
A'_1 &= \cosh\alpha \cdot A_1 - \sinh\alpha \cdot A_0 \\
A'_2 &= A_2 \\
A'_3 &= A_3
\end{aligned}
\tag{5-7b}
$$

The usefulness of α and its hyperbolic functions is not limited to transformations in one direction. For the pure Lorentz transformation [Eq. (5-3b)], we can simply express β and γ in terms of α, obtaining

$$
P(\theta, \alpha) =
\begin{pmatrix}
(\cosh\alpha - 1)\cos^2\theta + 1 & (\cosh\alpha - 1)\sin\theta\cos\theta & 0 & i\sinh\alpha\cos\theta \\
(\cosh\alpha - 1)\sin\theta\cos\theta & (\cosh\alpha - 1)\sin^2\theta + 1 & 0 & i\sinh\alpha\sin\theta \\
0 & 0 & 1 & 0 \\
-i\sinh\alpha\cos\theta & -i\sinh\alpha\sin\theta & 0 & \cosh\alpha
\end{pmatrix}
\tag{5-8}
$$

of which Eq. (5-7a) is the limiting case $\theta = 0$.

The hyperbolic functions of α have a direct kinematical interpretation in terms of *the motion in R of a particle at rest in R'*. Its velocity $\mathbf{v}(R')$ has the components

$$
\begin{aligned}
v_x &= (c\tanh\alpha)\cos\chi \\
v_y &= (c\tanh\alpha)\sin\chi \\
v_z &= 0
\end{aligned}
\tag{5-9}
$$

where χ is the angle in R between the velocity and the x axis. The y axis is in the plane of these two vectors. Its four-velocity, $\tilde{w}(R') = (\gamma\mathbf{v}, ic\gamma)$, thus has the components

$$
\begin{aligned}
w_1 &= c\sinh\alpha\cos\chi \\
w_2 &= c\sinh\alpha\sin\chi \\
w_3 &= 0 \\
w_4 &= ic\cosh\alpha \qquad w_0 = c\cosh\alpha
\end{aligned}
\tag{5-10}
$$

The transformation of α when the motion of the particle fixed in R' is referred to a different inertial frame, say S, is readily found from the four-vector transformation of $\tilde{w}(R')$. Let (Fig. 5-5) χ_R be the angle in R between the velocities of R' and S with respect to R, and χ_S be the angle in S between the velocities of R and R' with respect to S. The SLT Eq. (5-7b) from R to S applied to $\tilde{w}(R')$

$$w_S(R')_0 = \cosh \alpha_R(S) \cdot w_R(R')_0 - \sinh \alpha_R(S) \cdot w_R(R')_1$$

$$w_S(R')_1 = \cosh \alpha_R(S) \cdot w_R(R')_1 - \sinh \alpha_R(S) \cdot w_R(R')_0$$

$$w_S(R')_2 = w_R(R')_2$$

$$w_S(R')_3 = w_R(R')_3$$

gives, by Eq. (5-10),

$$\cosh \alpha_S(R') = \cosh \alpha_R(S) \cdot \cosh \alpha_R(R') - \sinh \alpha_R(S) \cdot \sinh \alpha_R(R') \cdot \cos \chi_R$$

$$-\sinh \alpha_S(R') \cos \chi_S = \cosh \alpha_R(S) \cdot \sinh \alpha_R(R') \cdot \cos \chi_R$$

$$- \sinh \alpha_R(S) \cdot \cosh \alpha_R(R') \tag{5-11}$$

$$\sinh \alpha_S(R') \sin \chi_S = \sinh \alpha_R(R') \sin \chi_R$$

Equations (5-11) tell how $\cosh \alpha$ and $\sinh \alpha$ transform under a change of basic reference frame (we describe the motion of R' relative to S rather than relative to R). The transformation is essentially that of the timelike and

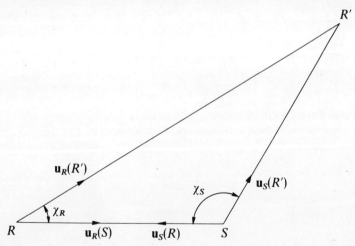

Figure 5-5.

spacelike components of the four-velocity or four-momentum of a particle at rest in R'.

The transformation equations, Eqs. (5-11), are, of course, essentially equivalent to the transformation equations for velocity, Eqs. (2-42), (2-43), and (5-4).

Composition of Lorentz Transformations in Different Directions

We consider the general case, depicted in Fig. 5-2. To save writing, we make the following notational simplification. We label a quantity with two indices simply by using the third:

$$\alpha_0 = |\alpha_1(2)| = |\alpha_2(1)|$$
$$\alpha_1 = |\alpha_2(3)| = |\alpha_3(2)|$$
$$\alpha_2 = |\alpha_0(1)| = |\alpha_1(0)|$$

and

$$u_0 = |u_2(1)| = |u_1(2)|, \quad \text{etc.}$$

The labeling of the velocity triangle is now as in Fig. 5-6. It agrees with the usual convention in trigonometry of pairing sides and opposite angles. Quotation marks are put on the labels to emphasize the symbolic character of the triangle.

In each reference frame there are two physically significant directions—the lines of motion of the other two reference frames. These define the angles at each vertex. The relations between the vertices are implicit in the velocity

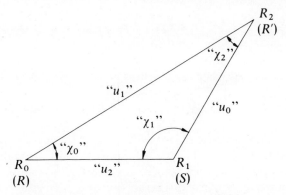

Figure 5-6. Diagrammatic representation of relative motion of the reference frames R_0, R_1, R_2.

transformation equations, Eqs. (5-11) or (5-4). In the new notation, Eqs. (5-11) read

$$\cosh \alpha_0 = \cosh \alpha_2 \cosh \alpha_1 - \sinh \alpha_2 \sinh \alpha_1 \cos \chi_0 \qquad (5\text{-}12\text{a})$$

$$-\sinh \alpha_0 \cos \chi_1 = \cosh \alpha_2 \sinh \alpha_1 \cos \chi_0 - \sinh \alpha_2 \cosh \alpha_1 \qquad (5\text{-}12\text{b})$$

$$\sinh \alpha_0 \sin \chi_1 = \sinh \alpha_1 \sin \chi_0 \qquad (5\text{-}12\text{c})$$

and Eqs. (5-4) are

$$\tanh^2 \alpha_1 = \frac{\left\{ \begin{matrix} \tanh^2 \alpha_0 + \tanh^2 \alpha_2 - 2 \tanh \alpha_0 \tanh \alpha_2 \cos \chi_1 \\ - \tanh^2 \alpha_0 \tanh^2 \alpha_2 \sin^2 \chi_1 \end{matrix} \right\}}{(1 - \tanh \alpha_0 \tanh \alpha_2 \cos \chi_1)^2}$$

$$(5\text{-}12\text{d})$$

$$\tan \chi_0 = \frac{1}{\cosh \alpha_2} \frac{\tanh \alpha_0 \sin \chi_1}{\tanh \alpha_2 - \tanh \alpha_0 \cos \chi_1} \qquad (5\text{-}12\text{e})$$

Each of these equations is the first of three, the others being obtained by cyclic permutation of the indices 0, 1, 2. They are simply ([57]; [58]) identities of hyperbolic trigonometry, obtained from those of Euclidean spherical trigonometry by treating the arc length as pure imaginary, and using

$$\cos(ix) = \cosh(x) \qquad \sin(ix) = i \sinh(x)$$

Figure 5-7 shows the hyperbolic triangle corresponding to the Eqs. (5-12). Thus, Eq. (5-12c) is simply the sine law

$$\frac{\sin \chi_0}{\sin(i\alpha_0)} = \frac{\sin \chi_1}{\sin(i\alpha_1)}$$

and Eq. (5-12a) is the cosine law of sides

$$\cos(i\alpha_0) = \cos(i\alpha_1) \cos(i\alpha_2) - \sin(i\alpha_1) \sin(i\alpha_2) \cos \chi_0$$

These two equations repeated around the triangle suffice for the derivation of the others. In effect, Eq. (5-12d) reduces to Eq. (5-12a) at the R_1 vertex. Equation (5-12e) follows from the sine law and Eq. (5-12b), and Eq. (5-12b) is a direct consequence of Eq. (5-12a) applied at R_0 and R_1.

Note that the sides are proportional to the velocity parameter α and not to the velocity u.

We shall need to evaluate the spherical defect ε, defined as

$$\varepsilon = \pi - (\chi_0 + \chi_1 + \chi_2) \tag{5-13}$$

In plane trigonometry, ε is zero. In spherical trigonometry, ε is negative and equal in absolute value to the area of the triangle (on unit sphere). In hyperbolic trigonometry, ε is positive. A useful expression in terms of three sides and an angle (compare, for example, [59], formulas 631 and 632) is

$$\sin\frac{\varepsilon}{2} = \left[\frac{(\cosh\alpha_2 - 1)(\cosh\alpha_0 - 1)}{2(\cosh\alpha_1 + 1)}\right]^{1/2} \sin\chi_1 \tag{5-14}$$

(There is, of course, one such formula for each vertex, obtained by cyclic advance of indices.)

The nonvanishing of ε means that the plane triangle of Fig. 5-6 is indeed only symbolic, for

$$\pi - \chi_1 \neq \chi_0 + \chi_2$$

One can legitimately draw vectors in one reference frame, but one may not take vectors from three reference frames, draw them in one plane, and expect sides and angles to link up in accordance with Euclidean geometry.

We proceed to calculate the effect of successive PLT's on the components of four-vectors.[2] We shall compare the indirect transformation from R_0

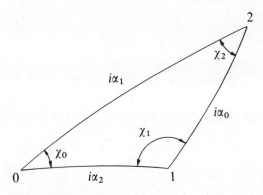

Figure 5-7. Hyperbolic triangle representing relative motion of R_0, R_1, and R_2.

---

[2] The following treatment of axis orientation, spin, and Thomas precession is based on unpublished notes by G. Ascoli.

to R_2 via R_1—a PLT from R_0 to R_1 followed by a PLT from R_1 to R_2—with the direct PLT from R_0 to R_2, starting, of course, from the same coordinate system in R_0. Denote the coordinate system reached in the two-step process by C_{012}, and that reached directly by C_{02}. Systems C_{012} and C_{02} are not necessarily identical. But we do know that for any four-vector $\tilde{A}$, its square A_μ^2 will have the same value in C_{012} and C_{02}, and so will the length squared of its spatial part A_m^2. Hence, C_{012} and C_{02} differ at most in the orientation of their space axes, thus by a rotation. We shall now prove that the space axes of C_{012} lag behind those of C_{02} by the angle ε—that is, in R_2 a rotation about the axis $\mathbf{u}_2(0) \times \mathbf{u}_2(1)$ by the angle ε brings the C_{012} axes into coincidence with those of C_{02}.

Start from a coordinate system C_0 in R_0 with (x_0, y_0) plane parallel to the plane of $\mathbf{u}_0(1)$ and $\mathbf{u}_0(2)$. Any x_0 axis orientation can be chosen in this plane. Suppose, for example, that it makes an angle of $-15°$ with $\mathbf{u}_0(1)$ (Fig. 5-8a). The PLT to R_1 leads to the coordinate system in R_1, C_{01}, shown

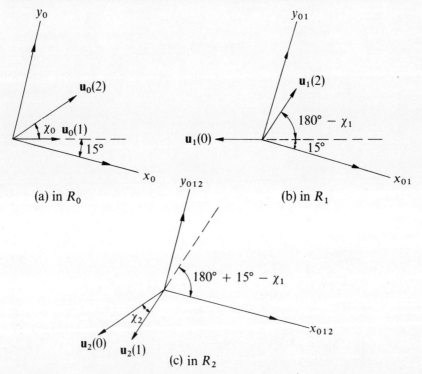

(a) in R_0

(b) in R_1

(c) in R_2

Figure 5-8. (a) The initial CS, C_0, in R_0, with axes x_0, y_0. In (b), we see the CS, C_{01}, in R_1 resulting from the PLT from C_0. Its axes are x_{01}, y_{01}. In (c), we see the CS, C_{012}, in R_2 resulting from the PLT from C_{01}. Its axes are x_{012}, y_{012}.

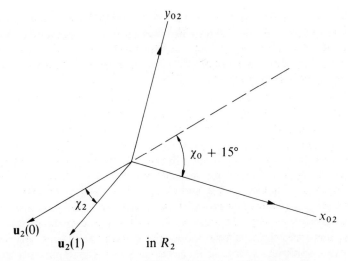

Figure 5-9.   The CS, C_{02}, resulting from the PLT from C_0.

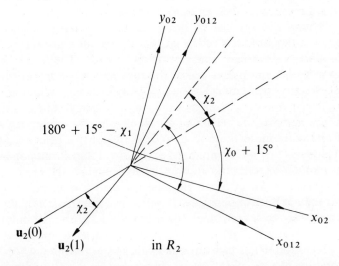

Figure 5-10.   Superposition of the CS's in R_2, C_{02} resulting from the direct PLT $R_0 \rightarrow R_2$ and C_{012} resulting from the succession of PLT's $R_0 \rightarrow R_1$, $R_1 \rightarrow R_2$. A rotation of C_{012} about its z_{012} axis by the angle ε brings its axes into coincidence with those of C_{02}.

in Fig. 5-8b. The x_{01} axis of C_{01} makes an angle of $-15°$ with $-\mathbf{u}_1(0)$. The PLT from C_{01} to the corresponding coordinate system in R_2, C_{012}, leads to the axis orientation in R_2 shown in Fig. 5-8c, the x_{012} axis making the angle $180° + 15° - \chi_1$ with $-\mathbf{u}_2(1)$. The direct PLT from R_0 to R_2 starting with the same C_0 leads, however, to the axis orientation in R_2 shown in Fig. 5-9, where the x_{02} axis makes the angle $\chi_0 + 15°$ with $-\mathbf{u}_2(0)$. The two figures in R_2, Figs. 5-8c and 5-9, are superposed in Fig. 5-10. The axis frame C_{02} obtained by direct transformation leads the axis frame C_{012} obtained by successive transformations by

$$(180° + 15° - \chi_1) - (\chi_2 + \chi_0 + 15°) = 180° - (\chi_1 + \chi_2 + \chi_0) = \varepsilon$$

The reader is urged to verify this result graphically by picking specific values for the angles. The figures were drawn with $\beta_2 = 1/2$, $\beta_0 = 3/4$, and $\chi_1 = 120°$, giving, by Eqs. (5-12a and c), $\chi_0 = 32.7°$ and $\chi_2 = 16.0°$.

The result just obtained shows that pure Lorentz transformations do not have the group property. Successive PLT's from a coordinate system in R_0 to R_1 and from the resulting coordinate system in R_1 to R_2 do not lead to the same coordinate system in R_2 as the direct PLT starting with the same coordinate system. If space rotations are included among the transformations considered, the group property is restored, for the result of successive PLT's, a PLT plus a rotation, is then an element of the group.

The "*rotation of axes due to successive Lorentz transformations*" has no effect on scalar products of the spacelike parts of four-vectors, for the scalar product is invariant under rotation of axes. The vector product continues to be the same pseudovector although its components are different. Every rotationally covariant equation (i.e., every equation written in correct vector or tensor notation) remains valid. In the example of the multistage reaction mentioned at the beginning of this chapter, the components in the Y_1^* rest frame of the momentum vectors of the different particles are different, depending on whether one computes them by direct transformation from the laboratory frame or indirectly by transforming first to the center of mass frame and thence to the Y_1^* frame. But one can analyze momentum relations (including angular distributions) just as well in one axis frame as in the other; it is only necessary to be consistent. In the kinematics of angular momentum, however, there is an effect of the defect ε on the spin orientation, known as the Thomas precession [60], which has significant physical consequences.

The rotation of axes effect can, of course, be described algebraically in terms of the transformation matrices for four-vector components. Our result is in matrix notation

$$\mathsf{P}(\chi_0 + \theta, \alpha_1) = \mathsf{Rot}_3(\varepsilon) \cdot \mathsf{P}(\pi - \chi_1 + \theta, \alpha_0) \cdot \mathsf{P}(\theta, \alpha_2) \qquad (5\text{-}15)$$

where θ is an arbitrary angle (15° in the preceding example). Multiplying both sides of Eq. (5-15) on the left by the inverse of $\mathsf{Rot}_3(\varepsilon)$—$\mathsf{Rot}_3(-\varepsilon)$— and on the right by the inverse of $\mathsf{P}(\chi_0 + \theta, \alpha_1)$—$\mathsf{P}(\pi + \chi_0 + \theta, \alpha_1)$—we obtain

$$\text{Rot}_3(-\varepsilon) = P(\pi - \chi_1 + \theta, \alpha_0) \cdot P(\theta, \alpha_2) \cdot P(\pi + \chi_0 + \theta, \alpha_1) \quad (5\text{-}16)$$

an equation giving the rotation matrix as a product of three PLT matrices. It means (Fig. 5-11) that if one starts with some coordinate system C_2 in R_2, transforms it by a PLT to R_0, transforms the coordinate system so obtained by a PLT to R_1, and finally transforms that coordinate system by a PLT to R_2, the resulting system C_{2012}, is behind C_2 by the angle ε. The reader can verify that the lag is zero in the parallel transformation case ($\chi_0 = 0$, $\chi_1 = \pi$, $\chi_2 = 0$), where the hyperbolic triangle has zero area.

Explicit calculation of the matrix elements is rather involved in the general case, and we limit ourselves in the next example to the important special case of perpendicular Lorentz transformations ($\chi_1 = \pi/2$).

EXERCISE Calculate the matrix elements of the right-hand side of Eq. (5-16) in the case $\chi_1 = \pi/2$. Show that they correspond to a rotation of the axis frame by the angle $-\varepsilon$ about the z_2 axis, with ε given by Eq. (5-14):

$$\sin\frac{\varepsilon}{2} = \left[\frac{(\cosh\alpha_2 - 1)(\cosh\alpha_0 - 1)}{2(\cosh\alpha_2 \cosh\alpha_0 + 1)}\right]^{1/2} \quad (5\text{-}17)$$

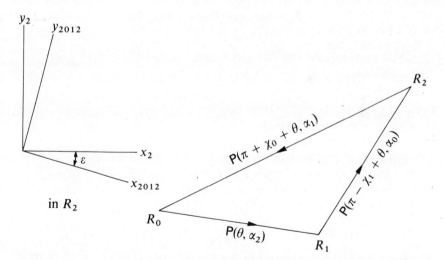

Figure 5-11. Cycle of PLT's. Starting from the CS x_2, y_2 in R_2, we make a succession of PLT's, $R_2 \to R_0$, $R_0 \to R_1$, $R_1 \to R_2$, ending in R_2 with the CS x_{2012}, y_{2012}.

SOLUTION   With no loss of generality, we set the arbitrary angle θ equal to zero. We can also save a good deal of unnecessary writing by omitting the third row and column of every matrix. The z components are not affected by any of the transformations considered, and one always has $0, 0, 1, 0$ in the third row and column. Thus,

$$P(\theta, \alpha) =$$

$$\begin{pmatrix} (\cosh \alpha - 1)\cos^2 \theta + 1 & (\cosh \alpha - 1)\sin \theta \cos \theta & i \sinh \alpha \cos \theta \\ (\cosh \alpha - 1)\sin \theta \cos \theta & (\cosh \alpha - 1)\sin^2 \theta + 1 & i \sinh \alpha \sin \theta \\ -i \sinh \alpha \cos \theta & -i \sinh \alpha \sin \theta & \cosh \alpha \end{pmatrix}$$

$$(5\text{-}8')$$

and

$$\mathsf{Rot}_3(\theta) = \begin{pmatrix} \cos \theta & \sin \theta & 0 \\ -\sin \theta & \cos \theta & 0 \\ 0 & 0 & 1 \end{pmatrix} \tag{5-1'}$$

We shall substitute from Eq. (5-8') in

$$P(\pi - \chi_1, \alpha_0) \cdot P(0, \alpha_2) \cdot P(\pi + \chi_0, \alpha_1)$$

With $\chi_1 = \pi/2$, the four-velocity transformation Eq. (5-12b) gives

$$\cos \chi_0 = \frac{\sinh \alpha_2 \cosh \alpha_1}{\cosh \alpha_2 \sinh \alpha_1}$$

Equation (5-12c) gives

$$\sin \chi_0 = \frac{\sinh \alpha_0}{\sinh \alpha_1}$$

Equation (5-12a) advanced to the R_1 vertex gives

$$\cosh \alpha_1 = \cosh \alpha_0 \cosh \alpha_2$$

After some tedious but straightforward algebra, the matrix product becomes, with these substitutions,

$$
\begin{pmatrix}
\dfrac{\cosh \alpha_2 + \cosh \alpha_0}{\cosh \alpha_2 \cosh \alpha_0 + 1} & \dfrac{-\sinh \alpha_2 \sinh \alpha_0}{\cosh \alpha_2 \cosh \alpha_0 + 1} & 0 \\[2ex]
\dfrac{\sinh \alpha_2 \sinh \alpha_0}{\cosh \alpha_2 \cosh \alpha_0 + 1} & \dfrac{\cosh \alpha_2 + \cosh \alpha_0}{\cosh \alpha_2 \cosh \alpha_0 + 1} & 0 \\[2ex]
0 & 0 & 1
\end{pmatrix}
$$

This matrix is just the one corresponding to a rotation of the axes about the z direction through the angle whose cosine is

$$
\frac{\cosh \alpha_2 + \cosh \alpha_0}{\cosh \alpha_2 \cosh \alpha_0 + 1}
$$

and whose sine is

$$
\frac{-\sinh \alpha_2 \sinh \alpha_0}{\cosh \alpha_2 \cosh \alpha_0 + 1}
$$

This angle is $-\varepsilon$, where ε is given by Eq. (5-17).   Q.E.D.

5-3. SPIN: KINEMATICS

In nonrelativistic mechanics, the *angular momentum* or *moment of momentum* of a system of particles is defined as the pseudovector

$$
\mathbf{L} = \sum_i {}^i\mathbf{r} \times {}^i\mathbf{p} \tag{5-18}
$$

The law of motion

$$
{}^i\mathbf{F} = \frac{d}{dt}({}^i\mathbf{p}) \tag{1-5}
$$

gives for the net torque

$$
\mathbf{N} = \sum_i {}^i\mathbf{r} \times {}^i\mathbf{F} = \sum_i {}^i\mathbf{r} \times \frac{d}{dt}({}^i\mathbf{p}) = \frac{d}{dt}\sum_i {}^i\mathbf{r} \times {}^i\mathbf{p}
$$

that is,

$$
\mathbf{N} = \frac{d\mathbf{L}}{dt} \tag{5-19}
$$

In words, the torque equals the rate of change of angular momentum. In the absence of torque, the angular momentum is a constant of the motion.

Inspection of Eq. (5-18) shows that, in general, $\mathbf{L}$ depends on the choice of origin of the coordinate system. A shift of origin from O to O' (Fig. 5-12) replaces $^i\mathbf{r}$ by $^i\mathbf{r} - \overrightarrow{OO'}$ but does not change $^i\mathbf{p}$. The angular momentum is independent of origin if and only if (abbreviated iff)

$$\overrightarrow{OO'} \times \sum_i {}^i\mathbf{p} = 0$$

that is, since $\overrightarrow{OO'}$ is arbitrary, iff

$$\mathbf{P} = \sum_i {}^i\mathbf{p} = 0$$

Only in the system's rest frame does the angular momentum have a unique value. It is this value that nature quantizes in multiples of $\hbar/2$.

Torques on microscopic systems arise from interaction of their electric or magnetic moments with external electromagnetic fields. The magnetic dipole moment $\mathbf{m}$ gives rise to a torque

$$\mathbf{N} = \mathbf{m} \times \mathscr{B} \tag{5-20}$$

tending to make the moment point along the field, and a force

$$\mathbf{F} = (\mathbf{m} \cdot \text{grad})\mathscr{B}$$

tending to drive the system to regions of stronger field. This force vanishes in a homogeneous magnetic field, whose only effect is then to produce

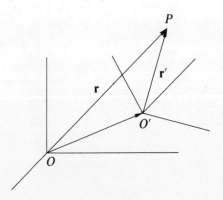

Figure 5-12.

a torque on the system. In the absence of magnetic field, the angular momentum is constant [Eq. (5-19)].

Relativistically, we define the *spin* **S** of a system as its angular momentum in the inertial frame in which its total momentum is zero (rest frame of the system as a whole). The angular momentum in this frame is not in general given by Eq. (5-18), because many of the known particles have *intrinsic* angular momentum that must be combined with the *orbital* **r** × **p** terms of Eq. (5-18). Rather it is defined by the law of motion [Eq. (5-19)]: It is a pseudovector [coinciding with **L** of Eq. (5-18) in the classical nonrelativistic limit] whose rate of change with time equals the applied torque.

$$\mathbf{N}^\dagger = \frac{d\mathbf{S}}{d\tau} \tag{5-21}$$

With this definition, elementary particles, such as the electron, have spin. The law of *conservation of angular momentum* of an isolated system is automatically maintained.[3] This is gratifyingly consistent, for that law can be shown to be equivalent to the *isotropy* of inertial frames.

The spin of a spacecraft is its angular momentum in its center of mass frame. The spin of an atom is its angular momentum, spin plus orbit, in its center of mass frame. The spin of one of its electrons or of the nucleus is *its* angular momentum in *its* rest frame.

Unlike momentum, which has definite components in each reference frame, angular momentum is defined only in one particular reference frame. It does not transform. Any statement about it refers to the rest frame as of that instant. *If we say that in the laboratory the spin of an electron makes the*

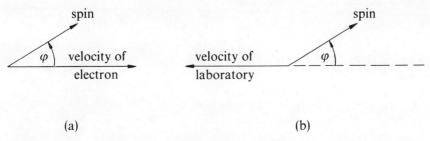

(a)   (b)

Figure 5-13.   Definition of spin direction. The vector marked "spin" is a unit vector in the direction of the spin. (a) Laboratory frame. (b) Electron frame.

---

[3] Saving the conservation law by postulating intrinsic angular momentum is not like adding another epicycle to save Ptolemaic cosmology. It is a *simplifying* assumption that permits coherent understanding of a host of diverse experimental observations on quantum states of atoms, nuclei, and free particles.

angle φ with its velocity (Fig. 5-13a), *we mean that in the electron's rest frame the spin makes this angle with the line of motion of the laboratory* (Fig. 5-13b).

It is thus erroneous to think of the spin direction as that of a weathervane or compass needle. The latter are material systems whose points have definite positions in each reference frame. In fact, the orientation of a rigid body is *different* in the laboratory and in its rest frame [Eqs. (2-30)].

Turning of Spin and Velocity in a Pure Lorentz Transformation

With this definition of spin direction there results a rotation of the spin when the system is observed from different frames. Let R_0 be the rest frame of the system, R_1 be one "laboratory" frame, R_2 another. Suppose that in R_0 (Fig. 5-14a), the angle between the spin and $\mathbf{u}_0(1)$ is φ. Its angle with

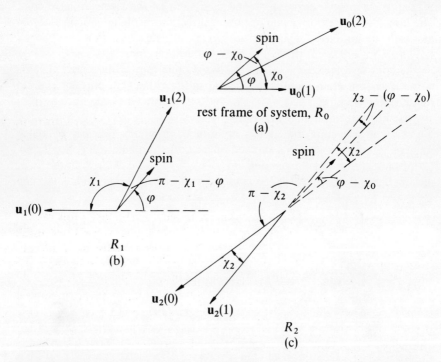

Figure 5-14. Rotation of the spin and velocity vectors in the pure Lorentz transformation from R_1 to R_2. R_0 is the rest frame in which the spin is defined. The system's velocity with respect to R_1 is $\mathbf{u}_1(0)$, with respect to R_2, $\mathbf{u}_2(0)$. Parts (a), (b), and (c) show the velocity and spin directions in R_0, R_1, and R_2, respectively. The drawings are for the case $\beta_2 = 1/2$, $\beta_0 = 3/4$, $\chi_1 = 120°$.

$\mathbf{u}_0(2)$ is then $\varphi - \chi_0$. In R_1, Fig. 5-14b, the spin by definition makes this same angle φ with $-\mathbf{u}_1(0)$, and thus the angle $(\pi - \chi_1 - \varphi)$ with the line of motion of R_2 in R_1. In R_2, Fig. 5-14c, the spin by definition makes the angle $\varphi - \chi_0$ with $-\mathbf{u}_2(0)$, and thus the angle $[\chi_2 - (\varphi - \chi_0)]$ with $-\mathbf{u}_2(1)$. Because of the spherical defect ε, the spin orientation in R_2 with respect to the direction of relative motion of R_2 and R_1 is not the same as in R_1. In effect, the difference is

$$(\chi_2 - \varphi + \chi_0) - (\pi - \chi_1 - \varphi) = \chi_2 + \chi_0 + \chi_1 - \pi = -\varepsilon$$

In the pure Lorentz transformation from R_1 to R_2, the spin gets turned through the angle $+\varepsilon$ so as to be more nearly parallel to the direction of relative motion of R_1 and R_2.

In the same transformation, the velocity vector of the system—$\mathbf{u}_1(0)$ in R_1 and $\mathbf{u}_2(0)$ in R_2—gets turned in the same sense through the angle

$$(\pi - \chi_2) - \chi_1 = \chi_0 + \varepsilon$$

There is thus a change in the angle between spin and velocity (or momentum); it is not invariant. The change is by the angle χ_0, the spin vector lagging behind the velocity vector.

For an extreme relativistic particle and a moderate transformation velocity ($\alpha_2 \gg 1$, $\alpha_1 \gg 1$, $\alpha_0 \sim 1$), we see from Eq. (5-14) at the R_0 vertex that

$$\chi_0 \ll \varepsilon$$

The spin vector almost keeps up with the momentum vector in this case.

We have tacitly assumed the spin to be in the plane of $\mathbf{u}_0(1)$ and $\mathbf{u}_0(2)$. In the general case, our considerations apply to the component of the spin in that plane. These results are purely kinematical. We are describing the same state in three different inertial reference frames.

Our definition of spin does not apply to a photon or neutrino, because a particle of speed c has no rest frame. It is possible to define spin for these particles by a limiting process ($\mu \to 0$). It turns out that the spin is either parallel or antiparallel to the velocity, in all frames [61].

Thomas Precession

Consider now a spinning system that is undergoing a sidewise acceleration —for example, an atomic electron orbiting about the nucleus (Fig. 5-15). Its rest frame is different at every instant. The spin direction in the laboratory must be constantly changing even in the absence of torque. This effect, which

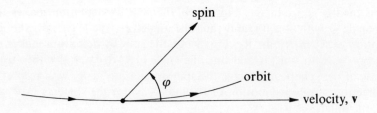

Figure 5-15.

was discovered and evaluated by L. H. Thomas in 1926 [60], shortly after the discovery of electron spin [62], is called the *Thomas precession*.

Let the component of the spin in the plane of the orbit make the angle φ with the velocity $\mathbf{v}$ (Fig. 5-15). Because of the acceleration $\mathbf{a}$, the velocity changes in the infinitesimal interval dt to $\mathbf{v} + \mathbf{a}\,dt$. The rotation of the velocity vector is (Fig. 5-16)

$$d\boldsymbol{\theta} = \frac{\mathbf{v} \times \mathbf{a}}{v^2}\,dt \tag{5-22}$$

We wish to calculate the rotation $d\varphi$ of the spin projection in the time dt.

Let R_0 be the atom's rest frame (laboratory), R_1 the inertial frame moving with velocity $\mathbf{v}$ relative to R_0, and R_2 the inertial frame moving with velocity $\mathbf{v} + \mathbf{a}\,dt$ relative to R_0. In the notation we have been using,

$$d\theta = \chi_0$$
$$\mathbf{v} = \mathbf{u}_0(1) \tag{5-23}$$
$$\mathbf{v} + \mathbf{a}\,dt = \mathbf{u}_0(2)$$

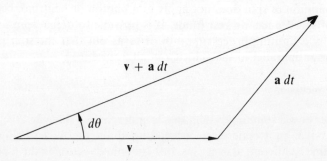

Figure 5-16.

In R_1, the electron experiences a force $e\mathscr{E}^{(1)}$ toward the nucleus. The velocity it acquires in the time interval dt_1 corresponding to dt in R_0 is $\mathbf{u}_1(2)$. In R_1, there is also a torque on the electron due to the interaction between the magnetic moment associated with its spin and the magnetic field of the moving nucleus (spin–orbit interaction).

When the electron is at rest in R_1 (at time t in R_0), its spin projection makes the angle φ with $\mathbf{u}_1(0)$, or the angle $\psi = \varphi - (\pi - \chi_1)$ with $\mathbf{u}_1(2)$ (Fig. 5-17a). In the interval dt_1, the torque changes the spin orientation. At the time $t + dt$ in R_0, when the electron is at rest in R_2 (Fig. 5-17b), the spin projection makes the angle $\psi + d\psi$ with the direction of relative motion of R_2 and R_1, $-\mathbf{u}_2(1)$, or $(\psi + d\psi + \chi_2)$ with the direction of relative motion of R_2 and R_0.

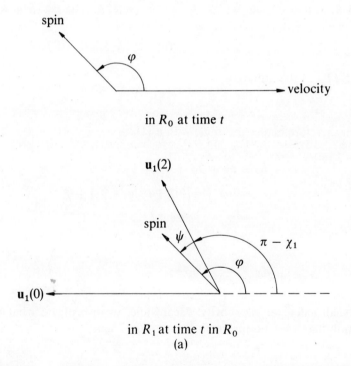

in R_0 at time t

in R_1 at time t in R_0
(a)

Figure 5-17. Rotation of spin and velocity vectors in accelerated motion. The laboratory frame is R_0, R_1 is the rest frame at time t in R_0, and R_2 is the rest frame at time $t + dt$ in R_0. The drawings are for the case $\gamma_1 = 2$, $\gamma_2 = 2$, $\chi_0 = 30°$, leading to $\varepsilon = 26.3°$. (a) The initial situation in R_0 and R_1. (b) The final situation in R_0 and R_2. In between, the spin was turned by the angle $d\psi$ in R_1, arbitrarily taken to be $35°$ in the drawing.

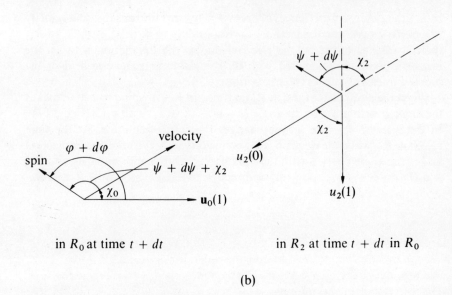

in R_0 at time $t + dt$ in R_2 at time $t + dt$ in R_0

(b)

Figure 5-17.   (Continued).

In R_0, it therefore makes the angle with $\mathbf{u}_0(1)$

$$\varphi + d\varphi = (\psi + d\psi + \chi_2) + \chi_0$$
$$= [(\varphi - \pi + \chi_1) + d\psi + \chi_2] + \chi_0$$
$$= \varphi - \varepsilon + d\psi$$

or

$$d\varphi = -\varepsilon + d\psi \tag{5-24a}$$

We shall calculate $d\psi$ shortly. Meanwhile, we investigate what would happen if there were no torque in $R_1 (d\psi = 0)$. Then

$$\boxed{d\varphi = -\varepsilon} \tag{5-24b}$$

This is the Thomas precession. While the velocity turns along the orbit by $d\theta$, the spin projection in the orbital plane turns in the opposite sense by ε

(Fig. 5-18). Equation (5-14) gives for the spherical defect

$$\sin\left(\frac{-d\varphi}{2}\right) = \left[\frac{(\cosh\alpha_1 - 1)(\cosh\alpha_2 - 1)}{2(\cosh\alpha_0 + 1)}\right]^{1/2} \sin(d\theta)$$

or, since $\alpha_0 \to 0$, $\alpha_1 \to \alpha_2$ (infinitesimal interval),

$$-d\varphi = (\gamma - 1)\, d\theta \qquad (5\text{-}25)$$

where

$$\gamma = \left(1 - \frac{v^2}{c^2}\right)^{-1/2}$$

Equation (5-25) is a simple, rigorous relation between the angle turned through by the spin projection and the angle turned through in the orbit. The turning of the spin is a relativistic effect, for $(\gamma - 1) \approx 0$ in the low velocity region. In the nonrelativistic approximation, the direction of the spin projection in the plane of the orbit is not affected by the particle's acceleration.

The pure Lorentz transformations ("boosts") considered here leave invariant the spin and velocity projections normal to the orbital plane. In effect, the rotation of axes due to successive PLT's is about this normal as axis.

If the acceleration is longitudinal ($\mathbf{a} \parallel \mathbf{v}$), $d\theta$ is zero, and therefore $d\varphi = 0$. Only transverse acceleration turns the spin.

Expressing $d\theta$ in terms of the acceleration and velocity [Eq. (5-22)],

$$d\varphi = -(\gamma - 1)\frac{|\mathbf{v} \times \mathbf{a}|}{v^2}\, dt = \frac{-\gamma^2}{\gamma + 1}\frac{|\mathbf{v} \times \mathbf{a}|}{c^2}\, dt \qquad (5\text{-}26a)$$

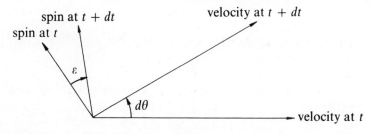

Figure 5-18.   Thomas precession, drawn for the case of zero torque in the rest frame of the particle ($d\psi = 0$). The drawing is made for $\gamma_1 = \gamma_2 = 2$, $\chi_0 = 30°$.

and the angular velocity of the Thomas precession, ω_T, is

$$\omega_T = \frac{-\gamma^2}{\gamma + 1} \frac{\mathbf{v} \times \mathbf{a}}{c^2} \qquad (5\text{-}26b)$$

For an atomic electron, $\gamma \approx 1$, and

$$\omega_T \approx -\frac{1}{2} \frac{\mathbf{v} \times \mathbf{a}}{c^2} \qquad (5\text{-}26c)$$

This equation can also be applied to a macroscopic body, such as one spinning in orbit about the earth, as in an experiment proposed by Schiff and others [63] to test general relativity. The Newtonian law of gravitation [Eq. (1-28)] gives

$$\mathbf{a} = -\frac{GM\mathbf{r}}{r^3}$$

where G is the gravitational constant, M the mass of the earth, and $\mathbf{r}$ the radius vector from the center of the earth to the satellite. Thus a naïve application of the kinematics of special relativity as expressed in Eq. (5-26c) leads us to expect the component of spin in the plane of the orbit to precess at the rate

$$\omega_T = \frac{GM}{2c^2 r^3} \mathbf{v} \times \mathbf{r}$$

The general relativity calculation gives, to a very good approximation [63], a precession frequency of

$$-3\frac{GM}{2c^2 r^3} \mathbf{v} \times \mathbf{r}$$

an effect three times as large and in the opposite sense.

Our evaluation of ε in Eq. (5-25) is rigorous; even when the torque on the spinning electron (or other system) is not negligible, Eq. (5-24a) gives, without approximation,

$$\boxed{d\varphi = d\psi - (\gamma - 1)\, d\theta} \qquad (5\text{-}27)$$

The determination of $d\psi$ involves the equation of motion of spin.

5-4. DYNAMICS OF SPIN

Let us now consider the effect of torque on spin.[4] We shall limit ourselves to the case of a particle with a magnetic dipole moment **m** in a microscopically homogeneous electromagnetic field.

All experiments are in agreement with Ampère's hypothesis that magnetism results only from the motion of electric charge. The magnetic dipole moment of a system is associated with the circulation of electric charge inside it, and thus with the system's angular momentum. For a system of charged particles without intrinsic magnetic moment,

$$\mathbf{m}_L = \sum_i \frac{e_i}{2c} {}^i\mathbf{r} \times {}^i\mathbf{v} = \sum \frac{e_i}{2m_i c} {}^i\mathbf{r} \times {}^i\mathbf{p}$$

In the nonrelativistic approximation ($m_i \approx \mu_i$), this expression becomes, for a system of identical particles,

$$\mathbf{m}_L \approx \frac{e_i}{2\mu_i c} \sum_i {}^i\mathbf{r} \times {}^i\mathbf{p} = \frac{e_i}{2\mu_i C} \mathbf{L} \tag{5-28}$$

The magnetic dipole moment (axial) vector is proportional to the orbital angular momentum (axial) vector. It is oppositely directed if the circulating charge is negative. Thus, the magnetic moment of an atom resulting from the revolving of the electrons about the nucleus is

$$\mathbf{m}_L \approx \frac{-e_0}{2\mu_e c} \mathbf{L}$$

where e_0 is the positive quantum of electric charge and μ_e is the electron's proper mass.

The proportionality of magnetic moment and angular momentum has been confirmed in "gyromagnetic" experiments on many different systems.[5]

---

[4] The reader might wonder why we bother to give a classical treatment of spin dynamics when we know that only a quantal description can be correct. The answer lies in the quantal theorem that the classical equation of motion of a dynamical variable is the quantal equation of motion of the mean value of that variable averaged over an ensemble of identical systems. The conclusions of our classical treatment will apply to averages over many identical particles prepared in the same way, like the electrons or muons in a beam or the valence electrons in a gas of atoms in a glow tube.

[5] One of the first two successful experiments was carried out in 1914–1915 by Einstein in collaboration with de Haas [64].

The constant of proportionality is one of the parameters characterizing the particular system. It is normally specified by giving the *gyromagnetic ratio or g factor*, defined by

$$\mathbf{m} = g\frac{e_0}{2\mu c}\mathbf{S} \qquad (5\text{-}29)$$

For a negative electron ($\mu = \mu_e, e = -e_0$), g is very nearly -2. The magnetic moment of an atom results from both electron orbital motion and electron spin; the g factor of the atom, with $\mu = \mu_e$, lies between -1 (pure electron orbit) and -2 (pure electron spin).

For the elementary particles, we use Eq. (5-29) with μ set equal to the particle's proper mass. The letter S is a positive multiple of $\hbar/2$, and g is then of order of magnitude unity. Some experimental values are given in Table 5-2. Note that for the electrons and muons, the g factor is very nearly 2; the exact values are within the estimated errors precisely those *predicted* by

Table 5-2
Spins and g Factors of Some Elementary Particles[a]

Name	Symbol	Spin (in units of $\hbar$)	g
Electron	e^-	$\frac{1}{2}$	$-2(1 + 1.1596 \times 10^{-3})$
Positron	e^+	$\frac{1}{2}$	$+2(1 + 1.17 \times 10^{-3})$
(+) muon	μ^+	$\frac{1}{2}$	$+2(1 + 1.16 \times 10^{-3})$
(−) muon	μ^-	$\frac{1}{2}$	$-2(1 + 1.166 \times 10^{-3})$
Pion	$\pi^\pm$	0	—
Proton	p^+	$\frac{1}{2}$	$+5.5855$
Neutron	n	$\frac{1}{2}$	-3.8262

[a] The adjective "elementary" is not meant to imply a belief that we have at last found the basic building blocks of the universe. It means merely that these systems are in the deepest layer yet reached in our explorations. The proton and neutron are certainly structured, the g factors being interpretable in terms of a virtual charged pion cloud circulating around a core.

relativistic quantum electrodynamics for a spin-1/2 particle. The symmetry between positive and negative charge is also impressive.

If $\mathscr{E}^\dagger$, $\mathscr{B}^\dagger$ are the electric and magnetic fields in the system's rest frame, there is a torque [Eq. (5-20)]

$$\mathbf{N}^\dagger = \frac{ge_0}{2\mu c}\mathbf{S} \times \mathscr{B}^\dagger \qquad (5\text{-}30)$$

and a force [Eq. (D-1)]

$$\mathbf{F}^\dagger = e\mathscr{E}^\dagger \tag{5-31}$$

acting on the charged system. [For a neutral system, such as a neutron or lithium atom, the expression in Eq. (5-31) vanishes, and the $(\mathbf{m} \cdot \text{grad})\,\mathscr{B}^\dagger$ force must be taken into account. For a charged system it is negligible.]

Our procedure will be to determine the motion in the rest frame and then use the relativistic transformation laws to convert the results to the laboratory frame.

The motion in the rest frame is determined by

$$\mathbf{N}^\dagger = \frac{d\mathbf{S}}{d\tau} \tag{5-21}$$

and

$$\mathbf{F}^\dagger = \frac{d\mathbf{p}^\dagger}{d\tau} \qquad (p^\dagger = 0) \tag{5-32}$$

Evidently the torque affects only the spin and the force affects only the momentum. The equations of motion are thus

$$\boxed{g\frac{e_0}{2\mu c}\mathbf{S} \times \mathscr{B}^\dagger = \frac{d\mathbf{S}}{d\tau}} \qquad \text{(rest frame)} \tag{5-33}$$

$$e\mathscr{E}^\dagger = \frac{d\mathbf{p}^\dagger}{d\tau} \qquad \text{(rest frame)} \tag{5-34}$$

It follows from Eq. (5-34) that the motion of the system as a whole in any frame is determined entirely by its charge, independent of magnetic dipole moment. This part of the motion has been treated in Chapter 3. We need now only consider the motion of the spin (boxed equation).

The general solution of Eq. (5-33) for a uniform field $\mathscr{B}^\dagger$ is regular precession around the field direction (Fig. 5-19) at the angular velocity

$$\omega_S = \frac{(N^\dagger\,d\tau/S\sin\zeta)}{d\tau} = \frac{(ge_0/2\mu c)S\mathscr{B}^\dagger\,\sin\zeta}{S\sin\zeta} = \frac{ge_0}{2\mu c}\mathscr{B}^\dagger \tag{5-35}$$

directed opposite to $\mathscr{B}^\dagger$ if $g > 0$. The polar angle ζ between spin and magnetic field is constant, and the magnitude S is constant. The azimuth changes linearly with time at the precession rate ω_S.

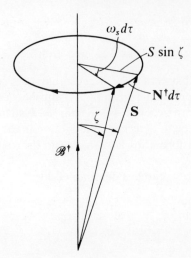

Figure 5-19.   Regular precession of spin in a uniform magnetic field. The spin vector **S** traces out a right circular cone with the direction of the magnetic field as axis. The figure is drawn for g positive.

For a system at rest or in rectilinear motion with respect to the laboratory, the dynamical problem has now been solved. The spin simply precesses at angular velocity ω_S.

For a transversely accelerated system, we have already worked out the transformation of momentum and spin direction in connection with the Thomas precession (Section 5-3). We considered (Fig. 5-15) a particle with spin, moving with velocity **v** and undergoing an acceleration **a**. In the laboratory frame R_0, we compared the situations at times t and $t + dt$. The inertial frame in which the particle is at rest at time t was called R_1; that in which it is at rest at time $t + dt$ was called R_2. Figure 5-15 shows the situation in the plane of the orbit in R_0. The angle between the projection of the spin in the orbital plane and the initial velocity is labelled φ. The turning of the velocity vector in dt is called $d\theta$ (Fig. 5-16). The angle between the spin projection and the line of relative motion of R_2 and R_1 is ψ in R_1 when R_1 is the rest frame (Fig. 5-17a) and $\psi + d\psi$ in R_2 when R_2 is the rest frame (Fig. 5-17b). We found that

$$d\varphi = d\psi - (\gamma - 1)\, d\theta \tag{5-27}$$

This simple equation relates the turning of the spin in the laboratory ($d\varphi$), the turning of the spin in the rest frame ($d\psi$), and the turning of the velocity (or momentum) in the laboratory ($d\theta$). We must now compute $d\psi$ and $d\theta$ from the dynamics of the situation.

Motion in a Transverse Magnetic Field. "g-2" Experiments

Introduce local Cartesian axes with the z direction along $\mathscr{B}$ and the x direction along $\mathbf{v}$ (Fig. 5-20a). Then a positively charged particle is accelerated in the $-y$ direction. We have already calculated the motion of the particle in the laboratory (Section 3-9) and know in particular [Eq. (3-55a)] that in the plane of the orbital element (Fig. 5-20b),

$$d\theta = \frac{e\mathscr{B}}{\mu\gamma c}\,dt \qquad (5\text{-}36)$$

We can quickly rederive this result in the present context.

In the initial rest frame R_1, the electromagnetic field is, according to Eqs. (2-55a),

$$\mathscr{E}^{(1)} = (0, -\gamma\beta\mathscr{B}, 0) \qquad \mathscr{B}^{(1)} = (0, 0, \gamma\mathscr{B})$$

In the time element $dt_1 (= d\tau = dt/\gamma)$, the particle acquires [Eq. (5-34)] a momentum $e\mathscr{E}^{(1)}\,dt_1$:

$$d\mathbf{p}^{(1)} = (0, -e\gamma\beta\mathscr{B}\,dt_1, 0)$$

and its energy becomes

$$\mu c^2 + dE^{(1)} = [(c\,dp^{(1)})^2 + \mu^2 c^4]^{1/2} = \mu c^2 + 0\,(dt_1{}^2)$$

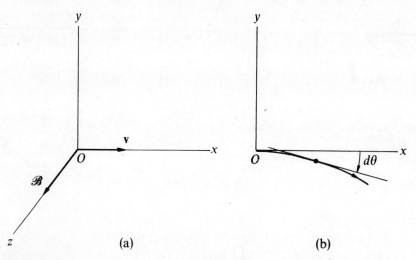

(a)   (b)

Figure 5-20.

Transforming the four-momentum to R_0 [Eqs. (3-23c)] yields

$$E = \gamma(\mu c^2 + 0) = \mu\gamma c^2$$

$$p_x = \gamma\left(0 + \frac{\beta}{c}\mu c^2\right) = \mu\beta\gamma c$$

$$dp_y = -e\gamma\beta\mathscr{B}\, dt_1$$

and

$$d\theta = \frac{-dp_y}{p_x} = \frac{(e\gamma\beta\mathscr{B}\, dt)/\gamma}{\mu\beta\gamma c} = \frac{e\mathscr{B}}{\mu\gamma c}\, dt \qquad (5\text{-}36)$$

In R_1, the spin precesses about the magnetic field ($\gamma\mathscr{B}$ in the z_1 direction) with angular velocity [Eq. (5-35)]

$$\omega_S = \frac{ge_0}{2\mu c}\gamma\mathscr{B}$$

Its projection in the x_1y_1 plane turns with this angular velocity, in the same sense for $g > 0$, as that in which θ increases for $e > 0$. As ψ is the azimuth of the spin in the orbital plane, measured from the y_1 axis (line of relative motion of R_2 and R_1),

$$d\psi = \omega_S\, dt_1 = \frac{ge_0}{2\mu c}\gamma\mathscr{B}\frac{dt}{\gamma} = \frac{g}{2}\frac{e_0\mathscr{B}}{\mu c}\, dt \qquad (5\text{-}37)$$

Substituting in Eq. (5-27) from (5-37), we obtain for the spin precession in the laboratory frame

$$d\varphi = \frac{g}{2}\frac{e_0\mathscr{B}}{\mu c}\, dt - (\gamma - 1)\, d\theta$$

or, by Eq. (5-36),

$$\frac{d\varphi}{dt} = \frac{g}{2}\left(\frac{e_0}{e}\right)\gamma\frac{d\theta}{dt} - (\gamma - 1)\frac{d\theta}{dt}$$

that is,

$$\frac{d\varphi}{dt} = \frac{d\theta}{dt}\left[1 + \gamma\left(\frac{g(e_0/e) - 2}{2}\right)\right] \qquad (5\text{-}38a)$$

Both φ and θ are measured from the initial direction of motion. The spin component in the direction of motion is proportional to $\cos(\varphi - \theta)$, called the *longitudinal polarization.* From Eq. (5-27), with Eqs. (5-36) and (5-37),

$$\frac{d(\varphi - \theta)}{dt} = \frac{e_0 \mathscr{B}}{\mu c}\left(\frac{g}{2} - \frac{e}{e_0}\right) \tag{5-38b}$$

For singly charged particles with g/e positive (see Table 5-2), the expression in parentheses is $(|g| - 2)/2$. If the absolute value of the g factor is 2, the

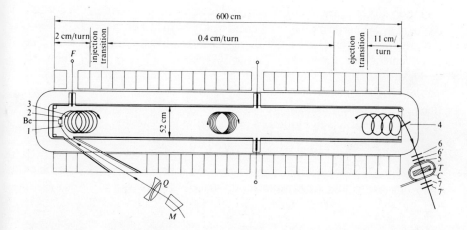

Figure 5-21. General view of the apparatus, showing magnet of pole surface $(600 \times 52)\,\text{cm}^2$. Muons, deflected by the bending magnet M and focused by the quadrupole pair Q, enter the magnet through a shielded channel. After slowing down in the beryllium moderator Be they describe many turns in the field. The quasicircular orbit is slowly displaced by the field gradient (2 cm/turn in the *injection* region, 0.4 cm/turn in the *storage* region, and 11 cm/turn in the final *ejection* region). Muons ejected from the magnet are stopped in target T of the polarization analyzer where the spin direction is determined by recording the decay electrons. Injected muons are indicated by the counter signature 123. Ejected muons by the signature $466'\,5\overline{7}$. Decay electrons by $66'\,4(\overline{77}')$ and $77'\,\overline{4}(\overline{66}')$. The time of flight of muons between counters 2 and 4 is recorded. [Figure and caption from G. Charpak, F. J. M. Farley, R. L. Garwin, T. Muller, J. C. Sens, and A. Zichichi, "The Anomalous Magnetic Moment of the Muon," *Nuovo Cimento* **37**, 1241 (1965).]

projection of the spin on the orbital plane turns at the same rate as the momentum (or velocity) vector. The spin component along the field direction is constant in the precession in R_1, and is also (Section 5-3) constant in R_0. The momentum and spin in the laboratory keep the same relative orientation; the longitudinal polarization is constant. Departures of $|g|$ from 2 manifest themselves in changes of spin orientation with respect to momentum.

"g–2" experiments have recently been carried out ([65]–[69]) in which the motion of the spin relative to the momentum in a known transverse field has been used to measure $|g| - 2$ to high precision. The particles are trapped in a storage magnet, and the difference between φ and θ builds up at a constant rate [Eq. (5-38b)] over many orbital revolutions. For negative electrons, the spin orientation has been determined by its effect on Coulomb scattering [65], whereas for positrons its effect on the direction of annihilation radiation has been used [66]. A fascinating account of the extremely difficult electron g–2 experiment has been given by Crane [67]. For muons ([68]; [69]), the angular distribution of the decay electrons locates the spin axis. The arrangement used for positive muons [68] is shown in Fig. 5-21. These experiments have given the g values for electrons and muons presented in Table 5-2. Their consistency with other experiments (e.g., [70]) that measure the spin precession frequency ω_S [Eq. (5-35)] gives strong support to the relativistic equations of motion as presented in this chapter.

A quantal derivation for spin-1/2 particles of the results of this section [Eqs. (5-36) and (5-38)] has been published by Mendlowitz and Case [71]. Schwinger [72] showed that for nonstrongly interacting spin-1/2 particles, the value of the "magnetic moment anomaly"

$$a = \frac{|g| - 2}{2}$$

is expected from quantum electrodynamics to be a power series in the fine structure constant α, with leading term $\alpha/2\pi$.

Motion in a Transverse Electric Field (Spin–Orbit Interaction)

This situation gives rise to the atomic spin–orbit interaction, where an electron moves essentially at right angles to the central Coulomb field (Fig. 5-22). Let us introduce local axes, the x axis parallel to $\mathbf{v}$ and the y axis parallel to the electric field $\mathscr{E}$. A negative singly charged particle ($e = -e_0$ for an electron) is accelerated in the $-y$ direction, its velocity turning at the rate

$$\frac{d\theta}{dt} = -\frac{dp_y/dt}{p_x} = \frac{e_0 \mathscr{E}}{\mu \gamma \beta c} \tag{5-39}$$

In the initial rest frame (R_1), the field is [Eqs. (2-55a)]

$$\mathscr{E}^{(1)} = (0, \gamma\mathscr{E}, 0) \qquad \mathscr{B}^{(1)} = (0, 0, -\gamma\beta\mathscr{E})$$

Again, the magnetic field is perpendicular to the plane of the orbit. The spin precesses about the z_1 direction with angular velocity [Eq. (5-35)]

$$\omega_S = \frac{ge_0}{2\mu c}\gamma\beta\mathscr{E} \tag{5-40}$$

For $g < 0$, as is the case for electrons, the precession is in the same sense as the acceleration. In the time $dt_1 = d\tau = dt/\gamma$ in R_1, the spin projection on the (x_1, y_1) plane turns by

$$d\psi = \omega_S \, dt_1 = \frac{|g|e_0}{2\mu c}\beta\mathscr{E} \, dt \tag{5-41}$$

and the kinematic relation Eq. (5-27) gives, with Eqs. (5-41) and (5-39),

$$\frac{d\varphi}{dt} = \frac{|g|e_0}{2\mu c}\beta\mathscr{E} - \frac{(\gamma - 1)e_0\mathscr{E}}{\mu\gamma\beta c}$$

or

$$\frac{d\varphi}{dt} = \frac{d\theta}{dt}\left[\left(\frac{|g|}{2} - 1\right)\gamma - \frac{|g|}{2\gamma} + 1\right] \tag{5-42}$$

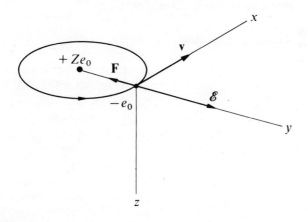

Figure 5-22.

The orbital angular velocity $d\theta/dt$ is proportional to the orbital angular momentum

$$\frac{d\theta}{dt} = \frac{L_z}{\mu\gamma r^2}$$

and we see that the spin precession frequency in the laboratory is therefore proportional to the orbital angular momentum. It can be shown that the spin precession alters the energy of the atom by an amount proportional to $d\varphi/dt$; the quantal calculation gives exact agreement with spectroscopic data. If we ignored the Thomas precession and assumed that the spin projection turned through the same angle in the laboratory frame as in the rest frame, we would expect the spin to precess at the rate $d\psi/dt$:

$$\left(\frac{d\varphi}{dt}\right)_{\text{naïve}} = \frac{d\psi}{dt} = \frac{|g|e_0}{2\mu c}\beta\mathscr{E} = \frac{d\theta}{dt}\frac{|g|}{2}\frac{(\gamma^2 - 1)}{\gamma}$$

Thus the ratio of rigorous precession rate to naïve is

$$\frac{(d\varphi/dt)}{(d\varphi/dt)_{\text{naïve}}} = 1 - \frac{\gamma}{(|g|/2)(\gamma + 1)}$$

which is about one-half, for $g \approx 2$ and $\gamma \approx 1$. In fact, the naïve calculation of the alkali doublet splittings made by the discoverers of electron spin [62] predicted effects twice as large as observed. This disagreement stimulated Thomas to discover his kinematical effect [60]. It then became clear that the relativistic spin–orbit interaction plus the relativistic momentum–velocity relation give results in quantitative agreement with the fine structure of the energy levels of atoms with one valence electron.

Equation of Motion of Four-Spin

When a particle is accelerated, it is at different times at rest in different inertial frames. In this case, integration of the simple differential equation of motion [Eq. (5-33)] is not a simple matter. One would like to have an equation of motion valid over a finite time in a single inertial frame.

In seeking such an equation, we shall be guided by the known dynamics in the rest frame and the known relativistic transformation laws. We have emphasized earlier that spin, being defined in a particular frame (the rest frame), does not transform in the usual sense. To form expressions with known transformation behavior, we need to introduce a four-tensor related to the spin. A convenient choice is a four- (pseudo-) vector $\tilde{\mathscr{S}}$ defined (see [73]; details in [74]) by the requirement that in the rest frame $R^\dagger$ its spacelike components are the components of the spin and its timelike component is zero:

$$\mathscr{S}^\dagger = \mathbf{S}$$
$$\mathscr{S}^\dagger_4 = 0 \qquad \mathscr{S}^\dagger_0 = 0 \tag{5-43}$$

We shall call it *four-spin*; when normalized by dividing by its invariant length, it is called the *polarization four-vector*. It is spacelike, and therefore in no frame does its spacelike part $\mathscr{S}$ vanish. Its invariant length squared is the square of the spin magnitude.

Recalling the components of the four-velocity in the rest frame,

$$\mathbf{w}^\dagger = \gamma \mathbf{v}^\dagger = 0$$
$$w^\dagger_4 = ic\gamma^\dagger = ic \quad \text{or} \quad w_0 = c$$

we see that $\mathscr{S}$ and $\tilde{w}$ are orthogonal:

$$\tilde{\mathscr{S}}\tilde{w} = \mathscr{S}^\dagger \cdot \mathbf{w}^\dagger + \mathscr{S}^\dagger_4 w^\dagger_4 = 0 \tag{5-44a}$$

Following [75], we deduce the equation of motion of $\tilde{\mathscr{S}}$ as follows. We know the equation of motion of $\mathscr{S}^\dagger$. Equation (5-33) reads

$$\frac{d\mathscr{S}^\dagger}{d\tau} = \frac{ge_0}{2\mu c}\mathscr{S}^\dagger \times \mathscr{B}^\dagger \tag{5-45a}$$

To find the equation of motion of $\mathscr{S}^\dagger_4$, we make use of the orthogonality property [Eq. (5-44a)] that relates $\mathscr{S}_4$ and $\mathscr{S}$:

$$\tilde{\mathscr{S}}\tilde{w} = \mathscr{S} \cdot \gamma \mathbf{v} + \mathscr{S}_4 ic\gamma = 0 \tag{5-44b}$$

so that

$$\mathscr{S}_4 = \frac{-1}{ic}\mathscr{S} \cdot \mathbf{v} \tag{5-46}$$

Differentiating with respect to τ, we obtain

$$\frac{d\mathscr{S}_4}{d\tau} = \frac{i}{c}\mathscr{S} \cdot \frac{d\mathbf{v}}{d\tau} + \frac{i}{c}\frac{d\mathscr{S}}{d\tau} \cdot \mathbf{v} \tag{5-47}$$

In the rest frame,

$$v^\dagger = 0$$

and

$$\frac{d\mathbf{v}^\dagger}{d\tau} = \frac{e\boldsymbol{\mathscr{E}}^\dagger}{\mu}$$

(the electromagnetic field is assumed to be homogeneous). The equation of motion of $\mathscr{S}_4$ [Eq. (5-47)] becomes, in this frame,

$$\frac{d\mathscr{S}_4^\dagger}{d\tau} = \frac{ie}{c\mu}\,\boldsymbol{\mathscr{S}}^\dagger \cdot \boldsymbol{\mathscr{E}}^\dagger \tag{5-45b}$$

We now have [Eqs. (5-45a and b)] the equations of motion of all the components of $\mathscr{S}^\dagger$. If we can put these equations in tensor notation we shall be able to transform them to any inertial frame. Recalling the definition [Eq. (2-54)] of the field tensor,

$$\mathscr{F}_{\mu\nu}^\dagger = \begin{pmatrix} 0 & \mathscr{B}_z^\dagger & -\mathscr{B}_y^\dagger & -i\mathscr{E}_x^\dagger \\ -\mathscr{B}_z^\dagger & 0 & \mathscr{B}_x^\dagger & -i\mathscr{E}_y^\dagger \\ \mathscr{B}_y^\dagger & -\mathscr{B}_x^\dagger & 0 & -i\mathscr{E}_z^\dagger \\ i\mathscr{E}_x^\dagger & i\mathscr{E}_y^\dagger & i\mathscr{E}_z^\dagger & 0 \end{pmatrix}$$

we see that

$$\mathscr{F}_{\mu\nu}^\dagger \mathscr{S}_\nu^\dagger = (\boldsymbol{\mathscr{S}}^\dagger \times \boldsymbol{\mathscr{B}}^\dagger, i\boldsymbol{\mathscr{S}}^\dagger \cdot \boldsymbol{\mathscr{E}}^\dagger)$$

$$\mathscr{S}_\mu^\dagger \mathscr{F}_{\mu\nu}^\dagger w_\nu^\dagger = c\boldsymbol{\mathscr{S}}^\dagger \cdot \boldsymbol{\mathscr{E}}^\dagger$$

Equations (5-45) can thus be written

$$\frac{d\mathscr{S}_\lambda^\dagger}{d\tau} = \frac{ge_0}{2\mu c}\left[\mathscr{F}_{\lambda\mu}^\dagger \mathscr{S}_\mu^\dagger - \frac{1}{c^2}(\mathscr{S}_\mu^\dagger \mathscr{F}_{\mu\nu}^\dagger w_\nu^\dagger)w_\lambda^\dagger \right] + \frac{e}{\mu c^3}(\mathscr{S}_\mu^\dagger \mathscr{F}_{\mu\nu}^\dagger w_\nu^\dagger)w_\lambda^\dagger \tag{5-48a}$$

The square bracket on the right-hand side is equal to $\mathscr{F}_{\lambda\mu}^\dagger \mathscr{S}_\mu^\dagger$ for $\lambda = 1, 2, 3$ and to zero for $\lambda = 4$. Thus, the first term of the right-hand side gives Eqs. (5-45a). The second term vanishes for $\lambda = 1, 2, 3$ and gives Eq. (5-45b) for $\lambda = 4$. Combining terms, we write Eq. (5-48a) in the form

$$\frac{d\mathscr{S}_\lambda^\dagger}{d\tau} = \frac{ge_0}{2\mu c}\mathscr{F}_{\lambda\mu}^\dagger \mathscr{S}_\mu^\dagger - \frac{e_0}{\mu c^3}\left(\frac{g}{2} - \frac{e}{e_0}\right)(\mathscr{S}_\mu^\dagger \mathscr{F}_{\mu\nu}^\dagger w_\nu^\dagger)w_\lambda^\dagger \tag{5-48b}$$

Equation (5-48) equates four-vector components. They transform alike; therefore, the equality must hold in every inertial frame

$$\frac{d\mathscr{S}_\lambda}{d\tau} = \frac{ge_0}{2\mu c}\mathscr{F}_{\lambda\mu}\mathscr{S}_\mu - \frac{e_0}{\mu c^3}\left(\frac{g}{2} - \frac{e}{e_0}\right)(\mathscr{S}_\mu\mathscr{F}_{\mu\nu}w_\nu)w_\lambda \qquad (5\text{-}49a)$$

or, in coordinate-free notation,

$$\frac{d\tilde{\mathscr{S}}}{d\tau} = \frac{ge_0}{2\mu c}\tilde{\tilde{\mathscr{F}}}\tilde{\mathscr{S}} - \frac{e_0}{\mu c^3}\left(\frac{g}{2} - \frac{e}{e_0}\right)(\tilde{\mathscr{S}}\tilde{\tilde{\mathscr{F}}}\tilde{w})\tilde{w} \qquad (5\text{-}49b)$$

This equation is the law of motion of four-spin for a particle in a homogeneous electromagnetic field.[6] It is manifestly covariant and can be used in any inertial frame. If $g = 2(e/e_0)$—this is very nearly the case for muons and electrons (Table 5-2)—the equation of motion has the very simple form

$$\frac{d\tilde{\mathscr{S}}}{d\tau} \approx \frac{e}{\mu c}\tilde{\tilde{\mathscr{F}}}\tilde{\mathscr{S}} \qquad (5\text{-}49c)$$

The magnetic moment anomaly a gives rise to the second term on the right-hand side of Eq. (5-49b), as well as slightly altering the coefficient in the first term.

The equation of motion for $\tilde{w}$ is, of course, Eq. (3-11c):

$$\frac{d\tilde{w}}{d\tau} = \frac{e}{\mu c}\tilde{\tilde{\mathscr{F}}}\tilde{w} \qquad (5\text{-}50)$$

This exact equation for $\tilde{w}$ is of the same form as the approximate equation [Eq. (5-49c)] for $\tilde{\mathscr{S}}$. If there is no magnetic moment anomaly, four-velocity and four-spin obey the same law of motion. In particular, then, the longitudinal polarization is constant.

From Eqs. (5-49b) and (5-50), we conclude that $\tilde{\mathscr{S}}^2$ and $\tilde{\mathscr{S}}\,\tilde{w}$ are constants of the motion.

---

[6] Homogeneous over the spatial extent of the particle. The assumption is that the gradient force $(\mathbf{m}\cdot\mathrm{grad})\,\mathscr{B}$ is negligible compared to the Lorentz force [Eq. (D-1)]. For the general equation of motion covering field inhomogeneities and also possible electric dipole moment, see [75].

In effect,

$$\frac{d\mathscr{S}^2}{d\tau} = 2\mathscr{S}\frac{d\mathscr{S}}{d\tau} = \frac{ge_0}{\mu c}(\tilde{\mathscr{S}}\tilde{\mathscr{F}}\tilde{\mathscr{S}}) - \frac{2e_0}{\mu c^3}\left(\frac{g}{2} - \frac{e}{e_0}\right)(\tilde{\mathscr{S}}\tilde{\mathscr{F}}\tilde{w})(\mathscr{S}\tilde{w})$$

The first term is zero because $\tilde{\mathscr{F}}$ is antisymmetric. The second term vanishes because $\tilde{\mathscr{S}}$ and $\tilde{w}$ are initially orthogonal [Eq. (5-44b)]. Similarly,

$$\frac{d(\tilde{\mathscr{S}}\tilde{w})}{d\tau} = \mathscr{S}\frac{d\tilde{w}}{d\tau} + \frac{d\tilde{\mathscr{S}}}{d\tau}\tilde{w} = \frac{e}{e_0}\frac{e_0}{\mu c}(\mathscr{S}\tilde{\mathscr{F}}\tilde{w}) + \frac{ge_0}{2\mu c}(\tilde{\mathscr{F}}\tilde{\mathscr{S}})\tilde{w}$$

$$- \frac{e^0}{\mu c^3}\left(\frac{g}{2} - \frac{e}{e_0}\right)(\mathscr{S}\tilde{\mathscr{F}}\tilde{w})\tilde{w}^2$$

Since

$$\tilde{w}^2 = -c^2$$

and

$$(\tilde{\mathscr{F}}\tilde{\mathscr{S}})\tilde{w} = \tilde{w}\tilde{\mathscr{F}}\tilde{\mathscr{S}} = -\tilde{\mathscr{S}}\tilde{\mathscr{F}}\tilde{w}$$

the right-hand side vanishes.

Since the square of $\tilde{\mathscr{S}}$ is constant in time, the spin magnitude is constant at its initial value. Because $\mathscr{S}\tilde{w}$ is initially zero, the four-spin remains orthogonal to the four-velocity. Thus $\tilde{\mathscr{S}}$ has the same properties at $\tau + d\tau$ that were postulated for it at τ. The equations of motion [Eqs. 5-49) and (5-50)] are consistent with the definition [Eq. (5-43)]. At any stage of the motion, the spin is found as the spacelike part of $\tilde{\mathscr{S}}$ in the rest frame as of that instant.

An integration procedure for Eqs. (5-49) has been indicated by Bargmann, Michel, and Telegdi [75], and expounded in detail by Hagedorn ([56], Section 9-4). For simple cases with constant homogeneous fields, such as those we considered in Section 5-4, there is really no advantage in using four-spin. The covariant equation of motion [Eq. (5-49)] comes into its own for numerical integration in the complicated fields used to transport and focus particle beams.

PROBLEMS

5-1. Consider the SLT $L_1(\alpha_2)$ from R_0 to R_1 followed by the SLT $L_2(\alpha_0)$ from R_1 to R_2, with $\alpha_2 = 1$, $\alpha_0 = 2$. Compare the resulting coordinate system with that obtained when the order of transformations is reversed: $L_2(\alpha_0)$ followed by $L_1(\alpha_2)$. Are they in the same reference frame? Make a sketch.

5-2. This problem is an exercise in handling successive transformations. The transition

$$\pi^+ + p \to \pi^+ + \Lambda + K^+$$

is observed in a bubble chamber. One wishes to analyze it as a sequence

$$\pi^+ + p \rightarrow Y_1^{*+} + K^+$$
$$\phantom{\pi^+ + p \rightarrow Y_1^{*+}} \hookrightarrow \Lambda + \pi^+$$

and would like to know in particular the invariant mass of the outgoing π^+ and Λ and the angle between the outgoing and incoming pion in the frame in which the momenta of the outgoing π^+ and Λ add to zero (rest frame of the Y_1^*).

From the measurements of track coordinates on several views of the chamber, we have computed the momenta of the tracks. Thence we have calculated the momenta in the overall center of mass frame. They are given, in spherical polar coordinates, in the following table:

Particle	$p_{\text{MeV}/c}$	$\cos \chi$	ψ
π_{in}^+	916	1	0
π_{out}^+	243	-0.4420	$18.8°$
Λ	427	-0.7634	$-83.9°$
K^+	534	0.8118	$139.1°$

Here χ is the polar angle measured from the z axis, ψ is the azimuth in the xy plane.

The following transformations are suggested:
(1) rotation about z axis, so $x'z'$ plane contains K^+ momentum;
(2) rotation about y axis, so z'' axis is parallel to K^+ momentum;
(3) SLT parallel to z'' axis to rest frame of Y_1^*.

Answers: $101°$, 1374 MeV.

5-3. In the disintegration of the charged pion into a muon and neutrino

$$\pi^+ \rightarrow \mu^+ + \nu_\mu \qquad \pi^- \rightarrow \mu^- + \bar{\nu}_\mu$$

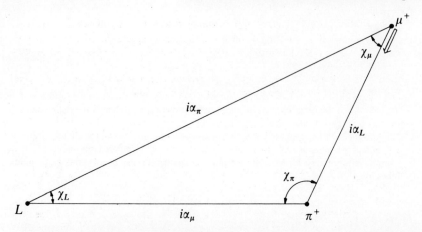

the muon is completely longitudinally polarized in the pion rest frame. In the positive charge case, for example, the muon spin is antiparallel to the momentum, as shown by the double arrow and loop in the sketch. It means simply that in the muon rest frame the muon's angular momentum is parallel to the pion momentum. The hyperbolic velocity triangle relating the pion rest frame, muon rest frame, and laboratory frame, is as shown, with $\chi_\mu (= \phi)$ giving the muon polarization in the laboratory frame (parallel, $\chi_\mu = \pi$; transverse, $\chi_\mu = \pi/2$; antiparallel, $\chi_\mu = 0$) and $\chi_L (= \theta)$ giving the muon angle of emission in the laboratory frame measured from the pion direction. The side $i\alpha_L$ is determined by the proper masses (Problem 4-3)

$$\alpha_L = \cosh^{-1} \gamma_L = \cosh^{-1}(1.0390)$$

Along with χ_π and the pion rapidity relative to the laboratory (α_μ), α_L determines the triangle.

Determine χ_μ as a function of χ_π and α_μ. Show that for a particular pion laboratory energy, complete transverse polarization is obtained at that χ_π for which the laboratory decay angle is greatest (Problem 3-8). (This is for a particular decay azimuth.)

Note that the longitudinal polarization $\mathscr{P}_\parallel$ of the positive muon in the laboratory is $-\cos.\chi_\mu$, and the transverse polarization $\mathscr{P}_\perp$ is $\sin \chi_\mu$. With a pencil beam of pions one can get any desired polarization by selecting the appropriate range of decay angles.

[For a careful derivation of the relation between muon spin orientation and center of mass decay angle, without use of four-spin or quantum mechanics, see G. Ascoli, Z. *Physik* **150**, 407 (1958).]

5-4. When cosmic protons and nuclei enter the earth's atmosphere, they interact with air nuclei, producing pions and other short-lived particles. Some of the pions decay in flight, giving rise to muons. Some of these muons survive to the bottom of the atmosphere, where they constitute, in fact, the bulk of the so-called cosmic radiation. A muon of a particular momentum at birth may be the decay product of parent pions of various energies and directions of motion. At one extreme (forward emission in pion frame, $\mathscr{P}_\parallel = \mp 1$ for $\pm$ charge), the pion energy $^\pi_L E$ is essentially equal to the muon energy $^\mu_L E$. At the other (backward emission, $\mathscr{P}_\parallel = \pm 1$), $^\pi_L E = 1.74 ^\mu_L E$.

Verify these limits for relativistic pions, using the results of Problem 4-3.

More slow pions are made than fast ones. The pion energy spectrum is of the form

$$f(^\pi_L E) \, d^\pi_L E = \text{const} \times (^\pi_L E)^{-n} \, d^\pi_L E \qquad n \approx 2.5$$

so that a given muon is more likely to result from a forward- than from a backward-decaying pion. Thus, a net longitudinal polarization is expected for muons of a particular sign and energy {I. I. Gol'dman, *Zh. Eksperim. i Teor. Fiz.* **34**, 1017 [1958] [*Soviet Phys.— JETP* **7** (**34**), 702 (1958)]}.

Derive Gol'dman's formula for the polarization of muons of given energy resulting from the decay of pions with the preceding energy distribution:

$$\mathscr{P}_{\|} = \mp\left\{\frac{1}{\beta} - \frac{n}{(n-1)}\frac{(1-\beta)}{\beta}\left[1 - \left(\frac{1-\beta}{1+\beta}\right)^{n-1}\right]\Big/\left[1 - \left(\frac{1-\beta}{1+\beta}\right)^{n}\right]\right\}$$

where $\beta = {}_{\pi}^{\mu}\beta = {}_{\mu}^{\pi}\beta = 0.270$ (Problem 4-3). Treat the process as occurring in vacuum, so that only the $\pi \to \mu$ decay probability needs to be taken into account (as in Section 2-7). Evaluate $|\mathscr{P}_{\|}|$ for $n = 2, 2.5, 3$.

The polarization of fast muons can be measured by bringing them to rest in an appropriate material and observing the angular distribution of their decay electrons. The slowing down does not depolarize appreciably. The distribution of the angle θ between the electron momentum and the muon direction can be shown [74] to be

$$f(\cos\theta)\,d(\cos\theta) = \tfrac{1}{2}(1 \pm a\mathscr{P}_{\|}\cos\theta)\,d(\cos\theta)$$

For the whole energy spectrum of decay electrons, $a = 1/3$; it can be calculated for any particular detection geometry. The measured angular distributions give values of $\mathscr{P}_{\|}$ in agreement with expectations (20–40%). [See H. V. Bradt and G. W. Clark, *Phys. Rev.* **132**, 1306 (1963), which contains references to earlier work.] A more sophisticated analysis of the dependence of $\mathscr{P}_{\|}$ on the pion spectrum near the top of the atmosphere has been published by V. S. Berezinskii [in *Proceedings of the International Conference on Cosmic Rays and the Earth Storm at Kyoto, 1961*, published in *J. Phys. Soc. Japan* **17**, Suppl. A–3 (1962), Vol. 3, p. 307].

5-5. Cosmic rays also make K mesons in the upper atmosphere, and the most common charged-K decay mode is

$$K^+ \to \mu^+ + \nu_\mu \qquad K^- \to \mu^- + \bar{\nu}_\mu$$

Like the pion, the K meson has zero spin, and the kinematics of this transition are exactly the same as for $\pi \to \mu + \nu_\mu$ (two preceding problems).

From the proper masses, compute ${}_{K}^{\mu}\beta = {}_{\mu}^{K}\beta$. For relativistic K mesons (${}_{L}^{K}\beta \to 1$), what range of K energies ${}_{L}^{K}E$ can contribute to the supply of muons of energy ${}_{L}^{\mu}E$?

If the atmospheric K mesons have a power-law spectrum like that of the pions, what muon polarization is expected for $n = 2$? $n = 3$?

The high values found ($> 90\%$) indicate that the muon polarization is a sensitive function of the K/π ratio at production as well as of the shape of the pion spectrum.

5-6. A particle with spin is moving parallel to a constant magnetic field. There is no electric field. Determine the spin precession frequency in the laboratory.

5-7. We would like to convert a beam of longitudinally polarized protons of momentum 5 GeV/c to transverse polarization. Since $|g| \neq 2$ (Table 5-2) we can do this by deflecting the protons in an appropriate transverse magnetic field. For how many seconds do the protons need to be deflected? With a 10^4-G field, how long should the magnet be? What is the deflection of the proton beam?

The same for protons of 0.5 GeV/c, 50 GeV/c, 500 GeV/c. (For protons, $e/\mu c = 9.58 \times 10^3$ rad sec^{-1} G^{-1}.)

5-8. If

$$\frac{|g| - 2}{2} = 1.16 \times 10^{-3}$$

what momentum should a muon have so that its spin turns 2π radians in one orbital revolution? The same for an electron.

Chapter 6
Principle of Equivalence. Motion in a Weak Gravitational Field

Special relativity is concerned only with transformations between members of a family of local inertial frames. It needs to be embedded in a broader theory, covering reference frames in arbitrary relative motion. Such a theory, *general relativity*, was worked out by Einstein in the period from 1907 to 1915 [76]. It turned out to be at the same time a theory of gravitation, containing Newton's inverse-square law [Eq. (1-8)] as an extremely close approximation. An essential part of the theory is the principle of equivalence [31]. This principle has been confirmed experimentally to high precision and can be regarded as solidly established. The rest of the theory, specifically the gravitational field equations, has not as yet been tested so decisively. It is in agreement with the available experimental data, but competing theories (for example, [77]) cannot be ruled out.[1]

General relativity is outside the scope of this book, and we can only urge the reader to go on to study it elsewhere.[2] But for describing phenomena in weak (that is, essentially homogeneous) gravitational fields, the principle of equivalence alone, used with special relativity, is sufficient.

6-1. THE PRINCIPLE OF EQUIVALENCE

This principle unifies gravitational and inertial forces. It states:

In an infinitesimal region of spacetime, two reference frames, Σ and Rho, are completely equivalent with regard to all physical phenomena provided that in Σ there is a uniform gravitational field **g** *while in Rho there is no gravitational field*

---

[1] General relativity is extremely attractive on aesthetic (and thus philosophic) grounds. As an anonymous editor [78] has remarked, "every lover of the beautiful must wish it to be true." Of course, aesthetic criteria go out the window as soon as a crucial experiment can be carried out.

[2] Some recommended books are Adler, Bazin, and Schiffer [79]; Landau and Lifshitz ([35], Chapters 10 and 11); Møller [29]; Bergmann [80]; and Einstein [81].

but the frame has translational acceleration $-\mathbf{g}$ *and velocity zero with respect to the inertial frame in which* Σ *is at rest.*

This principle elevates the fact (Chapter 1) of the indistinguishability of a translational acceleration and a uniform gravitational field into a law of nature. A gravitational field is equivalent to the corresponding acceleration of the reference frame. The effect of acceleration $\mathbf{a}$ of a frame with respect to an inertial frame in which it is instantaneously at rest is the same as that of adding a homogeneous gravitational field $\mathbf{g} = -\mathbf{a}$ to the other fields present. Alternatively, one can eliminate a uniform gravitational field $\mathbf{g}$ by replacing it with the equivalent acceleration of the laboratory $\mathbf{a} = -\mathbf{g}$ and making a purely kinematical calculation of the effect of this acceleration on the relative motion of the weightless body under consideration.

Note that the frames Σ and Rho are instantaneously at rest with respect to the same inertial frame. Any effects on measuring rods and clocks of velocity with respect to another inertial frame are the same in Σ and Rho.

The principle accomplishes a fusion of two apparently universal properties of physical systems—inertia (resistance to change of motion) and gravitation (attraction toward other bodies). They become inextricably intertwined. It suggestes an "explanation" for inertia: The inertial force $-m^{\dagger}\mathbf{a}^{\dagger}$ in the rest frame of the particle expresses the gravitational pull of the rest of the universe.

The main experimental basis for the principle of equivalence is the equal acceleration of all shielded bodies regardless of their weight or composition. This fact startled the pundits of Galileo's time and does not cease to impress. It has been confirmed in a series of experiments of ever higher precision, from Galileo through Newton and Bessel and Eötvös to Dicke and his associates [14]. The last find, comparing a piece of gold with a piece of aluminum, that the accelerations are equal to within one part in 10^{11}. Equal acceleration of all bodies is precisely the effect that would occur if there were no gravitational field and instead the laboratory were accelerated (Einstein's elevator).

Recent measurements of the gravitational red shift, to be described later, have confirmed that the principle of equivalence applies to photons as well as to macroscopic bodies. Now anything moving with constant velocity in an inertial frame, be it material particle or photon, follows a curved path in an accelerated frame. Therefore, a gravitational field deflects light. This acceleration of photons by gravity invalidates the assumption of constant light velocity made in special relativity. In a gravity-ridden frame, one cannot use light beams to survey the monuments and synchronize the clocks. One can, however, use these special-relativity procedures in an infinitesimal region of spacetime, and fortunately the gravitational field at the earth is so weak that such a region is physically quite large. One needs only to be clear about the order of magnitude of the various infinitesimals.

The principle of equivalence makes gravitation really universal; it acts on every kind of concentration of energy, not just on "bodies." (From here it is only a step further to geometrize gravitation, as Einstein did in general relativity: Describe gravitation not in terms of force but in terms of the metric of spacetime. The sources of the gravitational field alter the metric in such a way as to bend the natural path of a free particle.)

6-2. THE EFFECT OF A GRAVITATIONAL FIELD ON CLOCKS AND MEASURING RODS

The principle of equivalence permits us to calculate the effect of a gravitational field by purely kinematical considerations on the equivalent accelerated frame ([31]; also [82]). In these considerations, we must allow for the effects of acceleration and velocity on the measuring rods and clocks in the accelerated frame.

As we have already remarked in Section 2-5, correction must be made for specific effects of acceleration. A real measuring rod (such as a Michelson interferometer) may be bent or stretched depending on the way it is supported and its elastic constants. It will not do to use for a clock a pendulum clock whose rate depends on the effective gravitational field; a spring clock, on the other hand, is only affected by changes in the size and shape of its parts. An atomic clock is extremely insensitive to acceleration. The needed corrections can be determined in the laboratory by studying the effects of weight or acceleration. We assume from now on that they will have been made.

There are, in addition, effects of velocity that are independent of the type of measuring instrument used. We have learned in Section 2-5 about the Lorentz–Fitzgerald contraction of rods and the Einstein slowing down of clocks. Let R and R' be inertial frames with identical arrays of rods and clocks, duly surveyed and synchronized; let $\mathbf{u}$ be the velocity of R' relative to R. With respect to the rods and clocks of R, rods at rest in R' are shortened longitudinally by the factor $\gamma = (1 - u^2/c^2)^{-1/2}$; they are unchanged transversally; and clocks at rest in R' are slowed down by the factor γ. We now make the physical assumption that these statements also hold for the rods and clocks of an accelerated frame that is momentarily at rest with respect to R'. In other words, we shall calculate the behavior of accelerated rods and clocks by applying the formulas of special relativity to the rods and clocks of the comoving inertial frame.

There is really no alternative. A physicist in the noninertial frame knows that it is not inertial: a test body falls. So he knows that he cannot trust his rods and clocks. His only recourse is to calibrate his instruments against corresponding nearby ones in the inertial frame that is at rest with respect to his. His standard rod and clock are installed in the laboratory so as to be just starting to drop when observed.

Consider an accelerated frame Rho (Greek letters), and a sequence of inertial frames R, R', R'' (Latin letters) that are successively at rest with respect to Rho. All of the frames have identical arrays of monuments and clocks. In the inertial ones they have been surveyed with light beams and synchronized by light flashes, according to the procedures of Section 2-2. We are interested in events occurring in a small spacetime region of Rho (small spatial volume and short time interval). Pick some origin event in this region, to locate an origin of coordinates Θ and a zero time $\tau_\Theta = 0$ on the clock at the origin. Let the ξ axis be in the direction of acceleration (Fig. 6-1). The inertial frame R is at rest with respect to Rho at the origin event. Pick an origin O in R to coincide with Θ at this moment ($t = 0$), and choose x, y, z axes to coincide with the ξ, η, ζ axes. In the part of Rho under consideration, we graduate our measuring rods so as to agree with the matching ones in R, and we set our clocks to zero. As observed from R, the clocks of Rho then remain synchronized, because each one has the same motion relative to R. Two events that are simultaneous in R ($t_1 = t_2$) have equal readings of time (τ) on their local Rho clocks.

At a time τ_Θ later on the Θ clock, the inertial frame R' is at rest with respect to Rho. Choose axes $O'x'y'z'$ in R' to coincide with $Oxyz$ at $t = 0 = t'$; R and

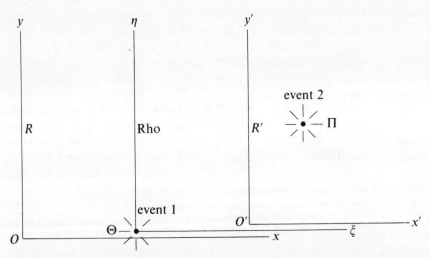

Figure 6-1. At $t = 0$ in the inertial frame R, the axes $Oxyz$ coincided with the axes $\Theta\xi\eta\zeta$ in the accelerated frame Rho. Rho has acceleration with respect to R of g in the x direction. Event 1 occurs at (infinitesimal) time τ_Θ on the clock at Θ. Event 2, simultaneous with it in the comoving inertial frame R', occurs at Π, where the clock reads τ.

R' are related by a SLT [Eq. (1-31)] with u given to first order in τ_Θ by

$$u \approx g\tau_\Theta$$

where g is the acceleration of Rho with respect to R at $t = 0$. Consider two neighboring events that are simultaneous in R'—number 1 at Θ when it is at rest in R' ($\xi_1 = 0, \eta_1 = 0, \zeta_1 = 0, \tau_1 = \tau_\Theta$), number 2 at $\Pi(\xi, \eta, \zeta, \tau)$. Because

$$t_2' - t_1' = \gamma\left[(t_2 - t_1) - \frac{u}{c^2}(x_2 - x_1)\right]$$

they are not simultaneous in R, but rather

$$t_2 - t_1 = \frac{u}{c^2}(x_2 - x_1)$$

Now the effects of nonzero velocity on the rods and clocks of Rho are of second order in τ_Θ [they depend on $\gamma(u)$], so to first order

$$t_2 \approx \tau \qquad t_1 \approx \tau_1 = \tau_\Theta$$
$$x_2 - x_1 \approx \xi_2 - \xi_1 = \xi$$

The equation expressing simultaneity in R' becomes

$$\tau - \tau_\Theta \approx \frac{g\tau_\Theta}{c^2}\xi$$

or

$$\tau = \tau_\Theta\left(1 + \frac{g\xi}{c^2}\right) \qquad\qquad (6\text{-}1a)$$

Equation (6-1a) is valid provided that

$$\tau \ll \frac{c}{g} \qquad \xi \ll \frac{c^2}{g} \qquad\qquad (6\text{-}1b)$$

It states that, comparing events that are simultaneous in the comoving inertial frame, the Rho clock reading is greater for the event located further in the $+\xi$ direction. The clock at Π runs faster than the clock at Θ by the fractional amount $g\xi/c^2$.

The same result is obtained if one compares the two clocks in Rho with essentially the same R clock as it passes near each Rho clock in turn. With

no loss of generality, we choose Π to be at ($\xi = -\alpha, \eta = 0, \zeta = 0$) in Rho.
We compare the clock at Θ with the master R clock at O when they coincide
(Fig. 6-2a) and the clock at Π with the same R clock later on when these two
coincide (Fig. 6-2b). At the first coincidence, clock Θ is at rest in R. At the
second, clock Π is moving with velocity

$$v \approx (2g\alpha)^{1/2}$$

The time dilation formula [Eq. (2-27)] gives for the number of ticks between
corresponding successive events near each clock

$$\frac{\Delta\tau_\Pi}{\Delta\tau_\Theta} = \frac{\Delta\tau_\Pi/\Delta\tau_O}{\Delta\tau_\Theta/\Delta\tau_O} = \frac{(1 - 2g\alpha/c^2)^{1/2}}{1} = 1 - \frac{g\alpha}{c^2} + \cdots$$

$$= 1 + \frac{g\xi}{c^2} + \cdots$$

(6-1c)

in agreement with Eqs. (6-1a and b).

We can apply the same argument to compare measuring rods at different ξ.
Comparing a longitudinal meter stick at Θ with one at Π by matching them
in turn against a standard meter nearby at O (Fig 6-2), we find from the
length contraction formula [Eqs. (2-30)] that the spatial distance between

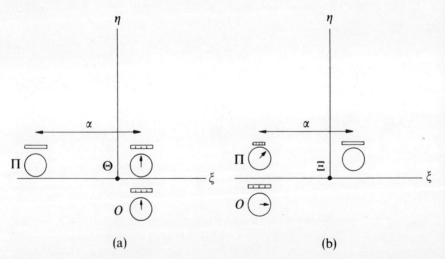

(a)   (b)

Figure 6-2.   (a) Comparison of clock and rod at Θ with master clock and
rod at O. (b) Comparison of clock and rod at Π with master clock and rod
at O.

corresponding events near each rod is

$$\frac{\Delta\xi_\Pi}{\Delta\xi_\Theta} = \frac{\Delta\xi_\Pi/\Delta x_O}{\Delta\xi_\Theta/\Delta x_O} = \frac{(1 - 2g\alpha/c^2)^{-1/2}}{1} = 1 + \frac{g\alpha}{c^2} + \cdots$$

or

$$\Delta\xi_\Pi = \Delta\xi_\Theta\left(1 - \frac{g\xi}{c^2} + \cdots\right) \tag{6-2}$$

Transverse meter sticks are unaffected by the acceleration.

The principle of equivalence now tells us that the results in Eqs. (6-1) and (6-2) also apply in the original unaccelerated frame Σ, with the homogeneous gravitational field of magnitude g pointing in the $-\xi$ direction. Here, $g\xi$ is the increase of gravitational potential from Θ to Π:

$$g\xi = \varphi_\Pi - \varphi_\Theta = \Delta\varphi$$

At higher gravitational potential, a clock runs faster; a meter stick parallel to the field is longer, one perpendicular to it is unchanged.

$$\tau_\Pi = \tau_\Theta\left(1 + \frac{\varphi_\Pi - \varphi_\Theta}{c^2}\right) \tag{6-3a}$$

$$\Delta\tau_\Pi = \Delta\tau_\Theta\left(1 + \frac{\varphi_\Pi - \varphi_\Theta}{c^2}\right) \tag{6-3b}$$

$$\Delta\xi_\Pi = \Delta\xi_\Theta\left(1 - \frac{\varphi_\Pi - \varphi_\Theta}{c^2}\right) \tag{6-4}$$

Equation (6-3b) was used in Section 2-7 in the discussion of the twin thought experiment.

We see that the presence of a gravitational field destroys the possibility of synchronizing clocks at different levels, unless we tamper with their rates, and destroys the possibility of surveying the monuments by assuming space to be Euclidian and light to travel in straight lines. The infinitesimal interval squared between neighboring events

$$ds^2 = c^2\,dt^2 - dx^2 - dy^2 - dz^2 \tag{2-24}$$

has, to a first approximation, the form

$$ds^2 = c^2\left(1 - \frac{2\,\Delta\varphi}{c^2}\right)d\tau^2 - \left(1 + \frac{2\,\Delta\varphi}{c^2}\right)d\xi^2 - d\eta^2 - d\zeta^2 \tag{6-5}$$

when expressed in terms of raw measurements in Σ. Space is neither isotropic nor homogeneous, and the time scale depends on position. Spacetime is not "flat."

6-3. THE GRAVITATIONAL RED SHIFT

When he derived the speeding up of a clock at a higher level in a gravitational field, Einstein [31] pointed out that it implies a shift toward the red for light climbing from sun to earth, of fractional magnitude 2×10^{-6}. In effect, the atom emitting the spectral line is a clock at low potential; the corresponding atom on earth, emitting the comparison line, is a clock at high potential. The wave train arriving at earth has a lower frequency in terms of the speeded-up clock; that is, it is shifted toward the red. By Eq. (6-3b),

$$-\frac{\Delta v}{v} = \frac{\Delta \varphi}{c^2} = \frac{1}{c^2} \int_{r_s}^{r_e} dr \left(-\frac{GM_s}{r^2} \right) = \frac{GM_s}{c^2} \frac{1}{r_s} = 2.1 \times 10^{-6}$$

where M_s is the mass of the sun, r_s is the radius of the chromosphere, and r_e is the distance of the earth from the center of the sun (the gravitational field of the earth is negligible). It is assumed here that for radial propagation the inhomogeneity of the field has no effect. Although the most recent astronomical data [83] agree with the prediction, there are a variety of perturbations that need to be taken into account. It is fortunate that a decisive terrestrial test has become possible.

Einstein [82] gave a variety of arguments for the red shift formula

$$-\frac{\Delta v}{v} = \frac{\Delta \varphi}{c^2} \tag{6-6}$$

One of the most convincing assumes only the Doppler effect formula of special relativity (Appendix F). It applies directly to the terrestrial tests using Mössbauer technique on vertically falling x rays. The illustrative numbers that we shall quote refer to the most precise of these experiments, carried out by Pound and his colleagues [84].

As before, let Rho be the upward-accelerated field-free frame equivalent to the real-life earth frame Σ. There is a laboratory A in the basement and another, B, on the roof at a height d (22.5 m) above A. An iron-57 Mössbauer source in B emits a sharp spectral line of frequency v_0 (3.47×10^{18} Hz) in B. One observes in A the attenuation of these x rays in an iron foil as a function of its vertical velocity. In Rho, we choose Cartesian axes with the ξ axis upward and origin at B (Fig. 6-3a). R is the inertial frame which is at rest in Rho when a particular wave crest leaves B in the downward direction. Choose axes $Oxyz$ in R to coincide with $\Theta\xi\eta\zeta$ at this instant, and let $t = 0$ then (Fig. 6-3b).

In R, the crest travels in the $-x$ direction with speed c. Laboratory A is accelerated in the x direction with acceleration g, from $x = -d$ at $t = 0$ to $x = -d + (1/2)gt^2$ at time t. (We can use nonrelativistic dynamics for A because $gd/c \ll c$. In fact, $gd/c^2 = 2.45 \times 10^{-15}$.) The crest meets A at

$$t \approx \frac{d}{c}$$

at which time A has an upward velocity in R of

$$gt = \frac{gd}{c}$$

Let R' be the inertial frame in which Rho (and thus A) is at rest at this event. Its velocity with respect to R is

$$u = \frac{gd}{c}$$

The period of the wave is very small compared with the time for an appreciable change in the velocity of A, and many crests reach A while its velocity is

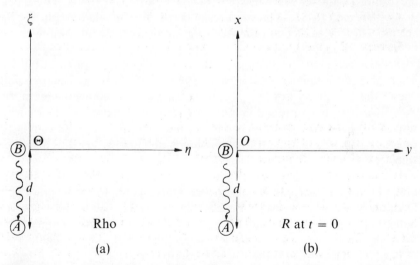

(a)   (b)

Figure 6-3. Blue shift of a falling photon. The roof laboratory B is where the gamma ray source is located. The basement laboratory A contains the absorber foil and detector. Axes $\Theta\xi\eta$ are fixed in the upward accelerated frame Rho. R with axes Oxy is an inertial frame at rest relative to Rho when a particular wave crest leaves B.

essentially gd/c. The Doppler effect formula for head-on approach [Eq. (F-8b)] gives, since $v_0 = v$,

$$v' = v_0\left(1 + \frac{gd}{c^2}\right) \qquad (6\text{-}7a)$$

or

$$v' = v_0\left(1 + \frac{\Delta\varphi}{c^2}\right) \qquad (6\text{-}7b)$$

where $\Delta\varphi = gd$ is the increase of gravitational potential from A to B in the frame Σ. This is the desired formula [Eq. (6-6)]. For falling light, the frequency is shifted toward the blue.

Regarding each crest as a tick of the nuclear clock in B, an observer in A will conclude that the clock in B is running faster than his identical clock in A by the fractional amount gd/c^2. Thus we have rederived the result [Eq. (6-3)] that a clock at a higher level runs faster.

The essential point in the preceding derivation of the gravitational red shift is that during the transit of the photon from source to absorber, the absorber acquires some velocity with respect to the inertial frame in which the source was at rest at the time of emission. The frequency shift is simply the resulting Doppler effect. There is, of course, a slowing down of the clock in A with respect to clocks in R as a result of the Einstein dilation [Eq. (2-27)], but this effect is of second order in (gd/c^2) and is completely swamped by the first-order Doppler effect.

In the experiment, use is made of the fact that the absorption in the foil is sensitive to a frequency shift between emitter and absorber. The blue shift due to the 22.5-m fall from roof (B) to basement (A) is compensated in the experiment by giving the foil in A the appropriate downward velocity with respect to the source gd/c (7.35×10^{-5} cm sec^{-1}). In fact, to an estimated precision of 0.8%, minimum transmission is obtained at just this absorber velocity.

Pound and Snider [84] have pointed out that the experiment is actually a direct test of the principle of equivalence. Only the photon absorption as a function of absorber velocity is measured, the result being that whatever happens to a photon in free fall through a distance d is exactly compensated by giving the absorber a velocity of g times the photon time of flight.

The "g" of Eq. (6-7a) is the g_{eff} of Section 1-5—earth gravity plus centrifugal and Coriolis inertial forces. The contribution of the latter is, however, smaller than the estimated error of the Boston experiment.

A direct test of the principle of equivalence for centrifugal force is provided by the experiments on the transverse Doppler effect (Section 2-7) that use rotary motion. These experiments ([25]; also [85] and [86]) study the frequency shift between source and absorber located on a radius of a rotating reference frame. The effect can be ascribed to their velocities in the (inertial) laboratory frame (transverse Doppler effect). It can equally well be described in the accelerated rotor frame as a gravitational red shift. The centrifugal field of force corresponds by the principle of equivalence to a radial gravitational field of strength $\rho\omega^2$, where ρ is the radial distance and ω is the angular velocity. Integrating between source at ρ_s and absorber at ρ_a (≈ 0), we find for the potential difference

$$\Delta\varphi = -\int_{\rho_s}^{\rho_a} d\rho\,\rho\omega^2 = -\frac{\omega^2}{2}[(\rho_a)^2 - (\rho_s)^2]$$

giving the frequency shift

$$\frac{\Delta\nu}{\nu} = -\frac{\Delta\varphi}{c^2} = \frac{\omega^2}{2c^2}[(\rho_a)^2 - (\rho_s)^2]$$

But the velocity in the laboratory of a point of the rotor is

$$v = \rho\omega$$

so that

$$\frac{\Delta\nu}{\nu} = \frac{1}{2}\left(\frac{v_a^2}{c^2} - \frac{v_s^2}{c^2}\right) \tag{6-8}$$

which is essentially the transverse Doppler effect formula [Eq. (2-34)]. The experiments agree with Eq. (6-8), the most precise [25] to within about 1%.

The consistency of the transverse Doppler effect formula and the gravitational red shift formula is not surprising, since both follow in a straightforward way from the interweaving of space and time in the Lorentz transformation. We have already noted this consistency in the discussion in Section 2-7 of the twin problem. There, the same result is obtained whether we calculate in the stationary twin's frame using special relativity or in the traveling twin's frame using special relativity and the principle of equivalence. Evidently the combination of special relativity and the principle of equivalence, as used in this chapter, is a logically consistent theory.

The gravitational frequency shift can be described in terms of a weight of the falling photon, given by

$$\left(\frac{\varepsilon}{c^2}\right)g$$

It follows from the quantal wave–particle relation (Section 4-9 first worked exercise, Appendix F)

$$\tilde{p} = \hbar\tilde{k} \tag{F-9c}$$

that the frequency increase Δv of Eq. (6-7a) corresponds to the energy increase

$$\Delta \varepsilon = h\,\Delta v = \frac{hvgd}{c^2} = \left(\frac{\varepsilon}{c^2}\right) g \cdot d$$

that is, the force $(\varepsilon/c^2)g$ times the distance d. It also corresponds to the momentum increase

$$\Delta \pi = \frac{\Delta \varepsilon}{c} = \left(\frac{\varepsilon}{c^2}\right) g \cdot \frac{d}{c}$$

which is the same force times the time of flight d/c. The force is the photon energy, divided by c^2, times the gravitational field strength. Electromagnetic energy has gravitational charge, just like the energy in bodies of nonzero proper mass.

6-4. FREE FALL OF A MATERIAL PARTICLE

The method just employed to analyze the free fall of a photon in a homogeneous gravitational field can be applied to the free fall of a material particle. Instead of the Doppler effect formula we use the velocity addition formulas [Eqs. (2-42)]. We replace the real laboratory frame Σ by its equivalent upward-accelerated field-free frame, Rho. The physicists in either frame know that their reference frame is not inertial (in it shielded bodies fall!), and they describe events by using time and position measurements in the instantaneously comoving inertial frames R and R'.

Motion Parallel to Field

Suppose the beam from an accelerator is bent straight upward, producing reversed cosmic rays (Fig. 6-4). Consider observations at two levels A and B in Rho, distance $\Delta \xi$ apart. Let R be the inertial frame at rest with respect to Rho when the particle is at A, and R' be the inertial frame at rest with respect to Rho when the particle is at B. Introduce parallel axes in Rho, R, and R' with $\Theta \xi$, Ox, $O'x'$ pointing up, and let $\tau_A = t = t' = 0$ when the particle is at A (τ is not proper time, but is local time in Rho with all clocks set to zero when R is at rest with respect to Rho). The velocity of the particle is not constant in Rho. At A at $\tau_A = 0$, it is $(v_A, 0, 0)$; at B at $\tau_B = \Delta \tau$ it is $(v_B = v_A - \Delta v, 0, 0)$. The downward acceleration of the particle in Rho is

$$a_{\parallel} = -a_{\xi} = \lim_{B \to A} \frac{\Delta v}{\Delta \tau} \tag{6-9}$$

In the (gravity-free) inertial frames R and R', the particle velocity is constant: $(v_A, 0, 0)$ in R; $(v_A - \Delta v, 0, 0)$ in R'. The velocity of R' with respect to R is, to a sufficient approximation, $g\,\Delta\tau$ in the x direction. The velocity transformation of special relativity, Eq. (2-42), reads

$$v'_x = \frac{v_x - u}{1 - uv_x/c^2}$$

that is,

$$v_A - \Delta v = \frac{v_A - g\,\Delta\tau}{1 - g\,\Delta\tau v_A/c^2}$$

$$= v_A - g\,\Delta\tau + g\,\Delta\tau\frac{v_A^{\ 2}}{c^2} + \cdots$$

$$= v_A - g\,\Delta\tau\frac{1 - v_A^{\ 2}}{c^2} + \cdots$$

Thus, the downward acceleration in Rho, Eq. (6-9), is

$$a_{\parallel} = \lim_{\Delta\tau \to 0}\left(\frac{\Delta v}{\Delta\tau}\right) = g\left(1 - \frac{v^2}{c^2}\right) = g\gamma^{-2} \qquad (6\text{-}10)$$

This is the general result of Ascoli, Eq. (1-42).

In speeding up from velocity 0 to velocity $\approx g\,\Delta\tau$ in R, Rho does suffer some alterations in its rods and clocks, but these effects are of second order in $\Delta\tau$. Equation (6-10) is exact.

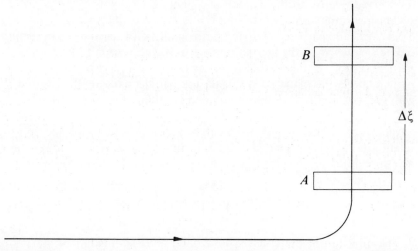

Figure 6-4.   Beam from an accelerator projected straight up. We calculate its slowing down in free fall between A and B.

The weight of the particle,

$$F_\xi = \frac{dp_\xi}{d\tau}$$

is, for longitudinal acceleration [first Eq. (3-4b)],

$$F_\xi = \mu\gamma^3 a_\xi = \mu\gamma^3(-g\gamma^{-2}) = -\mu\gamma g = -mg$$

The weight is proportional to the inertial mass m, or the energy mc^2. The gravitational field does not act on proper mass alone; all the energy in the particle contributes to its gravitational charge.

The inertial mass–energy relation and the principle of equivalence together imply that, in general, the gravitational field acts on all kinds of energy.

Equation (6-10) reduces to the familiar nonrelativistic expression

$$a_\parallel = g$$

for $v \ll c$. At the photon limit ($v \to c, \mu \to 0, \gamma = \infty$), the acceleration is zero, corresponding to constant velocity c. The change in photon speed with height is a higher order effect that can be neglected in a small region of spacetime.

Motion Perpendicular to Field

Now the beam is horizontal. We define the horizontal direction by the path of a light beam moving horizontally near the origin (i.e., grazing the liquid surface of a spirit level). There is nothing straighter. The free fall of the particle is observed relative to the free fall of a photon.

Consider observations at stations C and D in Rho, distance $\Delta\eta$ apart (Fig. 6-5a). At C, the particle is moving horizontally with speed v. As in the preceding case, let R be the inertial frame at rest with respect to Rho when the particle is at C, R' the inertial frame at rest with respect to Rho when the particle meets the plane D. We again use parallel axes in Rho, R, and R', with $\Theta\xi$, Ox, and $O'x'$ pointing up (along a plumb line at C), $\Theta\eta$, Oy, and $O'y'$ pointing in the horizontal direction, and $\tau_C = t = t' = 0$ when the particle is at C. The velocity of R' with respect to R is to a sufficient approximation $(g\,\Delta\eta/v, 0, 0)$. The constant velocity of the particle in R is $(0, v, 0)$. The special relativity velocity transformation [Eq. (2-42)] gives for its constant velocity in R'

$$\left(\frac{-g\,\Delta\eta}{v}, \left[1 - \left(\frac{g\,\Delta\eta}{vc} \right)^2 \right]^{1/2} v, 0 \right) \tag{6-11a}$$

The direction of the C-based horizontal at D is not that of the y' axis because of the fall of the light beam en route from C to D. Carrying out a similar calculation for the velocity transformation of a photon between an inertial frame S comoving with Rho when the photon leaves C, and one S' comoving with Rho when the photon reaches plane D, we find for the velocity components of the photon at D

$$\left(\frac{-g\,\Delta\eta}{c}, \left[1 - \left(\frac{g\,\Delta\eta}{c^2} \right)^2 \right]^{1/2} c, 0 \right)$$

The C horizontal at D is therefore obtained by rotating the $O'x'y'$ axes clockwise through the angle a (Fig. 6-5b), where

$$\tan a = \frac{g\,\Delta\eta/c^2}{[1 - (g\,\Delta\eta/c^2)^2]^{1/2}}$$

Projecting the components of the particle velocity at D [Eq. (6-11a)] onto the new vertical direction, we find for the vertical velocity component

$$\left(\frac{-g\,\Delta\eta}{v} \right) \cos a + \left\{ \left[1 - \left(\frac{g\,\Delta\eta}{vc} \right)^2 \right]^{1/2} v \right\} \sin a$$

$$= -\frac{g\,\Delta\eta}{v} \left\{ \left[1 - \left(\frac{g\,\Delta\eta}{c^2} \right)^2 \right]^{1/2} - \frac{v^2}{c^2} \left[1 - \left(\frac{g\,\Delta\eta}{vc} \right)^2 \right]^{1/2} \right\}$$

$$= -\frac{g\,\Delta\eta}{v} \left(1 - \frac{v^2}{c^2} \right) \left[1 + \frac{1}{8} \frac{v^2}{c^2} \left(\frac{g\,\Delta\eta}{vc} \right)^4 + \cdots \right]$$

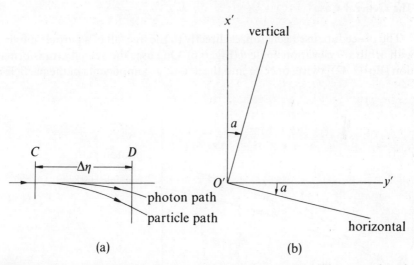

(a)   (b)

Figure 6-5.   Free fall of a horizontal particle beam. (a) The curved paths in Rho of a particle and of a photon used to define the horizontal direction. (b) The $O'x'y'$ axes in R' and the horizontal and vertical directions corresponding to the photon velocity when it meets the plane D.

The downward acceleration is the limit of this expression divided by the time ($\approx \Delta\eta/v$) as $\Delta\eta \to 0$:

$$a_\perp = \lim_{\Delta\eta \to 0} \frac{(g\,\Delta\eta/v)(1 - v^2/c^2)[1 + O(\Delta\eta)^4]}{(\Delta\eta/v)}$$

$$a_\perp = g\left(1 - \frac{v^2}{c^2}\right) = g\gamma^{-2} \tag{6-11b}$$

The horizontal component of the acceleration should be zero. In fact, the horizontal component at D of the particle's velocity in R' is

$$-\left(\frac{-g\,\Delta\eta}{v}\right)\sin a + \left\{\left[1 - \left[\frac{g\,\Delta\eta}{vc}\right]^2\right]^{1/2}v\right\}\cos a = v + O(\Delta\eta)^2$$

so that the horizontal acceleration

$$\lim_{\Delta\eta \to 0}\frac{v + O(\Delta\eta)^2 - v}{(\Delta\eta/v)} = 0$$

Again Ascoli's general expression [Eq. (1-42)] is confirmed.

The General Case

The procedure used generalizes directly to the free fall of a particle moving with arbitrary elevation angle θ (Fig. 6-6). One uses the velocity transformation [Eq. (2-42)] twice, once to find the x' and y' components of the particle's

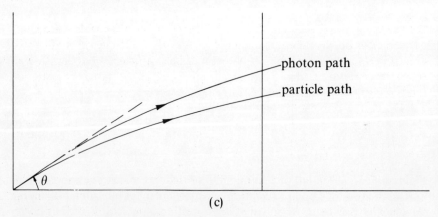

(c)

Figure 6-6.   The general case of free fall.

velocity in R' and again on a photon to find the amount of turning of the vertical. The result is once more the expression Eq. (1-42):

$$\mathbf{a} = \mathbf{g}\left(1 - \frac{v^2}{c^2}\right) = \mathbf{g}\gamma^{-2} \qquad (6\text{-}12)$$

Here g means the acceleration of a slowly moving body in the laboratory frame ($g \approx 980\,\text{cm sec}^{-2}$).

This result is due to G. Ascoli (unpublished), who gave a different proof. He has also derived it from the general relativity treatment of the motion of a particle of arbitrary velocity in a spherically symmetric gravitational field, showing that

$$g = \frac{GM/R^2}{(1 - 2GM/c^2R)^{1/2}}$$

where M is the mass of the earth and $2\pi R$ its circumference. The departure of g from the Newtonian numerator is only about one part in 10^9. Ascoli has also shown that in general relativity, the energy of the particle, $\mu c^2\gamma$, decreases by $\mu\gamma gh$ when the particle rises a small height h. In this sense, the weight is, as ever,

$$\mu\gamma g = mg = \left(\frac{E}{c^2}\right)g$$

The gravitational charge equals the inertial mass, or, to within a universal constant, the energy.

In terrestrial experiments involving forces other than gravitation, the accelerations occurring are much much greater than $980\,\text{cm sec}^{-2}$. As stated in Section 1-5, we can allow for the noninertial character of our earthbound reference frame by adding to each particle's nongravitational acceleration the very very slight correction term

$$\mathbf{g}\left(1 - \frac{v^2}{c^2}\right) \qquad (1\text{-}42)$$

By the same token, tremendously large equivalent gravitational fields exist in the rest frame of a microscopic particle during its collisions. Although these fields average to zero, it may be that it will not prove possible to develop a satisfactory quantal theory of fundamental particle dynamics in the framework of flat spacetime (special relativity). A theory unifying gravitation with the other interactions would then have to be found.

On this questioning note we conclude this book. Special relativistic particle mechanics is a mode of description whose limits of validity are still being probed.

PROBLEMS

6-1. Identical atomic clocks are placed on the ground and in an earth satellite moving in a circular orbit. The ticks of the orbiting clock are transmitted to the ground by radio, permitting comparison of the clock rates. For what height of the orbit above sea level are the clocks synchronous? [*Hint*: Consider both Doppler and gravity.]

6-2. Compare the straightness in a transverse gravitational field of a taut wire and a light beam. Show that the wire takes the shape of a catenary, and determine the dependence of its sag on the tension, density, and gravitational field strength. Look up the yield point and density of various solids and pick best material for the wire. Compare the smallest sag one can hope to realize on earth with that of a horizontal light beam.

Appendix A

Transformation Properties of Vectors. Tensors. Polar and Axial Vectors

ROTATION OF CARTESIAN AXES

Consider first rotation in two dimensions (Fig. A-1). The radius vector $\mathbf{r}$ from the origin to a point P of the plane can be written as a linear combination of the unit vectors $\mathbf{i}$ and $\mathbf{j}$ along the x and y axes respectively.

$$\mathbf{r} = x\mathbf{i} + y\mathbf{j} \tag{A-1a}$$

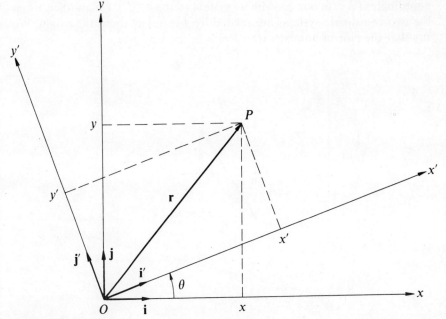

Figure A-1.

The numerical coefficients x and y are the components of $\mathbf{r}$ along the axes, Together they are the Cartesian coordinates (x, y) of the point P. Suppose the axes are rotated about the z axis by an angle θ. The radius vector $\mathbf{r}$ can also be expressed as a linear combination of the new unit vectors $\mathbf{i}'$ and $\mathbf{j}'$:

$$\mathbf{r} = x'\mathbf{i}' + y'\mathbf{j}' \qquad \text{(A-1b)}$$

corresponding to P having new Cartesian coordinates (x', y'). The relation between (x', y') and (x, y) is

$$\begin{aligned} x' &= \cos\theta \cdot x + \sin\theta \cdot y \\ y' &= -\sin\theta \cdot x + \cos\theta \cdot y \end{aligned} \qquad \text{(A-2a)}$$

The first of Eqs. (A-2a) is obtained by forming the scalar product of $\mathbf{r}$ with the unit vector $\mathbf{i}'$, using both Eqs. (A-1a) and (A-1b). The second equation is obtained by forming the scalar product with $\mathbf{j}'$.

$$\begin{aligned} x' &= (\mathbf{i}' \cdot \mathbf{i})x + (\mathbf{i}' \cdot \mathbf{j})y \\ y' &= (\mathbf{j}' \cdot \mathbf{i})x + (\mathbf{j}' \cdot \mathbf{j})y \end{aligned} \qquad \text{(A-2b)}$$

The pair of equations (A-2a) or (A-2b) describes the transformation of the coordinates (x, y) in one coordinate system to those (x', y') in another, when the two coordinate systems are related by a rotation about the origin. We say that the pair of numbers transforms in the way given.

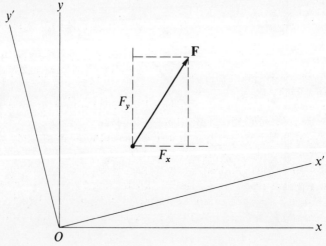

Figure A-2.

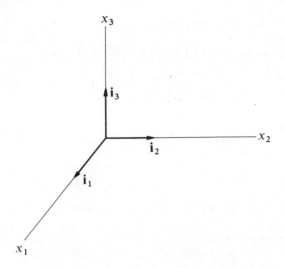

Figure A-3.

The same transformation law holds for the components of a velocity or a force or any *vector quantity*. If we have a force $\mathbf{F}$ acting on a particle P (Fig. A-2), its components (F_x, F_y) transform according to

$$F_{x'} = (\mathbf{i}' \cdot \mathbf{i})F_x + (\mathbf{i}' \cdot \mathbf{j})F_y$$

$$F_{y'} = (\mathbf{j}' \cdot \mathbf{i})F_x + (\mathbf{j}' \cdot \mathbf{j})F_y$$

The pair of numbers (F_x, F_y) has the same transformation behavior as the pair (x, y).

All pairs that have this transformation property specify vectors or vector quantities. The elements of the pair are the components of the vector in the coordinate system.

The components determine a directed line segment (Fig. A-2). Thus, a vector can be represented by an arrow.

The generalization to three or more dimensions is immediate. In three dimensions, we say that the triplet of numbers (A_1, A_2, A_3) associated with a Cartesian coordinate system represents a vector if it transforms under a rotation of the axes about the origin in the same way as the three components of a radius vector from the origin to a point in the space. Shifting from x, y, z to the algebraically more convenient x_1, x_2, x_3 (Fig. A-3) and calling

$$\mathbf{i}_1' \cdot \mathbf{i}_1 = a_{11} \qquad \mathbf{i}_1' \cdot \mathbf{i}_2 = a_{12} \qquad \text{etc}$$

the transformation law [Eq. (A-2b)] reads

$$x'_1 = a_{11}x_1 + a_{12}x_2 + a_{13}x_3$$

$$x'_2 = a_{21}x_1 + a_{22}x_2 + a_{23}x_3 \qquad (A\text{-}3)$$

$$x'_3 = a_{31}x_1 + a_{32}x_2 + a_{33}x_3$$

Using running indices,

$$x'_j = \sum_{k=1}^{3} a_{jk}x_k \qquad (j = 1, 2, 3) \qquad (A\text{-}3)$$

which is shortened still further by the Einstein summation convention (if an index appears twice in a product, the expression is to be summed over the whole range of the index) to

$$x'_j = a_{jk}x_k \qquad (j, k = 1, 2, 3) \qquad (A\text{-}3)$$

The numbers A_1, A_2, A_3 transform as

$$A'_j = a_{jk}A_k \qquad (j, k = 1, 2, 3) \qquad (1\text{-}10)$$

The numbers A_1, A_2, A_3 are called the components of the vector **A** in the $Ox_1x_2x_3$ coordinate system. The vector **A** corresponds to a directed line segment (arrow), whose projections on the axes are A_1, A_2, A_3 in one coordinate system, A'_1, A'_2, A'_3 in the other.

For a physical quantity to be a vector, it is not sufficient that it "have magnitude and direction"; it is also necessary that its algebra be the same as that of directed line segments. In particular, it must be possible to give an unambiguous meaning to the addition operation,

$$\mathbf{A} = \mathbf{B} + \mathbf{C} = \mathbf{C} + \mathbf{B}$$

corresponding to the law of geometric addition of Fig. A-4. Only then is it possible to assign components to the quantity, for the components are simply the numerical coefficients in the linear combination of basis vectors

$$\mathbf{A} = A_1\mathbf{i}_1 + A_2\mathbf{i}_2 + A_3\mathbf{i}_3 \qquad (A\text{-}4)$$

An example of a quantity that has magnitude, direction, and sense but fails to pass the algebra test is the finite rotation of a rigid body. The turning of a block of wood through some angle about an edge cannot be written as the sum of three rotations about three axes, for the resulting rotation depends on the order in which the three rotations are carried out. Yet the rotation in question can perfectly well be represented by an arrow along the edge, of length proportional to the angle turned through. Finite rotations are not vectors.

The 3×3 array of direction cosines a_{jk} is called the rotation matrix a. It specifies how the components of a vector change when the axes undergo the given rotation. Since the basis vectors are of unit length, the elements of a are the direction cosines of the angles between the new and old axes: for example,

$$a_{23} = \mathbf{i}'_2 \cdot \mathbf{i}_3 = \cos(Ox'_2, Ox_3)$$

Therefore, the sum of the squares of the elements of any row or column is one, and the sum of the products of corresponding elements of any two rows or columns is zero (the axes are orthogonal and remain orthogonal as they are turned). Using the Kronecker delta, defined by

$$\delta_{kl} = 0 \quad \text{if} \quad k \neq l$$
$$= 1 \quad \text{if} \quad k = l$$

the orthogonality of the unprimed axes is expressed by

$$a_{jk}a_{jl} = \delta_{kl} \qquad (j, k, l = 1, 2, 3)$$

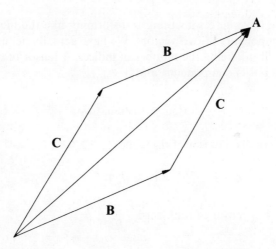

Figure A-4.

The orthogonality of the primed axes is expressed by

$$a_{jk}a_{lk} = \delta_{jl} \qquad (j, k, l = 1, 2, 3)$$

The rotation is an orthonormal transformation. It preserves the orthogonality and unit length of the basis vectors.

A quantity that is invariant under rotation is called a *scalar*. The scalar product deserves its name, for

$$\mathbf{A} \cdot \mathbf{B} = A_j B_j = A'_k B'_k$$

This result is evident geometrically from the fact that the scalar product can be written

$$A_j B_j = AB \cos(\mathbf{A}, \mathbf{B})$$

(the plain letter A stands for the length of the vector $\mathbf{A}$) and algebraically from the orthonormality of the transformation:

$$A'_k B'_k = a_{kl}A_l a_{km}B_m = \delta_{lm}A_l B_m = A_l B_l$$

The scalar product of a vector with itself is its square or length squared.

A *tensor* is something whose components in a particular coordinate system have two indices and transform as

$$T'_{jk} = a_{jl}a_{km}T_{lm} \tag{1-11c}$$

Evidently we have here something transforming like the nine products of the three components of two vectors. We have actually defined a tensor of rank two, the rank being the number of indices. A tensor of rank three has three indices on its components, and

$$T'_{jkl} = a_{jm}a_{kn}a_{lp}T_{mnp} \tag{1-11d}$$

A vector is actually a tensor of rank one

$$T'_j = a_{jk}T_k \tag{1-11b}$$

and a scalar is a tensor of rank zero

$$T' = T \tag{1-11a}$$

INVERSION OF CARTESIAN AXES

Reversing the sense of any of the axes leaves them perpendicular. The lengths, and, more generally, the scalar product of any two vectors

$$\mathbf{A} \cdot \mathbf{B} = A_1 B_1 + A_2 B_2 + A_3 B_3$$

are unchanged, because $(-1) \times (-1) = +1$. There is a change of handedness of the coordinate system if the number of axes reversed is odd. If curling the fingers of the right hand[1] from positive axis 1 to positive axis 2 makes the thumb point along positive axis 3, the coordinate system is called right handed (Fig. A-5a). An odd number of axis reversals makes the coordinate system anti-right handed, or, as we say, left handed. This is illustrated in Fig. A-5b, showing the effect of three reversals. It is impossible to bring a right-handed and a left-handed coordinate system into coincidence (congruence) by rigid rotation.

The coordinate transformation called *inversion* is defined as reversal of all the axes

$$x_1' = -x_1$$
$$x_2' = -x_2$$
$$x_3' = -x_3$$

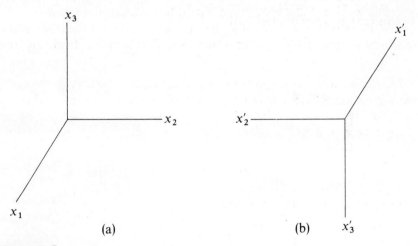

(a)   (b)

Figure A-5.

[1] The right hand is the hand opposite the heart. An example of *homo sapiens*, or, alternatively, a sample of right-handed or left-handed sugar is needed in the physicist's international bureau of weights and measures just as much as a lump of platinum called the kilogram, a krypton-86 glow tube for the meter, and a cesium-133 clock for the second.

or

$$x'_j = -x_j \qquad (j = 1, 2, 3)$$

It is, like rotation of axes, a linear homogeneous transformation

$$x'_j = a_{jk}x_k \qquad (j, k = 1, 2, 3)$$

but

$$a_{jk} = -\delta_{jk} \qquad (1\text{-}12)$$

The a_{jk} still have the meaning of direction cosines between new and old basis vectors. Evidently the transformation is orthonormal, but the determinant of the a_{jk} is -1 instead of $+1$.

$$|a| = \begin{vmatrix} -1 & 0 & 0 \\ 0 & -1 & 0 \\ 0 & 0 & -1 \end{vmatrix} = -1$$

The transformation from any Cartesian coordinate system to any other with the same origin and the same unit of length can be realized by combining an inversion (to change the handedness if need be) and a rigid rotation.

It is customary to extend the defining equations [Eqs. (1-11a–d)] for scalars, vectors, and tensors to cover inversion as well as rotation. Evidently, the

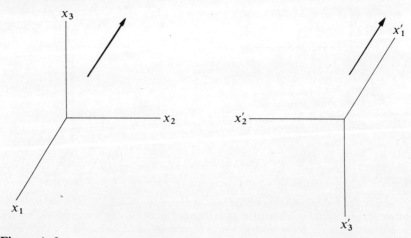

Figure A-6.

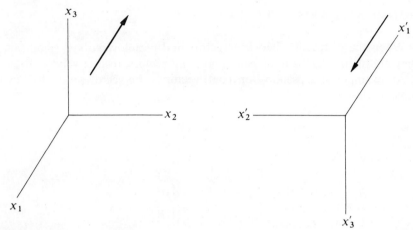

Figure A-7.

components of a vector are reversed in an inversion [as are those of a directed line segment (Fig. A-6)], whereas the components of tensors of even rank (scalars, tensors) are unchanged:

$$T'_{jk} = a_{jl}a_{km}T_{lm} = -a_{jl}T_{lk} = (-)(-)T_{jk} = T_{jk}$$

If a quantity transforms according to the preceding equations under rotation but with opposite sign under inversion, the prefix *pseudo* is applied to it. The numerical value of a pseudoscalar is unchanged by rotation of the axes, but its sign is reversed by inversion of the axes. The components of a pseudovector transform under rotation according to

$$A'_j = a_{jk}A_k \qquad (1\text{-}10)$$

but under inversion according to

$$A'_j = -a_{jk}A_k = -(-A_j) = A_j$$

A pseudovector can be represented by an arrow, but an inversion of axes requires reversal of the arrow representing the pseudovector (Fig. A-7). The components of a pseudotensor of even rank are reversed under inversion. Thus, under inversion,

$$P'_{jk} = -a_{jl}a_{km}P_{lm} = (-)(-)(-)P_{jk} = -P_{jk}$$

POLAR AND AXIAL VECTORS

When the distinction between vectors and pseudovectors is significant, they are often referred to as *polar* and *axial* vectors, respectively.

An example of a pseudo- (or axial) vector is the vector product of two (polar) vectors

$$\mathbf{C} = \mathbf{A} \times \mathbf{B}$$

defined by

$$C_1 = A_2 B_3 - A_3 B_2$$
$$C_2 = A_3 B_1 - A_1 B_3$$
$$C_3 = A_1 B_2 - A_2 B_1$$

or

$$C_j = \varepsilon_{jkl} A_k B_l$$

(Here ε_{jkl} is the Levi–Civita symbol, equal to zero whenever any two of the indices are equal and equal to $+1$ (-1) when *jkl* are an even (odd) permutation of 1, 2, 3.) Inasmuch as inversion reverses the sign of the components of both $\mathbf{A}$ and $\mathbf{B}$, the components of $\mathbf{C}$ are unchanged.

EXERCISE   Show that under a rotation the components of $\mathbf{C}$ transform according to the vector law.

SOLUTION   For example,

$$C_1' = A_2' B_3' - A_3' B_2' = a_{2j} A_j a_{3k} B_k - a_{3l} A_l a_{2m} B_m \qquad (j, k, l, m = 1, 2, 3)$$

Writing out this expression, we have

$$C_1' = (A_2 B_3 - A_3 B_2)(a_{22} a_{33} - a_{32} a_{23})$$
$$+ (A_3 B_1 - A_1 B_3)(a_{23} a_{31} - a_{33} a_{21})$$
$$+ (A_1 B_2 - A_2 B_1)(a_{21} a_{32} - a_{31} a_{22})$$
$$= C_1 \cdot \text{cofactor} (a_{11}) + C_2 \cdot \text{cofactor} (a_{12}) + C_3 \cdot \text{cofactor} (a_{13})$$

When the transformation is a rotation,

$$\text{cofactor} (a_{jk}) = a_{jk}$$

For example,

$$a_{12} = \mathbf{i}_1' \cdot \mathbf{i}_2 = (\mathbf{i}_2' \times \mathbf{i}_3') \cdot \mathbf{i}_2 = \begin{vmatrix} 0 & 1 & 0 \\ a_{21} & a_{22} & a_{23} \\ a_{31} & a_{32} & a_{33} \end{vmatrix} = \text{cofactor}(a_{12})$$

Thus,

$$C_1' = a_{11}C_1 + a_{12}C_2 + a_{13}C_3 \qquad \text{Q.E.D.}$$

The magnitude

$$C = (C_1^2 + C_2^2 + C_3^2)^{1/2} = AB\sin(\mathbf{A}, \mathbf{B})$$

is the area of the parallelogram $PQRSP$ determined by $\mathbf{A}$ and $\mathbf{B}$ (Fig. A-8). The order of the factors in the vector product specifies a sense of circulation around the parallelogram, as shown. In a right-handed coordinate system, the directed line segment of components C_1, C_2, C_3 is the arrow $\overrightarrow{PT}$. Its length equals the area of the parallelogram, it is perpendicular to it, and its sense is related to the sense of circulation by a right-hand rule. In a left-handed coordinate system, the directed line segment of components C_1, C_2, C_3 is the arrow $\overrightarrow{PU}$, equal and opposite to $\overrightarrow{PT}$. Its sense is related to the sense of circulation by a left-hand rule. The loop $PQRSP$ does not have a sidedness; there is no reason to single out one side of it over the other as the one on which to draw the arrow. The association of a directed line segment with a vector product involves an arbitrary handedness convention.

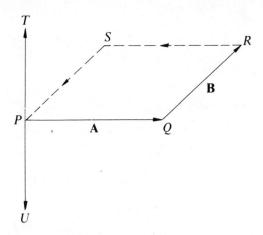

Figure A-8.

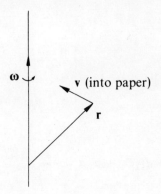

Figure A-9.

Our notation does not explicitly identify pseudotensors. A sure indication that a vector is axial is the presence of a hand rule in the specification of the associated directed line segment, as in the example just presented.

The vector product of a polar and an axial vector is a polar vector, apparent from the fact that a hand rule is used twice. An example is provided (Fig. A-9) by the velocity **v** of a point of a spinning rigid body,

$$\mathbf{v} = \boldsymbol{\omega} \times \mathbf{r}$$

where **r** is a radius vector from a point on the axis to the point in question and **ω** is the angular velocity. The angular velocity is an axial vector, because a hand rule is involved in specifying the associated arrow. The hand rule is used again in identifying the vector product of **ω** and **r** with an arrow.

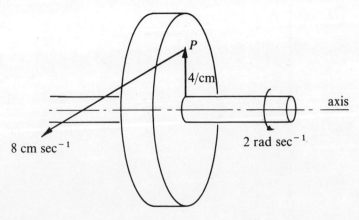

Figure A-10.

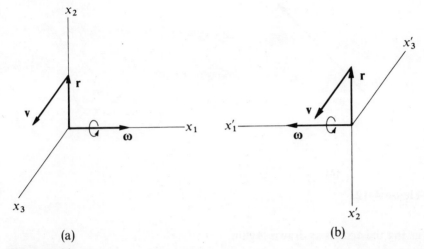

(a)   (b)

Figure A-11.

Suppose, for example, that P is a point 4 cm from the axis of a flywheel spinning about a fixed axis at $2 \, \text{rad sec}^{-1}$ (Fig. A-10). It is moving with a velocity of $8 \, \text{cm sec}^{-1}$, as shown. All this is independent of any choice of coordinate system. With the particular right-handed coordinate system selected in Fig. A-11a, we use the right-hand rule for determining the components of ω. The vector components are

$$x_1 = 0 \qquad v_1 = 0 \qquad \omega_1 = 2$$
$$x_2 = 4 \qquad v_2 = 0 \qquad \omega_2 = 0$$
$$x_3 = 0 \qquad v_3 = 8 \qquad \omega_3 = 0$$

giving the arrows as drawn. They satisfy

$$v_1 = \omega_2 x_3 - \omega_3 x_2$$
$$v_2 = \omega_3 x_1 - \omega_1 x_3$$
$$v_3 = \omega_1 x_2 - \omega_2 x_1$$

With the inverted coordinate system of Fig. A-11b, we use the left-hand rule for determining the components of ω. The components are

$$x_1' = 0 \qquad v_1' = 0 \qquad \omega_1' = 2$$
$$x_2' = -4 \qquad v_2' = 0 \qquad \omega_2' = 0$$
$$x_3' = 0 \qquad v_3' = -8 \qquad \omega_3' = 0$$

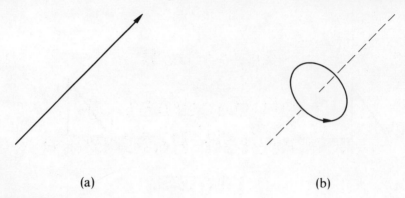

Figure A-12.

giving the arrows as drawn. Again,

$$v_1' = \omega_2' x_3' - \omega_3' x_2'$$

$$v_2' = \omega_3' x_1' - \omega_1' x_3'$$

$$v_3' = \omega_1' x_2' - \omega_2' x_1'$$

We see how the components of polar vectors reverse sign under inversion, whereas those of an axial vector do not.

Polar vectors correspond completely to directed line segments (Fig. A-12a). They are purely longitudinal in character, with no sidewise or circulatory aspect. Axial vectors obey the same algebra but have a different symmetry character, that of a circle with a sense of circulation (Fig. A-12b). They can be put in one–to–one correspondence with directed line segments by the use of a hand convention matching the handedness of the coordinate system.

Appendix B

Galilean Kinematics of Accelerated Reference Frames

ANALYSIS OF MOTION INTO TRANSLATION AND ROTATION

A translation of a rigid body is a bodily motion without change of aspect. Each point of the body has the same displacement. This displacement is a vector quantity specified in a coordinate system in R by three components. In an infinitesimal translation in time dt, the displacement vector is $\mathbf{u}\,dt$, where $\mathbf{u}$ is the instantaneous translational velocity.

A rotation of a rigid body is a displacement in which one point of the body remains in the same place; thus, it is pure pivoting. Every rotation has an axis—that is, a line of points whose position is unchanged (Euler's theorem[1]). An *infinitesimal* rotation in time dt can be specified by an axial vector $\mathbf{n}\,d\theta$, where $\mathbf{n}$ is a unit vector parallel to the axis. Its sense is given by a hand rule corresponding to the handedness of the axes used. The magnitude $d\theta$ is the angle turned through in radians. The angular velocity axial vector $\boldsymbol{\omega}$ is defined by

$$\boldsymbol{\omega} = \mathbf{n}\,\frac{d\theta}{dt}$$

In the infinitesimal rotation, the displacement of a point of the body is (Fig. B-1)

$$\mathbf{n}\,d\theta \times \mathbf{r} = \boldsymbol{\omega}\,dt \times \mathbf{r}$$

where $\mathbf{r}$ is the vector from a point on the axis to the point in question.

Consider now a perfectly general displacement of R' with respect to R. We shall show that the velocity $\mathbf{v}_{P'}$ of an arbitrary point P' of R' is

$$\mathbf{v}_{P'} = \mathbf{u} + \boldsymbol{\omega} \times \mathbf{r}'_{P'} \tag{B-1}$$

---

[1] For a proof of this intuitively obvious statement, see Lamb ([10], p. 3), Sommerfeld ([11], pp. 118–119), or Goldstein ([12], pp. 118–124).

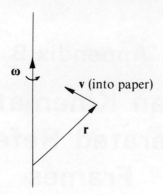

Figure B-1.

where **u** is the velocity of an arbitrary point chosen as origin in R' and $\mathbf{r}'_{P'}$ is the radius vector from that origin to the point P'.

Pick (Fig. B-2) axes fixed in R', $O'x'y'z'$, and consider the displacements of three points fixed in R': $O'(0,0,0)$, $A'(1,0,0)$, and $B'(0,1,0)$. These three noncollinear points determine the position of R' in R. They move to new positions in R marked O'', A'', and B''. We can obtain this general displacement by two steps. First, we translate R' by the vector $\overrightarrow{O'O''}$, by which O' gets to the right final position O'', A' to A'_1, and B' to B'_1. Then we pivot R' about O'' until the point at A'_1 gets to A'' and the point at B'_1 to B''. Evidently this pure translation $\overrightarrow{O'O''}$ followed by a pure rotation about O'' gets every point of R' from its initial to its final position. Specializing now to an infinitesimal

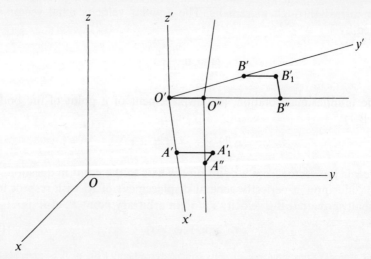

Figure B-2.

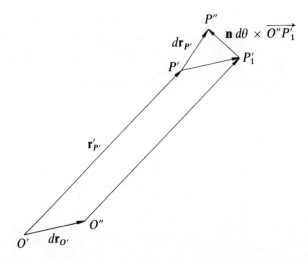

Figure B-3.

displacement in time dt, we can represent the rotation by a vector $\mathbf{n}\, d\theta = \boldsymbol{\omega}\, dt$. For an arbitrary point P' of R', the displacement to P'' is (Fig. B-3)

$$d\mathbf{r}_{P'} = \overrightarrow{O'O''} + \mathbf{n}\, d\theta \times \overrightarrow{O''P_1'}$$

Since $\overrightarrow{O''P_1'} = \overrightarrow{O'P'} = \mathbf{r}_{P'}'$, this equation becomes

$$d\mathbf{r}_{P'} = d\mathbf{r}_{O'} + \boldsymbol{\omega}\, dt \times \mathbf{r}_{P'}'$$

Using $\mathbf{v}_{P'}$ for the velocity of P' with respect to R and $\mathbf{u}$ for the velocity of O' with respect to R, we write it

$$\mathbf{v}_{P'}\, dt = \mathbf{u}\, dt + \boldsymbol{\omega}\, dt \times \mathbf{r}_{P'}'$$

or

$$\mathbf{v}_{P'} = \mathbf{u} + \boldsymbol{\omega} \times \mathbf{r}_{P'}' \qquad \text{Q.E.D.} \tag{B-1}$$

Although the translational velocity $\mathbf{u}$ depends on the choice of origin in R, the angular velocity $\boldsymbol{\omega}$ is the same for all choices of O'.

EXERCISE   Prove the preceding statement.

SOLUTION   ([13], p. 97.)

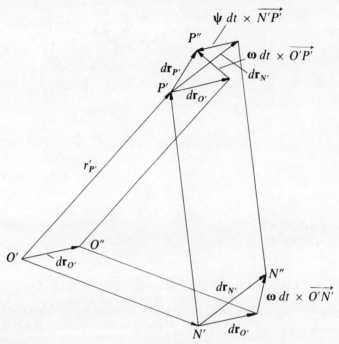

Figure B-4.

Try (Fig. B-4) an alternative origin N'. Suppose the angular velocity is then ψ.

$$d\mathbf{r}_{P'} = d\mathbf{r}_{O'} + \omega \, dt \times \overrightarrow{O'P'} = d\mathbf{r}_{N'} + \psi \, dt \times \overrightarrow{N'P'} \qquad \text{(B-2)}$$

As

$$\overrightarrow{O'P'} = \overrightarrow{O'N'} + \overrightarrow{N'P'}$$

the first equality [Eq. (B-2)] is

$$d\mathbf{r}_{P'} = d\mathbf{r}_{O'} + \omega \, dt \times \overrightarrow{O'N'} + \omega \, dt \times \overrightarrow{N'P'}$$

We can also apply Eq. (B-1) to the displacement of N'

$$d\mathbf{r}_{N'} = d\mathbf{r}_{O'} + \omega \, dt \times \overrightarrow{O'N'}$$

so that the second equality of Eq. (B-2) reads

$$d\mathbf{r}_{P'} = d\mathbf{r}_{O'} + \omega \, dt \times \overrightarrow{O'N'} + \psi \, dt \times \overrightarrow{N'P'}$$

Comparing the two equalities, we see that

$$\omega \, dt \times \overrightarrow{N'P'} = \psi \, dt \times \overrightarrow{N'P'}$$

Because N' and P' are arbitrary points of the moving space, we must have

$$\psi = \omega$$

The angular velocity of a rigid body is a property of the body independent of which point we choose as pivot point.

MOTION OF THE BASIS VECTORS

Equation (B-1) can be applied to the motion of any point fixed in R', in particular (Fig. B-5) to the extremities of the unit vectors $\mathbf{i}'$, $\mathbf{j}'$, and $\mathbf{k}'$. [These are just the points O', A', B' previously considered and $C'(0, 0, 1)$.] Thus,

$$d\mathbf{r}_{A'} = d\mathbf{r}_{O'} + \omega \, dt \times \mathbf{i}'$$

Forming

$$d\mathbf{i}' = d\mathbf{r}_{A'} - d\mathbf{r}_{O'}$$

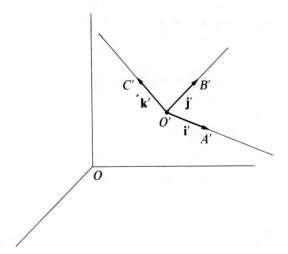

Figure B-5.

this equation gives

$$d\mathbf{i}' = \boldsymbol{\omega}\, dt \times \mathbf{i}'$$

or

$$\frac{d\mathbf{i}'}{dt} = \boldsymbol{\omega} \times \mathbf{i}' \qquad\qquad\qquad (\text{B-3a})$$

The same argument applied to B' and C' gives

$$\frac{d\mathbf{j}'}{dt} = \boldsymbol{\omega} \times \mathbf{j}' \qquad\qquad\qquad (\text{B-3b})$$

$$\frac{d\mathbf{k}'}{dt} = \boldsymbol{\omega} \times \mathbf{k}' \qquad\qquad\qquad (\text{B-3c})$$

We now have the tools for describing the motion of a particle with respect to R'.

TRANSFORMATION EQUATIONS FOR THE VELOCITY AND ACCELERATION OF A PARTICLE

Consider the situation shown in Fig. B-6. A particle P is in motion with respect to both R and R'. We assume the validity of the Galileo transformation:

$$t = t' \qquad\qquad\qquad (1\text{-}16)$$

$$\mathbf{r} = \overrightarrow{O\,O'} + \mathbf{r}' \qquad\qquad\qquad (1\text{-}7)$$

The positions and times refer to the same event—the particle's being at a certain place. Differentiating with respect to time, we have for the velocity of P with respect to R

$$\mathbf{v} = \frac{d\mathbf{r}}{dt} = \frac{d(\overrightarrow{O\,O'})}{dt} + \frac{d\mathbf{r}'}{dt} = \mathbf{u} + \frac{d\mathbf{r}'}{dt} \qquad\qquad (\text{B-4})$$

$\mathbf{u}$ being the velocity of O' in R. The last summand, $d\mathbf{r}'/dt$, is not the velocity of P with respect to R'. For example, if P were fixed in R', it would certainly have zero velocity with respect to R', but $\mathbf{r}'$ would not be constant in time because it would be changing direction as R' turned.

Returning to the general case of the unattached particle, let us express $\mathbf{r}'$ in terms of the unit vectors along the primed axes

$$\mathbf{r}' = \mathbf{i}'x' + \mathbf{j}'y' + \mathbf{k}'z' \qquad (1\text{-}15)$$

The velocity $\mathbf{v}'$ of P with respect to R' is that calculated on the assumption that $\mathbf{i}', \mathbf{j}'$, and $\mathbf{k}'$ are constant vectors:

$$\mathbf{v}' = \mathbf{i}'\frac{dx'}{dt'} + \mathbf{j}'\frac{dy'}{dt'} + \mathbf{k}'\frac{dz'}{dt'} = \mathbf{i}'\frac{dx'}{dt} + \mathbf{j}'\frac{dy'}{dt} + \mathbf{k}'\frac{dz'}{dt} \qquad (\text{B-5})$$

Thus, differentiation of Eq. (1-15) gives

$$\frac{d\mathbf{r}'}{dt} = \mathbf{v}' + \frac{d\mathbf{i}'}{dt}x' + \frac{d\mathbf{i}'}{dt}y' + \frac{d\mathbf{k}'}{dt}z'$$

By Eqs. (B-3),

$$\frac{d\mathbf{r}'}{dt} = \mathbf{v}' + \boldsymbol{\omega} \times \mathbf{i}'x' + \boldsymbol{\omega} \times \mathbf{j}'y' + \boldsymbol{\omega} \times \mathbf{k}'z'$$

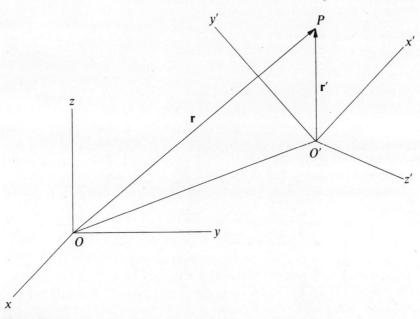

Figure B-6.

or, by Eq. (1-15),

$$\frac{d\mathbf{r}'}{dt} = \mathbf{v}' + \boldsymbol{\omega} \times \mathbf{r}' \qquad (B\text{-}6)$$

Introducing the symbol

$$\frac{d'}{dt}$$

for differentiation with respect to time with $\mathbf{i}'$, $\mathbf{j}'$, $\mathbf{k}'$ treated as constants,

$$\mathbf{v}' = \frac{d'}{dt}\mathbf{r}' = \frac{d}{dt}\mathbf{r}' - \boldsymbol{\omega} \times \mathbf{r}' \qquad (B\text{-}7a)$$

Any vector function of the time $\mathbf{f}(t)$ can be expressed in terms of the basis vectors $\mathbf{i}'$, $\mathbf{j}'$, $\mathbf{k}'$:

$$\mathbf{f} = \mathbf{i}'f_{x'} + \mathbf{j}'f_{y'} + \mathbf{k}'f_{z'}$$

Its rate of change with time is then

$$\frac{d\mathbf{f}}{dt} = \frac{d'\mathbf{f}}{dt} + \boldsymbol{\omega} \times \mathbf{f} \qquad (B\text{-}7b)$$

We have a general operator identity

$$\frac{d}{dt} = \frac{d'}{dt} + \boldsymbol{\omega}\times \qquad (B\text{-}7c)$$

Combining Eqs. (B-4) and (B-6) we have

$$\mathbf{v} = \mathbf{u} + \mathbf{v}' + \boldsymbol{\omega} \times \mathbf{r}' \qquad (1\text{-}34)$$

as the equation giving the velocity of P with respect to R in terms of the velocity and position of P with respect to R'.

Differentiating the velocity to get the acceleration, we find from Eq. (1-34)

$$\mathbf{a} = \frac{d\mathbf{v}}{dt} = \frac{d\mathbf{u}}{dt} + \frac{d}{dt}(\mathbf{v}' + \boldsymbol{\omega} \times \mathbf{r}')$$

$$= \frac{d\mathbf{u}}{dt} + \left(\frac{d'}{dt} + \boldsymbol{\omega} \times\right)(\mathbf{v}' + \boldsymbol{\omega} \times \mathbf{r}')$$

$$= \frac{d\mathbf{u}}{dt} + \frac{d'}{dt}\mathbf{v}' + \boldsymbol{\omega} \times \mathbf{v}' + \frac{d'}{dt}(\boldsymbol{\omega} \times \mathbf{r}') + \boldsymbol{\omega} \times (\boldsymbol{\omega} \times \mathbf{r}')$$

Noting that the second term on the right-hand side of the last equality is the acceleration $\mathbf{a}'$ of P with respect to R', and carrying out the differentiation of the fourth term, we have

$$\mathbf{a} = \frac{d\mathbf{u}}{dt} + \mathbf{a}' + 2\boldsymbol{\omega} \times \mathbf{v}' + \frac{d'\boldsymbol{\omega}}{dt} \times \mathbf{r}' + \boldsymbol{\omega} \times (\boldsymbol{\omega} \times \mathbf{r}')$$

Putting first the two terms on the right-hand side which vanish if P is attached to R', we get

$$\mathbf{a} = \mathbf{a}' + 2\boldsymbol{\omega} \times \mathbf{v}' + \boldsymbol{\omega} \times (\boldsymbol{\omega} \times \mathbf{r}') + \frac{d'\boldsymbol{\omega}}{dt} \times \mathbf{r}' + \frac{d\mathbf{u}}{dt} \qquad \text{(B-8)}$$
$$\quad\quad\quad\text{(Coriolis)}\quad\text{(centripetal)}\qquad\qquad\qquad\text{(translational)}$$

This equation gives the particle's acceleration with respect to R in terms of its acceleration, velocity, and position with respect to R'

On the right-hand side, the first term is the acceleration in R'. The second term, named after its discoverer Coriolis, was not noticed till about 1835. It manifests itself only when R' is rotating ($\omega \neq 0$) and the particle is moving with respect to $R'(v' \neq 0)$. It is sidewise in R', being perpendicular to $\mathbf{v}'$. It is also perpendicular to the angular velocity of R', $\boldsymbol{\omega}$.[2]

The last three terms do not involve the velocity of P with respect to R' and can therefore be called *entrainment* terms. They give the acceleration of a particle at rest in R'. The first entrainment term, called *centripetal*, is the familiar "v^2/ρ" acceleration of a particle toward the center of curvature of its orbit. In Fig. B-7, the velocity of a rotating particle P is $\boldsymbol{\omega} \times \mathbf{r}'$ directed into the page. The cross product of $\boldsymbol{\omega}$ with that velocity is the centripetal term. It is directed toward the axis and its magnitude is ω^2 times the distance of P from the axis, or $\omega^2 r \sin\theta' = v'^2/r' \sin\theta'$. The second entrainment term is proportional to the angular acceleration of R'. Note that in it one can replace d'/dt by d/dt, because in Eq. (B-7c), $\boldsymbol{\omega} \times \boldsymbol{\omega} = 0$. The third entrainment term is the acceleration of O' with respect to R. It results from a *translational acceleration* of R' without rotation.

Replacing d'/dt by d/dt, we obtain

$$\mathbf{a} = \mathbf{a}' + 2\boldsymbol{\omega} \times \mathbf{v}' + \boldsymbol{\omega} \times (\boldsymbol{\omega} \times \mathbf{r}') + \frac{d\boldsymbol{\omega}}{dt} \times \mathbf{r}' + \frac{d\mathbf{u}}{dt} \qquad \text{(1-35)}$$

---

[2] The formal similarity to the $\mathbf{v} \times \mathscr{B}$ sidewise force on a charged particle in a homogeneous magnetic field is the basis of Larmor's theorem on the effect of a weak external magnetic field on an axially symmetric system of charged particles. The Coriolis acceleration is important in meteorology in connection with the motion of air masses relative to the spinning earth. It is basic to Foucault's pendulum experiment (Appendix C), a very impressive purely dynamical demonstration of the rotation of the earth with respect to inertial frames.

The terms on the right-hand side after a' represent, from the point of view of R, the effect of the motion of R' under P.

The inverse transformation equations can be written down at once by interchanging primed and unprimed quantities and reversing the sign of u and ω.

$$\mathbf{v}' = -\mathbf{u} + \mathbf{v} - \omega \times \mathbf{r}$$

$$\mathbf{a}' = \mathbf{a} - 2\omega \times \mathbf{v} + \omega \times (\omega \times \mathbf{r}) - \frac{d\omega}{dt} \times \mathbf{r} - \frac{d\mathbf{u}}{dt}$$

In these equations, u is still the velocity of O' with respect to R, and ω is the angular velocity of R' with respect to R.

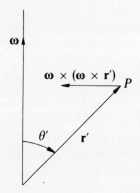

Figure B-7.

Appendix C

Detectable Effects of the Earth's Rotation with Respect to Inertial Frames

The study of observable effects of the earth's rotation with respect to inertial frames is of interest as showing the extent to which Newton's program for detecting absolute motion by observing inertial forces can be carried out. All "ether drift" experiments, designed to show an effect of the earth's *velocity* on terrestrial experiments, have given negative results. Efforts to show effects of the earth's *acceleration* have, on the other hand, given positive results as regards *rotation*. There is no doubt that the concept of rotation with respect to inertial frames has a definite operational meaning.

Let R be the fixed star frame and R' a frame attached to the earth's crust, with origin O' on the ground. As shown in Section 1-5, the equation of motion in R' for a particle of nonrelativistic velocity is

$$\mathbf{F}_{non} + m\mathbf{g}_{earth} - m\boldsymbol{\omega} \times [\boldsymbol{\omega} \times (\boldsymbol{\rho} + \mathbf{r}')] - m2(\boldsymbol{\omega} \times \mathbf{v}') = m\mathbf{a}' \quad (1\text{-}40)$$

The first term on the left-hand side represents physical forces other than gravitation; the second is the gravitational pull of the earth; the third is the centrifugal force ($\boldsymbol{\rho} + \mathbf{r}'$ is the vector from the center of mass of the earth to the particle); and the fourth is the Coriolis force. The effect of the earth's rotation is to alter the apparent earth gravity field to

$$\mathbf{g}_{eff} = \mathbf{g}_{earth} - \boldsymbol{\omega} \times [\boldsymbol{\omega} \times (\boldsymbol{\rho} + \mathbf{r}')] - 2(\boldsymbol{\omega} \times \mathbf{v}') \quad (1\text{-}41)$$

This, by Eq. (1-40), is the acceleration of a freely falling body ($F_{non} = 0$).

The flattening of the earth at the poles is consistent with the centrifugal (middle) term. Studies of the dependence of $\mathbf{g}_{eff}$ on position are complicated, however, by just the fact that $\mathbf{g}_{earth}$ varies with position in an a priori unknown way. The Coriolis term, however, provides an application of Eq. (1-40) that is unaffected by the earth's departure from spherical symmetry.

It is convenient to combine the first two terms

$$\mathbf{g}_{stat} = \mathbf{g}_{earth} - \mathbf{\omega} \times [\mathbf{\omega} \times (\mathbf{\rho} + \mathbf{r}')]$$

This part is independent of particle velocity and is essentially constant in the neighborhood of the laboratory. Its direction is given by a plumb line or spirit level.

The equation of motion of a freely falling body is

$$\mathbf{g}_{stat} - 2(\mathbf{\omega} \times \mathbf{v}') = \mathbf{a}' \qquad (C-1)$$

Choose axes in which $O'z'$ is parallel to $\mathbf{g}_{stat}$ and $O'x'$ points east (Fig. C-1). With ψ the angle shown (it is practically equal to the latitude), Eq. (C-1) reads

$$\frac{d'^2}{dt^2}z' = g_{stat} - 2\omega \cos \psi \frac{d'x'}{dt}$$

$$\frac{d'^2}{dt^2}y' = 2\omega \sin \psi \frac{d'x'}{dt} \qquad (C-2)$$

$$\frac{d'^2}{dt^2}x' = -2\omega \sin \psi \frac{d'y'}{dt} + 2\omega \cos \psi \frac{d'z'}{dt}$$

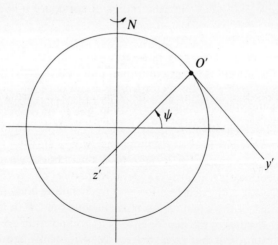

Figure C-1.   Coordinate system used for discussing projectile motion near O'.

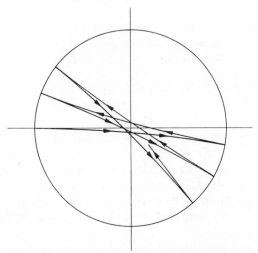

Figure C-2. Trace in the horizontal plane of the motion of the bob of Foucault's pendulum.

We have used here, as in Appendix B, the symbol d'/dt for differentiation with respect to time in the R' frame, $\mathbf{i'}$, $\mathbf{j'}$, and $\mathbf{k'}$ being regarded as constant.

For a particle moving horizontally ($d'z'/dt = 0$), the last two equations of Eqs. (C-2) show that the Coriolis acceleration is perpendicular to the velocity, to the right (i.e., clockwise, looking down from above) in the northern hemisphere and of magnitude $2\omega \sin \psi v'$. In principle one could determine the angular velocity of the earth (ω) by projecting a puck on a large horizontal frictionless table and measuring the radius of curvature ρ of its orbit:

$$\rho = \frac{v'^2}{a'} = \frac{v'}{2\omega \sin \psi}$$

Foucault's pendulum experiment is an elegant practical form of this thought experiment. A bob is suspended from a joint free to turn about the vertical. The bob oscillates back and forth in the horizontal plane and the repeated deflections to the right in the middle of the swing cumulate to give visible rotation of the plane of oscillation (Fig. C-2). It can be shown (e.g., [15], [16], [11]) that the plane rotates at the angular velocity

$$\lambda = \omega_F = \omega \sin \psi$$

The rate of rotation is proportional to the sine of the latitude, reaching $2\pi/24$ rad h^{-1} at the poles. This formula has been confirmed experimentally, showing that the angular velocity of the earth with respect to the fixed stars is approximately equal to its angular velocity with respect to inertial frames.

  More precise tests of this equality are provided by the functioning of gyro-compasses (pioneered by Foucault) and accelerometers used in inertial naviga-tion systems for missiles and spacecraft. No anomalies have been reported.

  An optical experiment to detect the rotation of the earth with respect to inertial frames was successfully carried out by Michelson and Gale [17]. It utilized an effect demonstrated by Sagnac [18], who constructed a two-beam interferometer in which the beams go around a loop of nonzero area in opposite senses before overlapping in the observation region, and showed that rotation of the whole apparatus gives rise to a fringe shift proportional to the angular velocity and the area of the loop. Michelson and Gale observed the fringe shift as between a moderate-sized loop and one of area 2.1×10^9 cm^2, obtaining a value for the earth's angular velocity correct to within 2.5%.

  We see that the distinction between rotating and nonrotating frames has a clear observational meaning. But the impossibility of distinguishing between translational acceleration of the reference frame and a homogeneous gravita-tional field is extremely well established, being based on the experimental evidence for the equal gravitational acceleration of all bodies [14].

Appendix D

Transformation Law of Charge and Electromagnetic Field[1]

The electric field $\mathscr{E}$ and magnetic field $\mathscr{B}$ are defined by the Lorentz expression for the force on a particle of charge e:

$$\mathbf{F} = e\mathscr{E} + \frac{e\mathbf{v}}{c} \times \mathscr{B} \qquad \text{(D-1)}$$

The unit of charge is defined as that charge which when confronted by an equal charge 1 cm away and at rest experiences a repulsive force of 1 dyn (Gaussian system of units). We determine e on this basis and map out $\mathscr{E}, \mathscr{B}$ by determining the force on the particle as a function of its position and velocity.

The same procedures hold in any other inertial frame. By the principle of relativity (first postulate), a particle found to have unit charge in R also has unit charge in R'. Otherwise there would be a physical basis for singling out a particular privileged reference frame. It follows that

$$e' = e \qquad \text{(D-2)}$$

The electric charge of a system is invariant.

The law of motion of the electromagnetic field is the Maxwell field equations. In vacuum for a region free of charge, they read

$$\text{(a)} \quad \frac{1}{c}\frac{\partial \mathscr{E}}{\partial t} = \text{curl}\,\mathscr{B} \qquad \text{(b)} \quad -\frac{1}{c}\frac{\partial \mathscr{B}}{\partial t} = \text{curl}\,\mathscr{E}$$

$$\text{(c)} \quad 0 = \text{div}\,\mathscr{B} \qquad \text{(d)} \quad 0 = \text{div}\,\mathscr{E} \qquad \text{(D-3)}$$

---

[1] We reproduce the derivation of Eq. (2-55a) given in Einstein's first paper ([2], Section 6).

Transforming the differential operators

$$\frac{1}{c}\frac{\partial}{\partial t}, \quad \frac{\partial}{\partial x}, \quad \frac{\partial}{\partial y}, \quad \frac{\partial}{\partial z}$$

that appear in these equations by the Lorentz transformation [Eq. (1-31b)], we obtain the following relations between rates of change in R and in R':

$$\frac{1}{c}\frac{\partial}{\partial t} = \gamma\frac{1}{c}\frac{\partial}{\partial t'} - \beta\gamma\frac{\partial}{\partial x'}$$

$$\frac{\partial}{\partial x} = -\beta\gamma\frac{1}{c}\frac{\partial}{\partial t'} + \gamma\frac{\partial}{\partial x'}$$

$$\frac{\partial}{\partial y} = \frac{\partial}{\partial y'} \tag{D-4}$$

$$\frac{\partial}{\partial z} = \frac{\partial}{\partial z'}$$

where, as always, $\beta = u/c$ and $\gamma = (1 - \beta^2)^{-1/2}$. The loop equation [Eq. (D-3a)] changes from

$$\frac{1}{c}\frac{\partial \mathscr{E}_x}{\partial t} = \frac{\partial}{\partial y}\mathscr{B}_z - \frac{\partial}{\partial z}\mathscr{B}_y$$

$$\frac{1}{c}\frac{\partial \mathscr{E}_y}{\partial t} = \frac{\partial}{\partial z}\mathscr{B}_x - \frac{\partial}{\partial x}\mathscr{B}_z$$

$$\frac{1}{c}\frac{\partial \mathscr{E}_z}{\partial t} = \frac{\partial}{\partial x}\mathscr{B}_y - \frac{\partial}{\partial y}\mathscr{B}_x$$

in R to

$$\frac{1}{c}\frac{\partial \mathscr{E}_x}{\partial t'} = \frac{\partial}{\partial y'}\gamma(\mathscr{B}_z - \beta\mathscr{E}_y) - \frac{\partial}{\partial z'}\gamma(\mathscr{B}_y + \beta\mathscr{E}_z)$$

$$\frac{1}{c}\frac{\partial}{\partial t'}\gamma(\mathscr{E}_y - \beta\mathscr{B}_z) = \frac{\partial}{\partial z'}\mathscr{B}_x - \frac{\partial}{\partial x'}\gamma(\mathscr{B}_z - \beta\mathscr{E}_y) \tag{D-5}$$

$$\frac{1}{c}\frac{\partial}{\partial t'}\gamma(\mathscr{E}_z + \beta\mathscr{B}_y) = \frac{\partial}{\partial x'}\gamma(\mathscr{B}_y + \beta\mathscr{E}_z) - \frac{\partial}{\partial y'}\mathscr{B}_x$$

in R', where the source equation [Eq. (D-3a)] has been used in rearranging the first equation. Similarly, the other loop equation [Eq. (D-3b)] changes from

$$-\frac{1}{c}\frac{\partial \mathscr{B}_x}{\partial t} = \frac{\partial}{\partial y}\mathscr{E}_z - \frac{\partial}{\partial z}\mathscr{E}_y$$

$$-\frac{1}{c}\frac{\partial \mathscr{B}_y}{\partial t} = \frac{\partial}{\partial z}\mathscr{E}_x - \frac{\partial}{\partial x}\mathscr{E}_z$$

$$-\frac{1}{c}\frac{\partial \mathscr{B}_z}{\partial t} = \frac{\partial}{\partial x}\mathscr{E}_y - \frac{\partial}{\partial y}\mathscr{E}_x$$

in R to

$$-\frac{1}{c}\frac{\partial \mathscr{B}_x}{\partial t'} = \frac{\partial}{\partial y'}\gamma(\mathscr{E}_z + \beta\mathscr{B}_y) - \frac{\partial}{\partial z'}\gamma(\mathscr{E}_y - \beta\mathscr{B}_z)$$

$$-\frac{1}{c}\frac{\partial}{\partial t'}\gamma(\mathscr{B}_y + \beta\mathscr{E}_z) = \frac{\partial}{\partial z'}\mathscr{E}_x - \frac{\partial}{\partial x'}\gamma(\mathscr{E}_z + \beta\mathscr{B}_y) \qquad \text{(D-6)}$$

$$-\frac{1}{c}\frac{\partial}{\partial t'}\gamma(\mathscr{B}_z - \beta\mathscr{E}_y) = \frac{\partial}{\partial x'}\gamma(\mathscr{E}_y - \beta\mathscr{B}_z) - \frac{\partial}{\partial y'}\mathscr{E}_x$$

in R', with help from Eq. (D-3c) in the first equation. The postulate of relativity requires that in R' the field equations have the same form and the same constant c as in R:

$$\text{(a)} \quad \frac{1}{c}\frac{\partial \mathscr{E}'}{\partial t'} = \text{curl } \mathscr{B}' \qquad \text{(b)} \quad -\frac{1}{c}\frac{\partial \mathscr{B}'}{\partial t'} = \text{curl } \mathscr{E}'$$

$$\text{(D-7)}$$

$$\text{(c)} \qquad 0 = \text{div } \mathscr{B}' \qquad \text{(d)} \qquad 0 = \text{div } \mathscr{E}'$$

It follows that the operands of Eqs. (D-5) and (D-6) must be equal, to within a factor depending only on u, to the corresponding primed components of the field in R'.

$$\mathscr{E}'_x = f(u)\mathscr{E}_x \qquad\qquad \mathscr{B}'_x = f(u)\mathscr{B}_x$$

$$\mathscr{E}'_y = f(u)\gamma(\mathscr{E}_y - \beta\mathscr{B}_z) \qquad \mathscr{B}'_y = f(u)\gamma(\mathscr{B}_y + \beta\mathscr{E}_z)$$

$$\mathscr{E}'_z = f(u)\gamma(\mathscr{E}_z + \beta\mathscr{B}_y) \qquad \mathscr{B}'_z = f(u)\gamma(\mathscr{B}_z - \beta\mathscr{E}_y)$$

We may assume that $f(u)$ is even. If, for example, $\mathscr{E} = 0$, $\mathscr{B}_x = 0$, $\mathscr{B}_y = 0$, then

$$\mathscr{E}'_y = -f(u)\frac{u}{c}\left(1 - \frac{u^2}{c^2}\right)^{-1/2}\mathscr{B}_z$$

Reversing u must reverse $\mathscr{E}'_y$, so that we must have

$$f(-u) = f(u)$$

Also, considering successive transformations $R \to R'$ with velocity u followed by $R' \to R''$ with velocity $-u$, we must have

$$f(-u)f(u) = 1$$

Hence,

$$f(u) = 1$$

and

$$
\begin{aligned}
\mathscr{E}'_x &= \mathscr{E}_x & \mathscr{B}'_x &= \mathscr{B}_x \\
\mathscr{E}'_y &= \gamma(\mathscr{E}_y - \beta\mathscr{B}_z) & \mathscr{B}'_y &= \gamma(\mathscr{B}_y + \beta\mathscr{E}_z) \\
\mathscr{E}'_z &= \gamma(\mathscr{E}_z + \beta\mathscr{B}_y) & \mathscr{B}'_z &= \gamma(\mathscr{B}_z - \beta\mathscr{E}_y)
\end{aligned}
\tag{2-55a}
$$

Rotations affect the components of $\mathscr{E}$ and $\mathscr{B}$ like those of any other vector, so that the transformation law under a general Lorentz transformation is

$$
\begin{aligned}
\mathscr{E}'_{\parallel} &= \mathscr{E}_{\parallel} & \mathscr{B}'_{\parallel} &= \mathscr{B}_{\parallel} \\
\mathscr{E}'_{\perp} &= \gamma\left(\mathscr{E} + \frac{\mathbf{u}}{c} \times \mathscr{B}\right)_{\perp} & \mathscr{B}'_{\perp} &= \gamma\left(\mathscr{B} - \frac{\mathbf{u}}{c} \times \mathscr{E}\right)_{\perp}
\end{aligned}
\tag{2-55b}
$$

where

$$
\begin{aligned}
\mathscr{E}' &= \mathscr{E}'_{\parallel} + \mathscr{E}'_{\perp} & \mathscr{B}' &= \mathscr{B}'_{\parallel} + \mathscr{B}'_{\perp} \\
\mathscr{E} &= \mathscr{E}_{\parallel} + \mathscr{E}_{\perp} & \mathscr{B} &= \mathscr{B}_{\parallel} + \mathscr{B}_{\perp}
\end{aligned}
$$

and $\parallel$ and $\perp$ refer to the direction of the velocity of R' with respect to R.

One readily verifies that the transformation Eqs. (2-55a) correspond to the transformation of the field-strength tensor

$$
\mathscr{F}_{\mu\nu} =
\begin{pmatrix}
0 & \mathscr{B}_z & -\mathscr{B}_y & -i\mathscr{E}_x \\
-\mathscr{B}_z & 0 & \mathscr{B}_x & -i\mathscr{E}_y \\
\mathscr{B}_y & -\mathscr{B}_x & 0 & -i\mathscr{E}_z \\
i\mathscr{E}_x & i\mathscr{E}_y & i\mathscr{E}_z & 0
\end{pmatrix}
\tag{2-54}
$$

under the special Lorentz transformation Eq. (1-31b), whose matrix is

$$a_{\mu\nu} = \begin{pmatrix} \gamma & 0 & 0 & i\beta\gamma \\ 0 & 1 & 0 & 0 \\ 0 & 0 & 1 & 0 \\ -i\beta\gamma & 0 & 0 & \gamma \end{pmatrix} \qquad (2\text{-}49c)$$

For example,

$$\mathcal{B}'_y = \mathcal{F}'_{31} = a_{3\lambda}a_{1\rho}\mathcal{F}_{\lambda\rho} = a_{33}a_{11}\mathcal{F}_{31} + a_{33}a_{14}\mathcal{F}_{34}$$

$$= \gamma\mathcal{B}_y + (i\beta\gamma)(-i\mathcal{E}_z) = \gamma(\mathcal{B}_y + \beta\mathcal{E}_z)$$

The antisymmetric four-tensor $\overset{\approx}{\mathcal{F}}$ has six independent components, specifying the electric and magnetic fields. The transformation equations show that $\mathcal{E}$ and $\mathcal{B}$ are inextricably intertwined. A field that is purely electric in one frame is both electric and magnetic in another. Useful invariants are

$$\mathcal{E} \cdot \mathcal{B} = \mathcal{E}' \cdot \mathcal{B}'$$

$$\mathcal{E}^2 - \mathcal{B}^2 = \mathcal{E}'^2 - \mathcal{B}'^2$$

Note also that the magnetic field is in the space–space part of the field tensor, the electric field in the space–time part. An inversion of axes does not change the components of $\mathcal{B}$, whereas it reverses those of $\mathcal{E}$. $\mathcal{B}$ is axial, $\mathcal{E}$ is polar.

The Maxwell equations [Eqs. (D-3a and d)] combine into one equation for the divergence of the field-strength tensor

$$\frac{\partial\mathcal{F}_{\lambda\mu}}{\partial x_\mu} = 0 \qquad (\lambda, \mu = 1, 2, 3, 4)$$

and the other two equations [Eqs. (D-3b and c)] combine into

$$\frac{\partial\mathcal{F}_{\lambda\mu}}{\partial x_\nu} + \frac{\partial\mathcal{F}_{\mu\nu}}{\partial x_\lambda} + \frac{\partial\mathcal{F}_{\nu\lambda}}{\partial x_\mu} = 0$$

where λ, μ, and ν are any three of the numbers 1, 2, 3, 4.

Appendix E

The Constant of Integration in the Expression for the Energy of a Particle

In Chapter 3 we showed that

$$\mathbf{p} = \mu \left(1 - \frac{v^2}{c^2}\right)^{-1/2} \mathbf{v} \tag{1-33}$$

and

$$dE = d\left[\mu c^2 \left(1 - \frac{v^2}{c^2}\right)^{-1/2} \right] \tag{3-13}$$

Integrating the last equation, we obtain

$$E = \mu c^2 \left(1 - \frac{v^2}{c^2}\right)^{-1/2} + A_0 \tag{E-1}$$

where A_0 is a constant independent of v. The purpose of this note is to show that A_0 must be set equal to zero.

We have seen [Eqs. (3-18)] that the expression Eq. (1-33) for $\mathbf{p}$ is the spacelike part of the four-momentum, and the first term of the right-hand side of Eq. (E-1) is the timelike part, to within a factor of c. With the x axis parallel to $\mathbf{v}$,

$$\mu \left(1 - \frac{v^2}{c^2}\right)^{-1/2} v = p_1$$

$$\mu c^2 \left(1 - \frac{v^2}{c^2}\right)^{-1/2} = c p_0$$

360

The principle of relativity requires that A_0 have the same transformation character as the other additive term in Eq. (E-1). Thus, A_0 must be the time-like component of some four-vector $\tilde{A}$ independent of v. In R', moving with respect to R in the x direction with velocity u ($\beta = u/c$, $\gamma = (1 - \beta^2)^{-1/2}$), the components are [Eqs. (2-49bb)],

$$p'_1 = \gamma(p_1 - \beta p_0) \qquad A'_1 = \gamma(A_1 - \beta A_0)$$

$$p'_0 = \gamma(p_0 - \beta p_1) \qquad A'_0 = \gamma(A_0 - \beta A_1)$$

Thus,

$$p'_x = p'_1 = \gamma\left[\mu\left(1 - \frac{v^2}{c^2}\right)^{-1/2} v - \beta\mu c\left(1 - \frac{v^2}{c^2}\right)^{-1/2}\right]$$

$$\text{(E-2)}$$

$$E' = cp'_0 + A'_0 = \gamma\left[\mu c^2\left(1 - \frac{v^2}{c^2}\right)^{-1/2} - \beta\mu c\left(1 - \frac{v^2}{c^2}\right)^{-1/2} v\right] + \gamma(A_0 - \beta A_1)$$

Let R' be the rest frame ($u = v$). Equations (E-2) reduce to

$$p'_x = 0$$

$$E' = \mu c^2 + \gamma(A_0 - \beta A_1)$$

$$\text{(E-3)}$$

The coefficient of γ in the second term of the second Eq. (E-3) must vanish. Otherwise, the energy in the particle's rest frame would be different for different velocities in R. Thus,

$$A_0 - \beta A_1 = 0$$

This leads to

$$\tilde{A}^2 = A_1{}^2 - A_0{}^2 = A_0{}^2\left(\frac{1}{\beta^2} - 1\right)$$

which is not invariant unless $A_0 = 0$.   Q.E.D.

Appendix F

Doppler Effect

We consider a simple harmonic plane wave in vacuum. In the inertial frame R its phase (the argument of the sine or cosine) is

$$\Phi = \mathbf{k} \cdot \mathbf{r} - \omega t \qquad \text{(F-1)}$$

where ω is called the circular frequency and $\mathbf{k}$ the circular wave vector. At a particular time, Φ has the same value throughout a plane perpendicular to $\mathbf{k}$, since (Fig. F-1) $\mathbf{k} \cdot \mathbf{r}$ is k times the distance from the origin to the plane. The value of Φ increases linearly with this distance, and changes by 2π on advancing a distance ("one wavelength")

$$\lambda = \frac{2\pi}{k} \qquad \text{(F-2a)}$$

in the direction of propagation. At a particular place, Φ decreases linearly with time, changing by 2π in a time ("one period")

$$T = \frac{2\pi}{\omega} \qquad \text{(F-2b)}$$

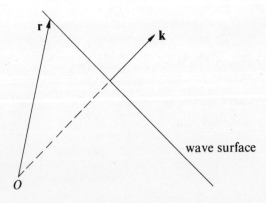

Figure F-1.

362

The frequency is, of course,

$$v = \frac{1}{T} = \frac{\omega}{2\pi} \tag{F-2c}$$

A plane of constant phase (wave surface) advances in the direction of **k** with speed ω/k; in effect, for a point moving in this way the increase of **k** · **r** balances the increase of ωt. Thus,

$$\frac{\omega}{k} = c \tag{F-3}$$

if the wave is electromagnetic.

Let us now look at this wave in another inertial frame, R', moving with constant velocity **u** with respect to R. The phase Φ must be invariant. No matter how the spacing and inclination of the wave surfaces may change in the transformation, a crest in R remains a crest in R'. Since **r** and t are essentially the spacelike and timelike parts of a four-vector $\tilde{r}$, **k** and ω must be the corresponding parts of a four-vector, the *wave four-vector* $\tilde{k}$. Then Φ is the scalar product of these four-vectors, and is invariant.

$$\Phi = \tilde{k}\tilde{r} = k_\mu r_\mu = \mathbf{k} \cdot \mathbf{r} - \omega t$$

The components of $\tilde{k}$ are

$$k_1 = k_x$$
$$k_2 = k_y$$
$$k_3 = k_z \tag{F-4}$$
$$k_4 = \frac{i\omega}{c} \quad \text{or} \quad k_0 = \frac{\omega}{c}$$

The invariant length squared of $\tilde{k}$ is

$$\tilde{k}\tilde{k} = k^2 - \frac{\omega^2}{c^2} \tag{F-5}$$

which, by Eq. (F-3), is zero if the wave is electromagnetic.

The wave four-vector transforms like any other four-vector. For a special Lorentz transformation (SLT) with $Ox \| \mathbf{u}$ [as in Eq. (1-31b)],

$$k_0' = \gamma(k_0 - \beta k_1)$$
$$k_1' = \gamma(k_1 - \beta k_0)$$
$$k_2' = k_2 \tag{F-6a}$$
$$k_3' = k_3$$

Specifying the direction of propagation by the polar angle χ with respect to the x axis and the azimuth ψ (as in Fig. 2-31), these become

$$\frac{\omega'}{c} = \gamma\left(\frac{\omega}{c} - \beta k \cos \chi\right)$$
$$k' \cos \chi' = \gamma\left(k \cos \chi - \beta\frac{\omega}{c}\right)$$
$$k' \sin \chi' \cos \psi' = k \sin \chi \cos \psi \tag{F-6b}$$
$$k' \sin \chi' \sin \psi' = k \sin \chi \sin \psi$$

Specializing to electromagnetic waves, where $\bar{k}$ is lightlike, the first Eq. (F-6) becomes

$$\omega' = \gamma\omega(1 - \beta \cos \chi) \tag{F-7a}$$

This equation gives the transformed frequency as a function of the angle between the direction of propagation and the direction of relative motion. It is the relativistic Doppler effect formula. In terms of ν instead of ω,

$$\nu' = \gamma\nu(1 - \beta \cos \chi) \tag{F-7b}$$

If the direction of propagation is transverse in $R(\chi = \pi/2)$,

$$\nu' = \gamma\nu$$

in agreement with Eq. (2-34) for the transverse Doppler effect. For head-on approach ($\chi = \pi$),

$$\nu' = \gamma\nu(1 + \beta)$$
$$= \nu\sqrt{\frac{1 + \beta}{1 - \beta}} \tag{F-8a}$$

and for small relative velocity,

$$v' = v(1 + \beta + \tfrac{1}{2}\beta^2 + \cdots)$$

or

$$v' \approx v\left(1 + \frac{u}{c}\right) \tag{F-8b}$$

The other Eqs. (F-6b) give the transformed direction of the wave

$$\psi' = \psi \tag{F-7c}$$

$$\tan \chi' = \frac{1}{\gamma} \frac{\sin \chi}{\cos \chi - \beta} \tag{F-7d}$$

These equations are just the aberration equations [Eqs. (2-45)] derived in the text for a particle moving with speed c.

We note in passing that the fundamental quantum equations

$$\mathbf{p} = \hbar\mathbf{k}\left(= \frac{h}{\lambda}\mathbf{n}\right) \tag{F-9a}$$

$$E = \hbar\omega(= h\nu) \tag{F-9b}$$

are one relativistically covariant equation

$$\tilde{p} = \hbar\tilde{k} \tag{F-9c}$$

applying not only to photons (Einstein) but also to material particles (De Broglie). It is interesting that already in his first 1905 paper on relativity ([2], Section 8), Einstein called attention to the covariance of "the energy and frequency of a light complex."

References

1. Is. Newton, *Principia* (London, 1686). A recent printing in English is *Sir Isaac Newton's Mathematical Principles of Natural Philosophy and His System of the World*, Motte's translation revised and edited by F. Cajori (University of California Press, Berkeley, 1962), 2 Vols., paperback.
2. A. Einstein, *Ann. Physik.* **17**, 891 (1905). This paper, as well as other important ones by Lorentz, Minkowski, and Einstein, is available in translation in *The Principle of Relativity* by A. Einstein and others, A. Sommerfeld, Ed. (Dover, New York, 1952), reprint, p. 35.
3. K. T. Bainbridge, *J. Franklin Inst.* **215**, 509 (1933).
4. C. W. Allen, *Astrophysical Quantities* (Athlone Press, London, 1963); A. Blaauw and M. Schmidt, *Galactic Structure* (University of Chicago Press, Chicago, 1965).
5. R. W. P. Drever, *Phil Mag.* **6**, 683 (1961).
6. Quoted in A. Einstein and L. Infeld, *The Evolution of Physics* (Simon and Schuster, New York, 1938), p. 58.
7. J. Larmor, *Aether and Matter* (Cambridge University Press, London, 1900).
8. H. A. Lorentz, *Amsterdam Proc.* **6**, 809 (1904). It is reprinted in *The Principle of Relativity* by A. Einstein and others, A. Sommerfeld, Ed. (Dover, New York, 1952), reprint, p. 11.
9. H. Poincaré, *Compt. Rend.* **140**, 1504 (1905); *Rend. Circ. Matem. Palermo* **21**, 1209 (1906).
10. H. Lamb, *Higher Mechanics* (Cambridge University Press, London, 1920).
11. A. Sommerfeld, *Mechanics* (Lectures on Theoretical Physics, Vol. 1), translated by M. O. Stern (Academic Press, New York, 1952).
12. H. Goldstein, *Classical Mechanics* (Addison-Wesley, Cambridge, Mass., 1950).
13. L. Landau and E. Lifshitz, *Mechanics*, translated by J. B. Sykes and J. S. Bell (Pergamon, Oxford, 1960).
14. P. G. Roll, R. Krotkov, and R. H. Dicke, *Ann. Phys.* (N.Y.) **26**, 442 (1964).
15. R. A. Becker, *Theoretical Mechanics* (McGraw-Hill, New York, 1954).
16. K. Symon, *Mechanics*, 2nd ed. (Addison-Wesley, Reading, Mass. 1960).
17. A. A. Michelson, *Astrophys. J.* **61**, 137 (1925); A. A. Michelson and H. G. Gale, assisted by Fred Pearson, *Astrophys. J.* **61**, 140 (1925). The experiment is described in A. A. Michelson, *Studies in Optics* (University of Chicago Press, Chicago, 1927), pp. 163–166.
18. E. J. Post, *Rev. Mod. Phys.* **39**, 475 (1967).
19. T. S. Jaseja, A. Javan, J. Murray, and C. H. Townes, *Phys. Rev.* **133**, A1221 (1964).
20. J. P. Cedarholm, G. F. Bland, B. L. Havens, and C. H. Townes, *Phys. Rev. Letters* **1**, 342 (1958); J. P. Cedarholm and C. H. Townes, *Nature* **184**, 1350 (1959).

21. K. C. Turner and H. A. Hill, *Phys. Rev.* **134**, B252 (1964).
22. T. Alväger, F. J. M. Farley, J. Kjellman, and I. Walling, *Phys. Letters* **12**, 260 (1964); T. Alväger, J. M. Bailey, F. J. M. Farley, J. Kjellman, and J. Wallin, *Arkiv Fysik* **31**, 145 (1966).
23. H. Minkowski, "Space and Time," address delivered at Cologne, Sept. 21, 1908. It is contained in *The Principle of Relativity* by A. Einstein and others, A. Sommerfeld Ed. (Dover, New York, 1952), reprint, p. 73.
24. H. E. Ives and G. R. Stilwell, *J. Opt. Soc. Am.* **28**, 215 (1938); **31**, 369 (1941). G. Otting, *Physik. Zeitschr.* **40**, 681 (1939). See also H. I. Mandelberg and L. Witten, *J. Opt. Soc. Am.* **52**, 529 (1962).
25. W. Kündig, *Phys. Rev.* **129**, 2371 (1963).
26. B. Rossi, K. Greisen, J. C. Stearns, D. Froman, and P. Koontz, *Phys. Rev.* **61**, 675 (1942).
27. D. S. Ayres, D. O. Caldwell, A. J. Greenberg, R. W. Kenney, R. J. Kurz, and B. F. Stearns, *Phys. Rev.* **157**, 1288 (1967); A. J. Greenberg, Thesis, Berkeley (1969).
28. M. v. Laue, *Relativitätstheorie*, Vol. 1, 7th ed. (Vieweg, Braunschweig, 1955).
29. C. Møller, *The Theory of Relativity* (Oxford University Press, London, 1952).
30. M. Born, *Einstein's Theory of Relativity*, revised edition with G. Leibfried and W. Biem (Dover, New York, 1962).
31. A. Einstein, *Jahrbuch der Radioaktivität und Elektronik* **4**, 411 (1907).
32. R. D. Sard, *Elec. Eng.* **66**, 61 (1947).
33. O. Heaviside, *Electrical Papers* (Macmillan, London, 1894), Vol. 2, p. 495.
34. D. Kerst, *Phys. Rev.* **60**, 47 (1941).
35. L. Landau and E. Lifshitz, *Classical Theory of Fields*, translated by M. H. Hamermesh (Addison-Wesley, Cambridge, Mass., 1951).
36. G. N. Lewis and R. C. Tolman, *Phil. Mag.* **18**, 510 (1909).
37. H. A. Lorentz, *The Theory of Electrons* (Dover, New York, 1954), reprint.
38. H. A. Lorentz, *Problems of Modern Physics* (Ginn, Boston, 1927); reprint, (Dover, New York, 1967).
39. Carl Störmer, *The Polar Aurora* (Oxford University Press, London, 1955); V. D. Hopper, *Cosmic Radiation and High Energy Interactions* (Logos, London, 1964), Chapter 5.
40. M. S. Livingston and J. P. Blewett, *Particle Accelerators* (McGraw-Hill, New York, 1962).
41. J. R. Pierce, *Theory and Design of Electron Beams*, 2nd ed. (Van Nostrand, New York, 1954).
42. J. D. Jackson, *Classical Electrodynamics* (Wiley, New York, 1962).
43. W. K. H. Panofsky and Melba Phillips, *Classical Electricity and Magnetism*, 2nd ed. (Addison-Wesley, Reading, Mass., 1962).
44. A. P. Banford, *The Transport of Charged Particle Beams* (Spon, London, 1966).
45. A. Sommerfeld, *Atomic Structure and Spectral Lines*, 3rd English ed., translated by H. L. Brose (Methuen, London, 1934).
46. C. L. Cowan, Jr., and F. Reines, *Phys. Rev.* **92**, 830 (1953); F. Reines and C. L. Cowan, Jr., *Phys. Rev.* **113**, 273 (1959).
47. A. Einstein, *Ann. Physik* **18**, 639 (1905). A translation into English is contained in *The Principle of Relativity* by A. Einstein and others, A. Sommerfeld, Ed. (Dover, New York, 1952), reprint, p. 69.

48. J. D. Cockcroft and E. T. S. Walton, *Proc. Roy. Soc.* **137**, 229 (1932).
49. K. T. Bainbridge, *Phys. Rev.* **44**, 123 (1933).
50. E. Feenberg and H. Primakoff, *Phys. Rev.* **73**, 449 (1948).
51. O. F. Kulikov, Y. Y. Telnov, E. L. Filippov, and M. N. Yakimenko, *Phys. Letters* **13**, 344 (1964).
52. C. Bemporad, R. H. Milburn, N. Tanaka, and M. Fotino, *Phys. Rev.* **138**, 1546 (1965).
53. O. I. Dahl, L. M. Hardy, R. I. Hess, J. Kirz, D. H. Miller, and J. A. Schwartz, *Phys. Rev.* **163**, 1430 (1967).
54. O. Chamberlain, E. Segrè, C. Wiegand, and T. Ypsilantis, *Phys. Rev.* **100**, 947 (1955).
55. V. L. Auslander, G. I. Budker, Ju. N. Pestov, V. A. Sidorov, A. N. Skrinsky, and A. G. Khabakpashev, *Phys. Letters* **25B**, 433 (1967).
56. R. Hagedorn, *Relativistic Kinematics* (Benjamin, New York, 1964).
57. A. Sommerfeld, *Physik. Zeitschr.* **10**, 826 (1909).
58. Ya. A. Smorodinskii in *Modern Aspects of Particle Physics*, A. I. Alikhanyan, Ed., Nor-Amberd Lectures, 1963 (Israel Program for Scientific Translations, Jerusalem, 1965).
59. B. O. Peirce, *A Short Table of Integrals* (Ginn, Boston, 1929).
60. L. H. Thomas, *Nature* **117**, 514 (1926); *Phil. Mag.* **3**, 1 (1927).
61. E. P. Wigner, *Rev. Mod. Phys.* **29**, 255 (1957).
62. G. Uhlenbeck and S. Goudsmit, *Naturwiss.* **13**, 953 (1925); *Nature* **117**, 264 (1926).
63. L. I. Schiff, *Proc. Natl. Acad. Sci. U. S.* **46**, 871 (1960).
64. A. Einstein and W. J. de Haas, *Verhandl. Deut. Phys. Ges.* **17**, 152 (1915); S. J. Barnett, *Rev. Mod. Phys.* **7**, 129 (1935).
65. D. T. Wilkinson and H. R. Crane, *Phys. Rev.* **130**, 852 (1963).
66. A. Rich and H. R. Crane, *Phys. Rev. Letters* **17**, 271 (1966).
67. H. R. Crane, *Scientific American* **218**, 72 (1968).
68. G. Charpak, F. J. M. Farley, R. L. Garwin, T. Muller, J. C. Sens, and A. Zichichi, *Nuovo Cimento* **37**, 1241 (1965).
69. J. Bailey, W. Bartl, G. von Bochmann, R. Brown, F. J. M. Farley, H. Jöstlein, E. Picasso, and R. W. Williams, *Phys. Letters* **28B**, 287 (1968).
70. D. P. Hutchinson, J. Menes, G. Shapiro, and A. M. Patlach, *Phys. Rev.* **131**, 1351 (1963); G. M. Bingham, *Nuovo Cimento* **27**, 1352 (1963).
71. H. Mendlowitz and K. M. Case, *Phys. Rev.* **97**, 33 (1955).
72. J. Schwinger, *Phys. Rev.* **76**, 790 (1949); **82**, 664 (1951).
73. L. Michel and A. S. Wightman, *Phys. Rev.* **98**, 1190 (1955).
74. J. M. Fowler, H. Primakoff, and R. D. Sard, *Nuovo Cimento* (10) **9**, 1027 (1958).
75. V. Bargmann, L. Michel, and V. L. Telegdi, *Phys. Rev. Letters* **2**, 435 (1959).
76. A. Einstein, *Ann. Physik* **49**, 769 (1916). An English translation is contained in *The Principle of Relativity* by A. Einstein and others, A. Sommerfeld, Ed. (Dover, New York, 1951), reprint, p. 111.
77. C. Brans and R. H. Dicke, *Phys. Rev.* **124**, 925 (1961).
78. H. A. Lorentz, *The Einstein Theory of Relativity* (Brentano's, New York, 1920).
79. R. Adler, M. Bazin, and M. Schiffer, *Introduction to General Relativity* (McGraw-Hill, New York, 1965).
80. P. Bergmann, *Introduction to the Theory of Relativity* (Prentice-Hall, New York, 1947).

81. A. Einstein, *The Meaning of Relativity*, 5th ed. (Princeton University Press, Princeton, 1955).

82. A. Einstein, *Ann. Physik* **35**, 898 (1911). An English translation is contained in *The Principle of Relativity* by A. Einstein and others, A. Sommerfeld, Ed. (Dover, New York, 1952), reprint, p. 99.

83. J. Brandt, Princeton Thesis, 1962, as reported by R. H. Dicke, *The Theoretical Significance of Experimental Relativity* (Gordon and Breach, New York, 1964), p. 25.

84. R. V. Pound and J. L. Snider, *Phys. Rev.* **140**, B788 (1965).

85. H. J. Hay, J. P. Schiffer, T. E. Cranshaw, and P. A. Egelstaff, *Phys. Rev. Letters* **4**, 165 (1960).

86. D. C. Champeney, G. R. Isaak, and A. M. Khan, *Nature* **198**, 1186 (1963).

87. W. Pauli, Jr., *Relativitätstheorie* (Teubner, Leipzig, 1921). It is available in English as *Theory of Relativity*, translated by G. Field with supplementary notes by the author (Pergamon, New York, 1958).

Index

Aberration, 108, 365
Absolute space (*see* Space)
Accelerated reference frame (*see* Reference frame)
Acceleration
  centripetal, 349
  constant transverse, 127–130
  Coriolis, 349
  four- (*see* Four-acceleration)
  of gravity, 16, 45, 351
  transformation of (*see* Transformation)
  translational, 349
  uniform longitudinal, 124–127, 189
Action at a distance (*see* Force)
Ampère, A. M., 291
Angular momentum, 25, 281–284
  conservation of (*see* Conservation)
  (*see also* Spin)
Angular velocity, 341–345
Annihilation (*see* Transition)
Argonne, 100
Ascoli, G., 45, 306, 325
Axial vector (*see* Vector)

Bargmann, V., 304
Berezinskii, V. S., 307
Berkeley, 248
Bessel, F. W., 310
Binary collision (*see* Transition)
Bohr, N., 190
Boost (*see* Lorentz transformation)
Born, M., 102
Bradt, H. V., 307
Brookhaven, 140, 247

Case, K. M., 298
Causality, 77, 108

Center of mass frame, 209
Center of momentum frame (*see* Center of mass frame)
Centrifugal force (*see* Force, inertial)
CERN, 52, 248
Charge
  electric, 355
  gravitational, 9, 41, 320, 322
  invariance of, 355
  transformation of (*see* Transformation)
Charged particle (*see* Equation of motion; Electromagnetic field)
Clark, G. W., 307
Clashing beams, 248
Clock, 6
  effect of gravitational field on (*see* Gravitation)
  rate of moving, 81, 94
  specific effect of acceleration on, 81, 311
  synchronization, 7, 55–57
Colliding beams, 248
Collision (*see* Scattering)
Compton scattering, 239–241, 254
Conservation
  of angular momentum, 283
  of energy, 13, 214–216
  of four-momentum, 8, 11, 214–216
  of momentum, 8, 151, 214–216
Coordinate system, 5
  Cartesian, 5
  right handed, 23, 333
Coordinate transformation, 16
  inversion, 23, 333–335
  rotation, 20, 327–332
  shift of origin, 17

Coriolis acceleration (*see* Acceleration)
Coriolis force (*see* Force, inertial)
Cosmic rays, 97, 99, 187, 250, 253, 306–307
Covariance, 16–17
  with respect to displacement of origin, 17
  with respect to Galileo transformation, 28, 30–33
  with respect to inversion, 25
  with respect to Lorentz transformation, 36, 140
  with respect to rotation of axes, 17
  with respect to uniform translation, 48
Crane, H. R., 298
Cross section, 177

De Broglie, L., 365
Decay in flight, 96–100, 179, 198, 199, 252, 306
de Haas, W. J., 291
Dicke, R. H., 51, 310
Distribution function, 2, 171–177
Doppler effect, 362–365
  transverse, 52, 95, 318–319, 364
Drever, R. W. P., 17

Earthbound reference frame (*see* Reference frame)
Ehrenfest, P., 190
Einstein, A., 1, 27, 36, 48, 49, 136, 138, 140, 142, 291, 316, 365
Einstein dilation (*see* Time, dilation)
Einstein's elevator, 310
Elastic scattering (*see* Scattering)
Electromagnetic field, 355
  invariants, 359
  of a point charge moving with constant velocity, 197
  tensor, 115, 358
  transformation of (*see* Transformation)
Electromagnetic interaction (*see* Interaction)
Electron theory, 151
Electron volt, 205

Energy, 8, 12, 154, 159, 209
  conservation of (*see* Conservation)
  constant of integration in, 155, 211, 360–361
  inertia of, 10, 11, 158, 211–212
  kinetic, 157
  -momentum relation, 166
  rest, 13, 163, 217–218
  in rest frame, 212
  transfer, 253
  transformation of (*see* Transformation)
  velocity relation, 154, 155
Eötvös, R. V., 310
Equation of motion, 142
  charged particle, 136–141
  covariant, for charged particle, 152–154
  general covariant, 166–168
  nonrelativistic, 32
  of spin (*see* Spin)
Equivalence principle (*see* Principle of equivalence)
Ether, 36, 49, 54
Ether drift, 17, 51, 351
Event, 13, 111
  interval between (*see* Interval)
  location, in space and time, 28, 37, 54–57

Fine-structure constant, 193
Fitzgerald, G. F., 82
Fitzgerald–Lorentz contraction (*see* Length)
Fixed stars, 15
Force, 11, 141
  on charged particle, 137
  equal and opposite pair, 8, 12
  four (*see* Four-force)
  gravitational, 14
  inertial, 14, 39–41
  instantaneous action at a distance, 33
  Minkowski (*see* Four-force)
  transformation of (*see* Transformation)
Foucault, L., 354
Four-acceleration, 121–123

Four-force, 167
Four-momentum, 8, 11, 13, 136, 159–166
    of a complex system, 207–208
    conservation of (*see* Conservation)
    right triangle, 165, 212
    transfer, 236, 242
Four-scalar, 112
Four-spin, 301
    equation of motion, 300–304
Four-tensor, 112, 115
Four-vector, 112
    lightlike, 114
    spacelike, 114
    timelike, 114
Four-velocity, 119–121
    equation of motion, 303
    of system, 213
Freely falling laboratory, 14, 41

Gale, H. G., 354
Galilean invariance (*see* Invariance)
Galileo, G., 27, 310
Galileo transformation, 28, 48–49, 65, 346
General relativity, 3, 17, 309
g factor, 292
Gol'dman, I. I., 307
Gravitation, 309–311
    effect of field on clock, 102, 311–314
    effect of field on measuring rod, 314–315
    free fall of a material particle, 320–325
    and inertia, 40, 310
    Newton's law of, 20, 33, 142
    weak gravitational field, 309
    weightlessness, 41
Gravitational charge (*see* Charge)
Gravitational deflection of light, 310
Gravitational field strength, 41
Gravitational red shift, 316–320
Gyromagnetic ratio, 292

Hagedorn, R., 304
Heitler, W., 254
Helmholtz, H. van, 34
Horizontal, 322
Hydrogenic atom, 189

Hyperbolic triangle, 274
Hyperbolic motion (*see* Acceleration, uniform longitudinal)
Hyperbolic trigonometry, 274

Inelastic scattering (*see* Scattering)
Inertia (*see* Mass, inertial; Gravitation, and inertia)
Inertia of energy (*see* Energy)
Inertial force (*see* Force)
Inertial frame (*see* Reference frame)
Inertial mass (*see* Mass)
Instantaneous action at a distance (*see* Force)
Interaction
    Coulomb, 33, 143
    electromagnetic, 34
    gravitational (*see* Gravitation)
    hadronic, 151
    weak, 151
Interval, 75–80
    space, 75
    spacelike, 76
    squared, 71, 76
    time, 75
    timelike, 76
Invariance, 16
    Galilean, 33
    Lorentz, 4, 68, 111
    rotational, 20, 68, 111
Invariant mass (*see* Mass)
Inversion (*see* Coordinate transformation)
Ives, H. E., 95

Klein-Nishina formula, 254
Kronecker delta, 23, 46, 331

Larmor, J., 36
Laser, 241, 249, 254
Laue, M. v., 102
Law of motion, 7
    (*see also* Equation of motion)
Leibniz, G. W., 13
Length, 5, 82
    contraction, 82, 91
    proper, 82
Leprince-Ringuet, L., 253

Levi-Civita symbol, 25, 46
Lewis, G. N., 146
Lhéritier, M., 253
Lifetime (*see* Transition probability)
Light cone, 78
Line element
  in space, 26
  in spacetime, 71
Lorentz contraction (*see* Length)
Lorentz, H. A., 36, 82, 151
Lorentz transformation, 36–38, 58–75
  boost, 74
  collimation due to, 109–111
  composition of,
    in different directions, 273–281
    parallel, 266–270
  general, 73–74
  inversion as, 74
  Minkowski graph (*see* Minkowski's
    hyperbolic graph)
  pure (PLT), 74
  rotation as, 73
  without rotation, 74
  special (SLT), 73, 113
  successive, 258

Magnetic dipole moment, 282, 291
Magnetic moment anomaly, 298
Magnetic rigidity, 187
Maser, 51
Mass, 9–11, 141
  gravitational, 9, 212
  inertial, 9, 30. 141, 149, 210, 212
  invariant, 10, 163, 212
  longitudinal, 138
  nonconservation of proper, 217, 231
  proper, 10, 137, 141, 163, 164, 212
  transverse, 138
Mass–energy relation, 11, 158, 164,
  210, 217, 234
Matrix
  inversion, 23
  multiplication, 258
  pure Lorentz transformation (PLT),
    260–261
  rotation, 22, 259, 331
  spacetime transformation, 258–261

Matrix *continued*
  special Lorentz transformation (SLT),
    115, 260
Maxwell field equations, 355, 359
Mendlowitz, H., 298
Mercury, 196
Michel, L., 304
Michelson, A. A., 354
Michelson–Morley experiment, 49, 51
Minkowski, H., 86, 111
Minkowski force (*see* Four-force)
Minkowski's hyperbolic graph, 86–94
Moment of momentum (*see* Angular
  momentum)
Momentum, 7, 140, 159
  angular (*see* Angular momentum)
  of complex system, 208
  conservation of (*see* Conservation)
  distribution, 168
  four (*see* Four-momentum)
  from orbit in magnetic field, 186, 206
  space, 168
  transfer, 238
  transformation of (*see* Transforma-
    tion)
Momentum–velocity relation, 38, 141,
  146, 155, 213
  (Newtonian), 38
Mössbauer effect, 51, 95
Muon polarization (*see* Polarization)

Neutrino beam, 179–181
Newton, I., 1, 13, 27, 34, 310
Nonconservation of proper mass (*see*
  Mass)
Notation, xix
Novosibirsk, 249
Nuclear transmutation, 202

Particle, 3, 203
  charged, field due to (*see* Electro-
    magnetic field)
  in constant electromagnetic field,
    182
  in constant homogeneous electric
    field, 187–189

Particle *continued*
  in constant homogeneous field, $\mathscr{E}$
    and $\mathscr{B}$ parallel, 200
  in Coulomb field, 189–195
  in magnetic field, 182–187
  charged particles moving with same
    velocity, 143–145
  complex, 203
  creation and destruction, 202–203
  equation of motion (*see* Equation of
    motion)
  free fall (*see* Gravitation)
  Table, xviii
Perihelion advance, 195–197
Phase, 362
Pinch effect, 145
Planck, M., 138
PLT (*see* Lorentz transformation)
Poincaré, H., 36, 260
Polarization
  of cosmic-ray muons, 307
  four-vector, 301
  longitudinal, 297, 303
  of muon, from K decay, 307
    from pion decay, 306
Polar vector (*see* Vector)
Pound, R. V., 316, 318
Primakoff, H., 4
Principle of equivalence, 15, 102–104,
    309–311, 318
Probability density, 172
Proper length (*see* Length)
Proper mass (*see* Mass)
Proper time (*see* Time)
Proper velocity (*see* Four-velocity)
Pseudoscalar (*see* Scalar)
Pseudotensor (*see* Tensor)
Pseudovector (*see* Vector)

Quantum mechanics, 2, 96, 190, 203,
    282, 291

Radiative transition (*see* Transition)
Rapidity (*see* Velocity)
Reference frame, 4–5, 54
  accelerated, 39, 102, 309, 341–350
  detectable effects of rotation of (*see*
    Rotation)

Reference frame
  earthbound, 38, 41–45, 325, 351–354
  inertial, 13–16, 38, 43, 351
  laboratory, 232
  local inertial, 15
  rest frame, 78, 80, 163
    of the system, 209
  rotating, 354
    (*see also* Reference frame, acceler-
    ated)
Relativistic kinematics, 250
Relativistic units, 204–207
  practical, 205
Relativistic velocity (*see* Four-velocity)
Rest energy (*see* Energy)
Rest frame (*see* Reference frame)
Rest mass (*see* Mass, proper)
Rigid body, 4, 341
Rod
  change of length (*see* Length)
  change of inclination, 84–85
  effect of gravitational field on (*see*
    Gravitation)
  specific effect of acceleration on, 311
Rossi, B., 97
Rotation
  of axes due to successive Lorentz
    transformations, 278
  of Cartesian axes (*see* Coordinate
    transformation)
  detectable effects of, 27–28, 351–354
  finite, 330
  infinitesimal, 341
  matrix, 22, 331
  of rigid body, 341
Rutherford scattering experiment, 235

Sagnac, G., 354
Scalar, 20, 332
  product, 22, 114, 332
  pseudo- 24, 335
Scattering, 203, 231–250
  angle, 236
  elastic, 203, 235–241, 253
  inelastic, 203
    the general case, 247–250
    two-body final state, 241–246

Schiff, L. I., 290
Serpukhov, 247
Simultaneity
  absolute, 65
  relativity of, 50, 89
SLT (*see* Lorentz transformation)
Snider, J. L., 318
Sommerfeld, A., 190, 196
Space, 4
  absolute, 13, 36
  Euclidean, 5–6, 55
  homogeneity of, 8, 17–20
  isotropy of, 17, 20–23
Spacetime, 111–118, 310
Special relativity, 1, 48
  postulates of, 1, 48–54
  principle of, 26–27, 36, 48
Speed (*see* Velocity)
Spherical defect, 275
Spin, 25, 283
  of elementary particles, 292
  "g-2" experiment, 298
  motion of, 293–304
    in transverse electric field, 298–300
    in transverse magnetic field, 295–298
  -orbit interaction, 298–300
  polarization, 297
  Thomas precession, 285–290, 300
  turning of, in pure Lorentz transformation, 284–285
Stanford, 45, 159, 250, 254, 256
Stillwell, G. R., 95
System, 203

Table of Particles, xviii
Telegdi, V. L., 304
Tensor, 22, 332
  electromagnetic field (*see* Electromagnetic field)
  four (*see* Four-tensor)
  pseudo-, 24, 335
Thomas, L. H., 286, 300
Thomas precession (*see* Spin)
Threshold, 241, 247

Time, 7
  dilation, 81, 93, 94
  laboratory, 81
  proper, 80, 118
  (*see also* Clock)
Tolman, R. C., 146
Torque, 281–282
Transformation
  of acceleration, 123–124, 134
    (Galilean), 39, 348–350
  of charge, 136, 355
  of coordinates (*see* Coordinate transformation)
  of distribution function, 172, 176–179
  of electromagnetic field, 116, 136, 355–359
  of energy, 160
  of force, 12, 142, 161–162
  Galileo (*see* Galileo transformation)
  Lorentz (*see* Lorentz transformation)
  of momentum, 160
  of phase, 363
  properties of particle parameters, 4
  properties of pseudoquantities, 24
  properties of tensors, 22
  properties of vectors, 21, 327–330
  of velocity, 104–111, 169–170
    (Galilean), 39, 346–348
  parameter α, 272

Transition, 203
  one-body final state, 233–235
  probability, 96
  radiative, 202, 218–224
  threshold for (*see* Threshold)
  two-body final state,
    both of zero proper mass, 225–227
    general case, 227–231
    one has zero proper mass, 224
  two-body initial state (binary collision), 231–250
Translation, 26, 341
Triplet production, 249–250
Twin thought experiment, 101–104
Two-photon annihilation (*see* Transition, two-body final state, both of zero proper mass)

Vector, 20, 329–331
    axial, 24, 336–340
    four (*see* Four-vector)
    polar, 24, 336–340
    product, 25, 336–340
    pseudo- 24, 335
    transformation properties (*see* Transformation)
Velocity
    four (*see* Four-velocity)
    Galilean addition of, 30, 106
    of light, 53, 158
    parameter ("rapidity") α, 268, 270
    separator, 200
    of system, 209
    transformation of (*see* Transformation)

Velocity
    triangle, 264–266, 273, 306
    turning of, in pure Lorentz transformation, 285

Wave, 362
    four-vector, 240, 363
    frequency, 240, 362
    phase, 362
    surface, 363
    velocity, 363
    wavelength, 240, 362
Wave-particle relation, 240, 365
Weathervane, 262
Weston, 248
Work, 12, 155
World line, 87, 111